AF355970

Waste Management Practices in Developing Countries

Waste Management Practices in Developing Countries

Editor

Linda Godfrey

MDPI • Basel • Beijing • Wuhan • Barcelona • Belgrade • Manchester • Tokyo • Cluj • Tianjin

Editor
Linda Godfrey
Council for Scientific and
Industrial Research (CSIR)
South Africa
North-West University
South Africa

Editorial Office
MDPI
St. Alban-Anlage 66
4052 Basel, Switzerland

This is a reprint of articles from the Special Issue published online in the open access journal *Recycling* (ISSN 2313-4321) (available at: https://www.mdpi.com/journal/recycling/special_issues/Developing_Countries).

For citation purposes, cite each article independently as indicated on the article page online and as indicated below:

LastName, A.A.; LastName, B.B.; LastName, C.C. Article Title. *Journal Name* **Year**, *Volume Number*, Page Range.

ISBN 978-3-0365-0592-3 (Hbk)
ISBN 978-3-0365-0593-0 (PDF)

Cover image courtesy of Linda Godfrey.

Contents

About the Editor

Linda Godfrey (Prof.) is a Principal Scientist at the Council for Scientific and Industrial Research (CSIR) and Extraordinary Professor at North-West University in South Africa, and holds a Ph.D. in Engineering from the University of KwaZulu-Natal. With over 20 years of sector experience, she manages the Waste Research Development and Innovation (RDI) Roadmap Implementation Unit on behalf of the Department of Science and Innovation, a unit tasked with implementing South Africa's 10-year Waste RDI Roadmap. She has provided strategic input to a number of local, regional and international waste and circular economy initiatives for the United Nations, European Union, World Bank, South African Government Departments, Academy of Sciences, International Solid Waste Association, universities and businesses. She lectures internationally on solid waste management in developing countries, including the social, economic and environmental opportunities of "waste" within a circular economy context. She has published extensively in the field.

Preface to "Waste Management Practices in Developing Countries"

Developing countries face many challenges in the management of waste. These include inadequate waste collection systems, the dumping and burning of waste in open spaces, the operation of uncontrolled or controlled dumpsites, and limited to no waste recycling or recovery. The "leakage" of waste into the environment has a direct negative impact on human and environmental health. At the same time, valuable resources are lost to local economies through the disposal of end-of-life products to land, resulting in negative social and economic impacts. The Global Waste Management Outlook (UNEP, 2015) called on developing countries to stop the uncontrolled dumping and burning of waste; to bring hazardous waste under control; to focus on the prevention of waste; and to develop feedback loops by maximizing recycling and integrating existing small-scale entrepreneurial recyclers into mainstream waste management. The International Solid Waste Association (ISWA, 2014) called for the closure and rehabilitation of dumpsites as a top priority to ensure health and environmental protection.

This book provides insights into waste management practices in developing countries. It highlights the application of research and innovation in finding locally appropriate solutions to improved waste management. The chapters have been selected with a focus on organic waste beneficiation; the role of government and policy interventions; citizen behaviour in driving waste recycling; and the safe management of hazardous waste—particularly healthcare risk waste. Organic waste such as food, garden, and agricultural waste is a relatively large percentage of the waste stream generated in developing countries. The disposal of this waste to land often results in the generation of leachate, impacting water resources; generates odours and attracts vermin; and through its aerobic or anaerobic digestion, generates greenhouses gases, with the potential to negatively impact climates. Yet, organic waste is easily beneficiated, through various technologies such as composting, anaerobic digestion, or more advanced biorefinery technologies. The separation of organic waste at its source and its diversion to beneficiation technologies creates numerous environmental, social, and economic opportunities, and as such, must be considered by developing countries.

All spheres of government have an important role to play in the improved management of waste, through appropriate policy development and implementation. Effective service delivery to all citizens must be a priority, including the consideration of partnerships with the private sector, where this will lead to more effective and efficient integrated waste management.

Recognising that responsible waste management starts with us, as citizens and consumers, improved education and awareness is important in supporting behaviour change towards more sustainable waste management practices, including eliminating littering and illegal dumping, and driving greater recycling behaviour.

Finally, it is important for waste practitioners in developing countries to share and publish their work, through rigorous peer-review processes, so that others may learn from their insights, thereby creating global communities of practice aimed at improving waste management in developing countries.

We trust that the chapters published here will provide such insights to assist communities in bringing waste under control while also unlocking the opportunities of waste as resource.

Linda Godfrey
Editor

 *recycling*

Article

Cassava Waste Management and Biogas Generation Potential in Selected Local Government Areas in Ogun State, Nigeria

David O. Olukanni * and Tope O. Olatunji

Department of Civil Engineering, College of Engineering, Covenant University, Ota 121101, Ogun State, Nigeria; engrolatunji@yahoo.com
* Correspondence: david.olukanni@covenantuniversity.edu.ng

Received: 5 September 2018; Accepted: 12 December 2018; Published: 14 December 2018

Abstract: Agricultural products such as cassava produce huge amounts of waste when processed into consumable goods. The waste generated is generally considered to contribute largely to environmental pollution. This study therefore investigates the waste management practice that is adopted by cassava processors in Ogun State, Nigeria. Five local government areas (LGAs) dominant in processing cassava were selected for the study on the basis of spatial location distribution, landmass, and population. The survey involved the use of structured questionnaires administered to cassava processors of the selected LGAs. The Statistical Package for Social Sciences (SPSS) software application and descriptive statistics were used for data analysis. Results of the analysis show that the majority (70%) of the cassava processors are females. Cassava peel constitutes 10% of the waste produced, of which 91% is heaped at refuse dumps in most communities. Results also reveal that 86.3% of cassava residues are used for animal feeds. Other findings show that the peels, when dried, are used as biofuel for cooking and there is a significant potential for biogas production. From the data captured from respondents during the study, most processors are willing to pay for an improved waste management system. The study therefore recommends the proper waste management of cassava waste to minimize environmental pollution.

Keywords: solid waste management; environmental pollution; agricultural waste; cassava waste; biogas generation; sustainable technology

---

1. Introduction

Solid waste management is the most pressing environmental challenge faced by urban and rural areas of Nigeria, with a population exceeding 170 million people. Among several wastes generated by this huge population is agricultural waste. Improper handling of agricultural waste has raised a significant challenge in the past decades. In 2016, agriculture contributed 19.17% to the gross domestic product (GDP) of Nigeria and it also generated large amount of waste materials. Nigeria is involved in growing and producing many food crops. One of such crops is cassava, a starchy staple food crop which has the ability to resist drought and diseases. In 2012, the production of cassava worldwide was estimated at over 260 million tonnes, with Nigeria being the largest producer, contributing over 20% of the global production [1,2]. In Nigeria, cassava is mostly produced and processed by small-scale farmers at the family or village level. Cassava provides a reliable and inexpensive source of carbohydrates for people in Sub-Saharan Africa, especially in Nigeria, where its production, processing, and consumption is most predominant and significant on a global scale [3,4]. It also provides different job opportunities for both men and women from the production stage until it gets to the final stage. There are indications that the domestic demand for cassava, particularly as a staple food, tends to outweigh the demands of the industrial sector. As farmers are unable to meet their demand, some industries are now engaging in the direct production of their cassava requirements.

Globally, 60% of the cassava produced is mostly used for consumption in numerous forms by humans, while the animal food industry uses about 33% of the world production. The remaining 7% is used by industries to produce products such as textiles, paper, organic acids, flavor and aroma compounds, and cassava bagasse [5]. Three main types of residues are generated during the industrial processing of cassava: peels, solids, and wastewater. These wastes are poor in protein content, but their residues are very rich in carbohydrate and are generated in large amounts during the production of 'garri' and cassava flour from the tubers. The cost associated with the handling and disposal of these wastes constitutes a huge financial burden to the cassava-processing industries in most rural regions of the country. As a result of this challenge, most rural cassava processors choose to dispose the cassava-processing wastes generated into the environment. These wastes have been identified to be toxic to the environment [2,4,6].

The technology of processing cassava roots predominantly includes peeling, grating, dewatering, fermenting, drying, frying, etc. The type and composition of the waste depend on the processing method and type of technology used [7]. In most cassava-processing communities, several tonnes of cassava peels are generated as a waste product from the processing activity and are generally considered to contribute largely to environmental pollution [8]. With an expected increase in cassava production, it is also expected that waste generation will continue to rise. Even though cassava peels can be used as feed for livestock, the quantities generated and the remoteness of many of the communities where processing takes place leave behind a lot of waste, which is burnt or left to rot, with many environmental consequences [9]. Tonukari et al. [10] presented a report of a cassava starch production center which produces 100 tons of tubers per day, with an output of about 47 tons of byproducts. This output may cause environmental problems when abandoned in the surroundings of processing plants or carelessly disposed. The basic form of cassava flour production comprises sorting, weighing, peeling, washing, grating, machine/milling, detoxification, dewatering, granulation, drying milling, sieving, and packaging [11,12].

Management of cassava waste varies across several processing centers in the country, and over 55% of waste generated from its processing is disposed in dump sites. This implies that a great number of cassava processors do not get benefit from the waste they produce [13]. The majority of the cassava peels in Nigeria are abandoned close to the processing site, while some are used for landfilling or burnt. This approach causes a serious threat to the environment and a health hazard to processors and communities [2]. Oparaku et al. [14], from their experiment, expressed that cassava wastes can be used as a biogas substrate, either as a standalone raw material or in combination with livestock manure. Attempts have been made by various researchers to produce products such as organic acid, flavor and aroma compounds, methane and hydrogen gas, enzymes, ethanol, lactic acid, biosurfactant, polyhydroxyalkanoate, essential oils, xanthan gum, and fertilizer from cassava bagasse, peels, and wastewater [2,5,12,13,15,16].

Furthermore, prior studies on cassava waste management [2,4,13] focused on different aspects of cassava waste management; however, there is still a dearth in the literature of studies that combine the potential in the reuse of cassava waste, gender composition of cassava processors in Nigeria, and also factors affecting the willingness to pay for cassava waste management in Ogun State. According to Echebiri and Edaba [17], there is a high positive correlation between the increase of cassava production and the estimated demand for the commodity. It was also found that the waste disposal habits of the people, corruption, work attitude, and inadequate plants and equipment, among others militate against effective waste management in Nigeria [18]. From the foregoing, there is a need for better management and utilization of these waste residues through a better waste management system. It is against this backdrop that this research investigates the potential for an integrated cassava waste management strategy with a focus on Ogun State, Nigeria. Furthermore, this study examined the waste management systems presently in use by cassava processors and their willingness to pay for a value-added solid waste management system.

The main objective of this study is to investigate the cassava waste management methods in Ogun State. Other specific objectives of the study are to:

i.   find out the method of waste disposal that is adopted by cassava processors in the selected local government areas,

ii. find out what cassava residues are used for in the selected cassava-processing factories,

iii. find out the factors that influence processors' willingness to pay for an improved waste management system, and

iv. investigate if cassava wastes generated in the selected local government areas have the potential for the generation of biofuel.

2. Materials and Methods

This study focuses on five local government areas (LGAs) (Yewa North, Odeda, Ijebu North, Ijebu East, and Remo North LGAs) (Figure 1). These local government areas are dominant in processing cassava. A total of 500 questionnaires were administered to selected cassava processors, with 100 questionnaires in each of the LGAs. Figure 2 shows the typical cassava crop harvest while Figure 3 (a–c) shows the activities in one of the cassava-processing factories. In line with Omilani et al. [2], a survey research design was used for this study as it was appropriate because the nature of the research requires the investigation of the opinions and experiences of a group of people by asking them questions.

2.1. Sample Size and Sample Technique

A good representation of the population was chosen from each local government area where cassava processing is prominent for proper evaluation and analysis. The questionnaire was constructed to provide precise and accurate answers through closed-ended questions. The questions were derived from the statement of problem, research questions, research objectives, and hypothesis for testing. Section A dealt with personal data of the respondents, while section B addressed hypothetical questions. The questionnaires were administered to participants that were educated, while those that were not literate had the questionnaires read and interpreted to them in order to get their responses. All the questionnaires administered were retrieved. A structured interview was also used to elicit information from the respondents.

2.2. Reliability of Instruments

The reliability test utilized in this research is Cronbach's alpha reliability test. A result obtained for a sample should be of a reliability of 0.70 or even higher before the research instrument can be used. This study makes use of tables, percentages, and various statistical techniques in the presentation and analysis of the data collected at the significance level of 95%; that is, at the 5% error limit. The data generated through the questionnaire were analyzed through the aid of a computer application, Statistical Packages for Social Sciences (SPSS). In specific terms, the frequency distribution, simple percentage, and mean were deployed in the data analysis.

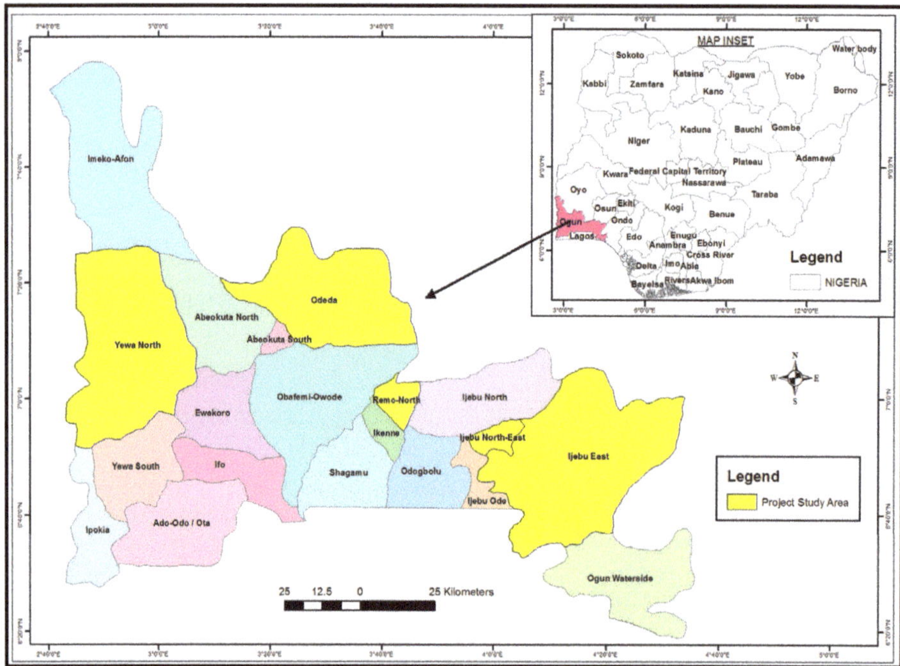

Figure 1. Map of Ogun State showing the study areas in yellow.

Figure 2. Typical Cassava crop harvest.

Figure 3. Activities in the Cassava Processing Factory. (**a**) Cassava peeling section; (**b**) Cassava grinding machine in operation; (**c**) Cassava peel dump site.

3. Results and Discussion

Figure 4 shows that the majority of the cassava processors are females, comprising over 70%. This, according to Popescu et al. [19] in their study titled "Managers' gender and SMEs production", implies that the productivity level of local cassava processors is expected to be higher, although this was not explicitly tested for in this study.

Figure 5 shows the age range of the respondents, with 14.2% of the respondents being between the ages of 20 and 29 years, 34% between the age of 30 and 39 years, and 30.5% and 20% aged between 40 and 49 years and 50 years and above, respectively. From the result in Table 1, it is observed that a great number of cassava processors are in the age range of 30–49 years.

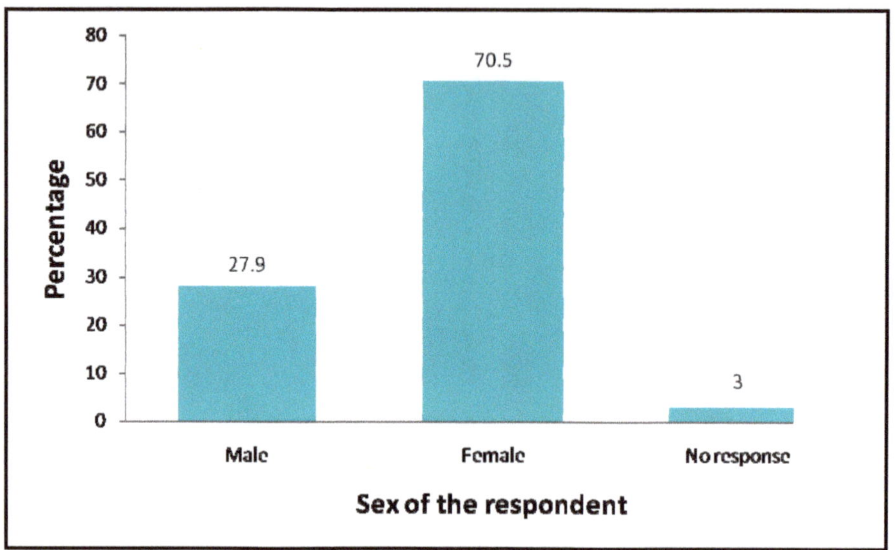

Figure 4. Gender percentage of respondents.

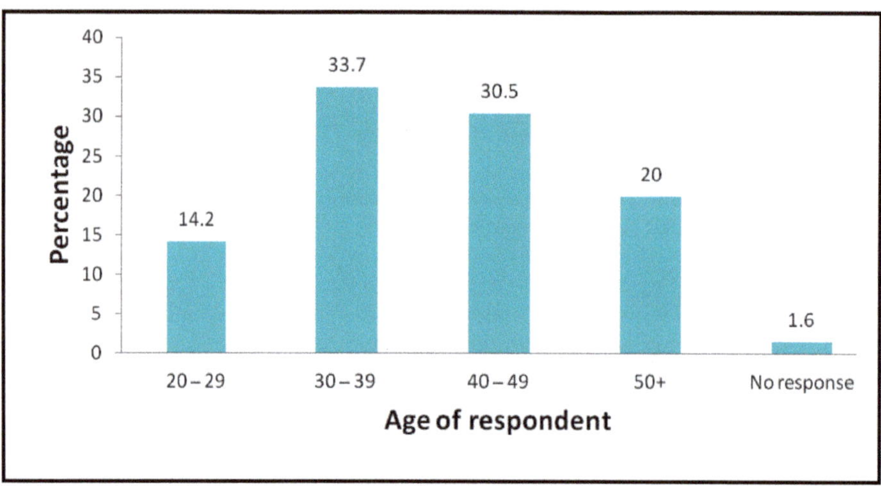

Figure 5. Age variation of respondents.

Figure 6 shows that the majority of cassava processors are married women who work very hard to earn a living in order to take care of their families.

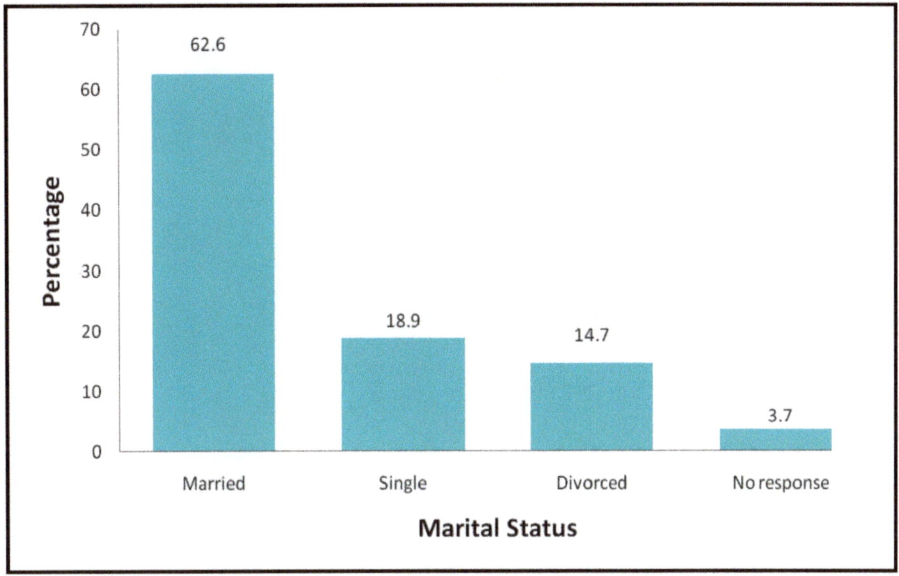

Figure 6. Marital status of respondents.

Figure 7 shows that 46.2% of respondents have at least a primary or secondary education. However, the results show that the cassava sector does provide a livelihood opportunity for people with no schooling, with 17.9% of respondents having never attended school.

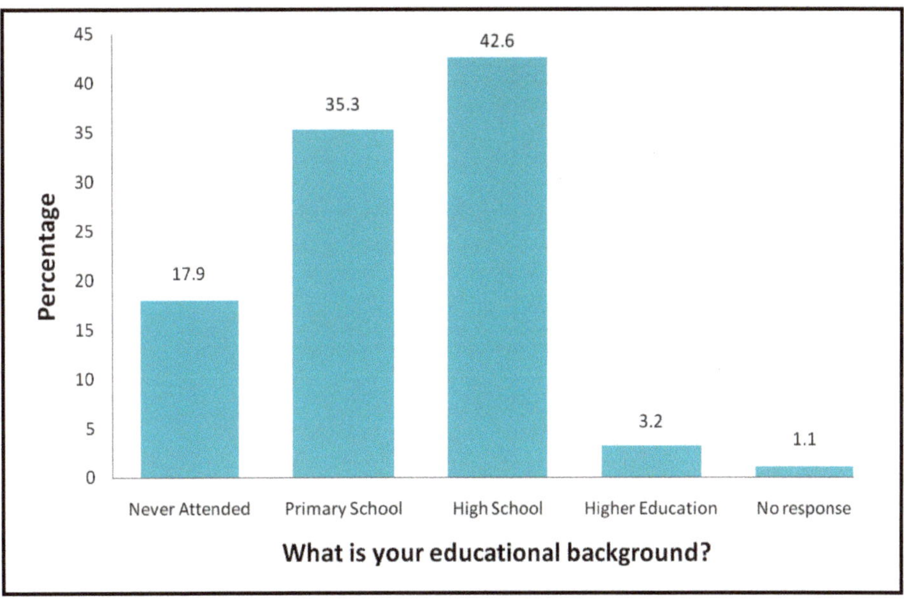

Figure 7. Education level of respondents.

Figure 8 shows that 55.3% of respondents have more than 10 years of experience and 34.2% have more than 5 years of experience in cassava processing. However, the results show that cassava processing does provide new livelihood opportunities for locals, with 9.5% of respondents having joined this sector in the last 5 years.

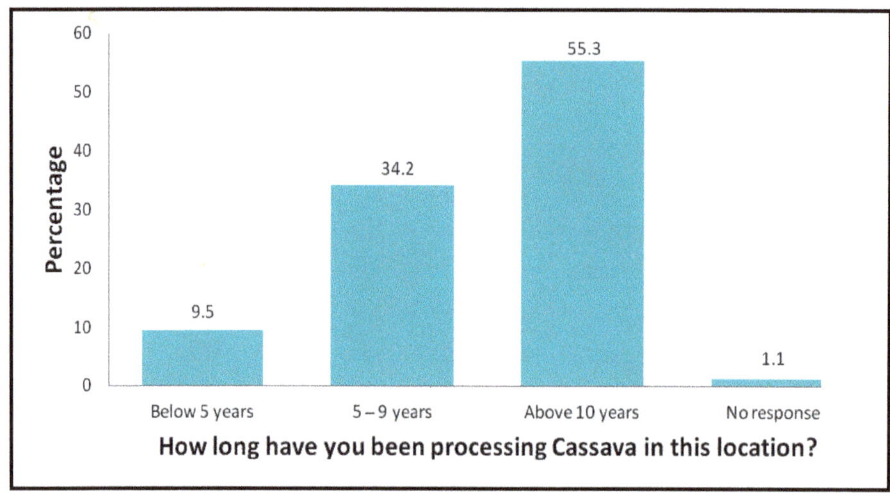

Figure 8. Number of years respondents have worked in cassava processing.

Table 1. Cassava-processing operations in the selected local government areas (LGAs).

	Major Type of Solid Waste Produced from the Cassava-Processing Operation	Percent	Valid Percent	Cumulative Percent
	Cassava peel and cassava pomace	86.8	86.8	86.8
	Cassava peel	10.5	10.5	97.3
Valid	Cassava pomace	1.6	1.6	98.9
	No response	1.1	1.1	100.0
	Total	100.0	100.0	
	How are the cassava solid wastes disposed?			
Valid	Cassava waste dump site	90.5	90.5	90.5
	Burnt near the factory	4.2	4.2	94.7
	No response	5.3	5.3	100.0
	Total	100.0	100.0	
	What do you use cassava residues for?			
Valid	Animal feeds	97.3	97.3	97.3
	Fertilizer	1.6	1.6	98.9
	No response	1.1	1.1	100.0
	Total	100.0	100.0	
	Willingness to pay for an improved waste management system			
Valid	Yes	74.7 {F = 87.3; M = 12.7}	74.7	74.7
	No	24.2 {F = 17.4; M = 82.6}	24.2	98.9
	No response	1.1	1.1	100.0
	Total	100.0	100.0	

Table 1 shows that over 86% of the respondents confirmed that cassava peels and cassava pomace are the major solid wastes generated, with 10.5% indicating that cassava peel is the major solid waste they produce. However, 1.6% indicated that cassava pomace is the major solid waste they produce. This result indicates that the majority of the cassava-processing units are involved in the production of 'garri' and 'fufu' [20]. Results obtained by Coker et al. [21] showed that the percentage and composition of solid waste (peels and bagasse) and liquid waste generated during cassava processing depends on the nature of the final product. The study conducted by Niringiye and Omortor [22] was on factors influencing willingness to pay (WTP) for waste management. They found that the age of the respondents has a negative and significant effect on WTP for waste management in Kampala city in Uganda. Coker et al. [21] conducted a study that focused on evaluating the cassava production activities in six selected cassava-processing sites in Ibadan city. Results showed that the percentage and composition of solid waste (peels and bagasse) and liquid waste generated during cassava processing depends on the nature of the final product. Irene and Richard [23], in their study, focused on the types of waste generated by cassava-processing plants. The survey showed that the wastes generated were cassava peels, fibrous material, chaff, wash water, and liquor. The study did not investigate the methods of waste management adopted.

Table 1 shows that the majority (90.5%) of the cassava processors dump the cassava waste generated at the cassava waste dump site. The table also indicates that 4.2% of the respondents burn the cassava solid waste generated near the factory, which is a major source of environmental pollution. Table 1 indicates that 97.3% of the respondents that use the cassava residue use it for animal feed or give to those that use them as animal feed. However, 1.6% of respondents that use the cassava residue use it as fertilizer. Table 1 shows that the majority (74.7%) of the respondents are willing to pay for an improved waste management system. The result further shows that 87% of the respondents who showed interest in paying for an improved waste management system are female. This is in line with the result of Omilani et al. [8]. Their result showed that the majority (68.73%) of the respondents who are female cassava processors were willing to pay for a value-added waste management system. The study compared the level of environmental pollution between small-scale cassava-processing firms and large-scale cassava firms. Investigations conducted confirmed that small-scale cassava processing affects the environment more than large-scale processing. Awunyo-Victor et al. [24] further revealed that the significant factors determining households' willingness to pay for improved solid waste management (collection and disposal) are the posted cost of the service, age, educational level, household size, and household monthly expenditure. In addition, cost sharing of waste management is affected by family income. From the study of Oyegbami et al. [25] on the awareness of occupational and environmental hazards associated with cassava processing in south-western Nigeria, it was established that cassava processors were aware of occupational and environmental hazards associated with cassava processing.

Biogas Production Potential

Table 2 shows the average cassava production capacity for factories in each of the LGAs. From 1000 kg of cassava, 300 kg of peel could be produced. Therefore, the weight of cassava peel in each LGA would be the equivalent of the (weight of cassava produced in kg × 300/1000). According to Wantanee and Rodtong [26], 1000 kg of dry cassava tubers could produce 497.01 L of biogas and 1000 kg of fresh cassava tubers could produce 235.1 L of biogas. Therefore, the weight in kg of cassava peel will produce (weight in kg × 497.01/1000) = volume in L of biogas.

Table 2. Biogas production potential from cassava peel.

Local Government (5 Factories)	Yewa North	Odeda	Ijebu North East	Ijebu East	Remo North
Cassava production per day (kg)	7500	10,000	12,000	8750	7500
Quantity of cassava peel per day (kg)	2250	3000	3600	2625	2250
Quantity of biogas (L/day)	(2250 kg × 497.01) = 1,118.273	(3000 kg × 497.01) = 1,491.030	(3600 kg × 497.01) = 1,789.236	(2625 kg × 497.01) = 1,304.651	(2250 kg × 497.01) = 1,118.273

4. Conclusions

The study revealed that there is no proper waste disposal method adopted by cassava processors in the selected LGAs in Ogun State. This is because the wastes generated are not properly disposed of. There was no significant difference in the waste disposal methods as adopted by the LGAs. This is because the majorities of the respondents either use the residues from the cassava processing for feeding their livestock or use them as fertilizer. It is therefore evident that the respondents are not exposed to the income-generating potentials of the cassava wastes. This was based on the responses on the questionnaire served and the focus group discussion. Most of them are local people with limited education.

The study concludes that the major waste is from 'garri' and 'fufu'-processing production. Though the cassava processors have an awareness of the health hazards associated with improper waste management, they prefer to dump the wastes at dump hills or burn them near the factories. The study concludes that majority of the cassava processors are female. Therefore, gender is a major determining factor as it influences the processors' willingness to pay. Also, the majority of the cassava processors do not generate income from the waste generated from the cassava residues.

Finally, the study concludes that based on the quantity of cassava peels generated which are dumped and burnt by the cassava processors in the selected LGAs, there is a huge potential for biogas production from these wastes generated. This biogas can be used as a substitute for the firewood which is currently in use by all the cassava processors.

5. Recommendations

From the findings, the study therefore recommends the following:

i.   That proper waste disposal methods should be adopted by the cassava processors in Ogun State to minimize pollution and reduce health risks.

ii.   There is a need for awareness of the income generation potential of cassava waste among the cassava processors in Ogun State.

iii.   It is highly recommended that the waste water generated from cassava processing undergoes proper treatment before it is discharged.

iv.   Based on the biogas generation potential of the cassava-processing factories in Ogun State, it is important that this resource be harnessed.

Author Contributions: Conceptualization of the research work came from D.O.O. and T.O.O.; and the methodology adopted was carried out by the authors: D.O.O. and T.O.O. All the authors played a contributing role in the use of the software, and the validation of the work was carefully checked by D.O.O. The formal analysis and investigation was done by D.O.O. Resources and data curation was carried out by D.O.O., and T.O.O. While the writing of original draft preparation was done by D.O.O. Writing-review editing was done by D.O.O. Visualization, supervision and project administration was done by D.O.O.

Funding: This research received no external funding.

Acknowledgments: The authors are grateful to the management of Covenant University for providing an enabling environment in doing the report of this research work. We also appreciate the members of the communities in the selected local government areas for making available information that led to the success of this work and giving access to their communities.

Conflicts of Interest: The authors declare no conflict of interest.

References

1. Asante-Pok, A. *Analysis of Incentives and Disincentives for Cassava in Nigeria*; FAO Technical Notes Series; FAO: Rome, Italy, 2013.

2. Omilani, O.; Abass, A.; Okoruwa, V.O. Willingness to Pay for Value-Added Solid Waste Management System among Cassava Processors in Nigeria. 2015. Available online: www.tropentag.de/2015/proceedings/node515.html (accessed on 22 May 2018).

3. Westby, A. Cassava Utilization, Storage and S mall-scale Processing. In *Cassava Biology, Production and Utilization*; Hillock, R., Thresh, J., Bellotti, A.C., Eds.; CABI Publishing: Wallingford, UK, 2008.

4. Olukanni, D.O.; Agunwamba, J.C.; Abalogu, R.U. Interaction between suspended and settled solid particles in cassava waste water. *Sci. Res. Essay* **2013**, *8*, 414–424.

5. Pandey, A.; Soccol, C.R.; Nigam, P.; Soccol, V.T.; Vandenberghe, L.P.; Mohan, R. Biotechnological potential of agro-industrial residues. II: Cassava bagasse. *Bioresour. Technol.* **2000**, *74*, 81–87. [CrossRef]

6. Barros, F.; Ana, P.D.; Júnio, C.; Gláucia, M. Potential Uses of Cassava Wastewater in Biotechnological Processes. In *Agriculture Issues and Policies: Cassava Farming, Uses and Economic Impact*; Pace, C.M., Ed.; NOVA Science Publishers Inc.: New York, NY, USA, 2012; Chapter 2; pp. 33–54. ISBN 978-1-61209655-1.

7. Osunbitan, J.A. Short term effects of Cassava processing waste water on some chemical properties of loamy sand soil in Nigeria. *J. Soil Sci. Environ. Manag.* **2012**, *3*, 164–171.

8. FAO (Food and Agricultural Organization) of United Nations. *Annual Statistics*; FAO: Rome, Italy, 2004.

9. Eze, J.I. Converting Cassava (Manihot spp) Waste from Gari Processing Industry to Energy and Bio-Fertilizer. *Glob. J. Res. Eng.* **2010**, *10*, 113–117.

10. Tonukari, N.J.; Ezedom, T.; Enuma, C.C.; Sakpa, S.O.; Avwioroko, O.J.; Eraga, L.; Odiyoma, E. White Gold: Cassava as an Industrial Base. *Am. J. Plant Sci.* **2015**, *6*, 972–979. [CrossRef]

11. Dada, A.D.; Afolabi, O.O.; Siyanbola, W.O. Harnessing Science and Technology for Cassava Productivity and Food Security in Nigeria. In Proceedings of the Portland International Conference on Management of Engineering & Technology, Portland, OR, USA, 2–6 August 2009.

12. Aro, S.O.; Aletor, V.A.; Tewe, O.O.; Agbede, J.O. Nutritional potentials of cassava tuber wastes: A case study of a cassava starch processing factory in south-western Nigeria. *Livest. Res. Rural Dev.* **2010**, *22*, 1–11.

13. Sackey, I.S.; Bani, R.J. Survey of waste management practices in cassava processing to gari in selected districts of Ghana. *J. Food Agric. Environ.* **2007**, *5*, 325–328.

14. Oparaku, N.F.; Ofomatah, A.C.; Okoroigwe, E.C. Biodigestion of cassava peels blended with pig dung for methane generation. *Afr. J. Biotechnol.* **2013**, *12*, 5956–5961. [CrossRef]

15. Olukanni, D.O.; Aremu, O.D. Provisional Evaluation of Composting as Priority Option for Sustainable Waste Management in South-West Nigeria. *Pollution* **2017**, *3*, 417–428.

16. Oresanya, O.O.; Olukanni, D.O. The evolution of solid waste management in Lagos State, Nigeria. A glance at the world. *Waste Manag.* **2017**, *66*, 1–3.

17. Echebiri, R.N.; Edaba, M.E. Production and Utilization of Cassava in Nigeria: Prospects for Food Security and Infant Nutrition. *Prod. Agric. Technol.* **2008**, *4*, 38–52.

18. Onu, B.; Surendran, S.S.; Price, T. Impact of Inadequate Urban Planning on Municipal Solid Waste Management in the Niger Delta Region of Nigeria. *J. Sustain. Dev.* **2014**, *7*, 27–45. [CrossRef]

19. Popescu, A.M.; Alecxandrina, D.; Popescu, T. The impact of gender difference at Romanian small and medium enterprises (SME) management level, analyzed by organizational citizenship behavior. *Eur. J. Hum. Soc. Sci.* **2013**, *27*, 1390–1406. [CrossRef]

20. Adebayo, K.; Anyanwu, A.C.; Osiyale, A.O. Perception of Environmental Issues by Cassava Processors in Ogun State, Nigeria—Implications for Environmental Extension Education. *J. Ext. Syst.* **2003**, *19*, 103–112.

21. Coker, A.; Achi, C.; Sridhar, M.K. Utilization of Cassava Processing Waste As A Viable And Sustainable Strategy For Meeting Cassava Processing Energy Needs: Case Study From Ibadan City, Nigeria. In Proceedings of the Thirtieth International Conference on Solid Waste Technology and Management, Philadelphia, PA, USA, 3–6 April 2010.

22. Niringiye, A.; Omortor G., D. Determinants of Willingness to Pay for Solid Waste Management in Kampala City. *Curr. Res. J. Econ. Theory* **2010**, *2*, 119–122.

23. Olukanni, D.O.; Adeleke, J.O.; Aremu, D.D. A Review of Local Factors affecting Solid Waste Collection in Nigeria. *Pollution* **2016**, *2*, 339–356.

24. Awunyo-Vitor, D.; Ishak, S.; Jasaw, G.S. Urban Households' Willingness to Pay for Improved Solid Waste Disposal Services in Kumasi Metropolis, Ghana. *Urban Stud. Res.* **2013**, *2013*, 2–8. [CrossRef]

25. Oyegbami, A.; Oboh, G.; Omueti, O. Cassava processors' awareness of occupational and environmental hazards associated with cassava processing in South-Western Nigeria. *Afr. J. Food Agric. Dev.* **2010**, *10*, 2177–2186.
26. Wantanee, A.; Rodtong, S. Laboratory Scale Experiments for Biogas Production from Cassava Tubers. Presented at the Joint International Conference on "Sustainable Energy and Environment", Hua Hin, Thailand, 1–3 December 2004.

Article

A Voluntary Delivery Point in Reverse Supply Chain for Waste Cooking Oil: An Action Plan for Participation of a Public-School in the State of Rio De Janeiro, Brazil

Luana dos Santos Ferreira [1], **Aldara da Silva César** [2,*], **Marco Antonio Conejero** [1] **and Ricardo César da Silva Guabiroba**

[1] Department of Business and Public Management, Fluminense Federal University, Volta Redonda RJ 27213-145, Brazil; lusferreira@globo.com (L.d.S.F.); marcoac@id.uff.br (M.A.C.); ricardocesar@id.uff.br (R.C.d.S.G.)

[2] Grupo de Análise e Sistemas Agroindustriais Agribusiness, Engineering Department, GASA, Fluminense Federal University, Volta Redonda RJ 27.255-125, Brazil

* Correspondence: aldaracesar@id.uff.br

Received: 13 August 2018; Accepted: 5 October 2018; Published: 10 October 2018

Abstract: Improper disposal of waste cooking oil into sewer systems is harmful to the environment. Through the selective collection, this highly polluting residue can be handled in a less harmful way. The present study presents an action plan for a public school in the Region of Médio Paraíba Fluminense of the state of Rio de Janeiro, Brazil to serve as a voluntary delivery point in a reverse logistics chain for waste cooking oil. A case study method with semi-structured interviews was carried out with agents who are part of the chain, including the government, commercial residue generators, collectors, the biodiesel production industry, the community, and teachers and students of the public school. Even though the reverse supply chain for waste cooking oil in the region lacks structure, this study showed that the actors were interested in correctly disposing of waste cooking oil through partnerships and agreements. In addition to the environmental benefits, environmental education actions in public schools, such as the one in this study, can help raise student awareness of issues relative to citizenship and social responsibility and promote employment and income opportunities for recyclable material collector cooperatives and industries that use waste cooking oil as raw material.

Keywords: reverse logistics; used cooking oil; selective collection; school; environmental education

---

1. Introduction

The circular economy concept is a new business model focused on the management of discarded products and materials that hold promise for reducing their volume, contributing to the economy and the environment. This approach is increasingly seen as a solution to a number of challenges, such as waste generation, food waste, resource scarcity, and the sustainability of economic benefits [1].

Stahel [2] suggested that a circular economy would minimize waste by reducing the amount of waste; reusing what can be reused; recycling what cannot be reused; recovering materials or energy from what cannot be reduced, reused, or recycled. This process is becoming increasingly important as the amount of waste is growing even faster than the rate of urbanization [3]. Rathi [4] argued that rapid population growth and industrialization degrade the urban environment and place a heavy burden on natural resources, which undermines sustainable and equitable development. Inefficient management and disposal of solid waste is an obvious cause of environmental degradation in most developing countries.

One type of urban solid waste is waste cooking oil (WCO), which can cause environmental damage if discarded improperly [5]. It has been estimated that one liter of cooking oil pollutes 20,000 L of water, the same amount the average person consumes in about 14 years [6]. Waste cooking oil can obstruct sewage pipelines, while also retaining other solid waste. Sewage blockages increase the pressure in pipelines and can lead to sewage leakage into the soil. Furthermore, the collection of WCO can help reduce damage to wastewater treatment plants.

In Brazil, the National Solid Waste Policy (PNRS), established by Federal Law n. 12.305, on 2nd August 2010, sets forth rules for collecting, disposing of, and managing post-consumption products [7]. The PNRS is considered to be the legal framework that establishes the framework for the reduction, reuse, and recycling of food waste. It promotes a sense of sustainability management and, above all, assigns responsibilities appropriately between the public sector and waste generators about the waste generated [7]. Waste cooking oil is generated daily by households, restaurants, and the food service sector in general; however, this policy does not include WCO. The legislation that addresses the issue, Bill n. 2074 of 19th September 2007, is currently shelved [8]. It proposes that gas stations, supermarkets, companies, and other similar establishments that sell or distribute cooking oil maintain facilities designated for collection of WCO.

In Brazil, an estimated 6.5 million L of oil are collected for recycling, which is only 2.5% of what is produced [6]. This low collection rate is explained, in part, by the scarcity of collection points [9], which raises the costs of collection [10] and by lack of awareness-raising campaigns aimed at the population [11].

When recycled, WCO can be used as a raw material for several production chains, such as biodiesel, soap, animal feed, paint, asphalt, and putty [9]. According to César [12], Brazil has great potential to produce WCO biodiesel on a commercial scale.

Post-consumption reverse logistics are the reverse flow of a fraction of the by-products and materials that arise from the disposal of products after their original intended use, in order to return them to the production cycle in some way [13]. When WCO is returned to the production cycle, its life cycle does not end after the cooking process [14]. According to Zucatto [15], despite the several potential uses of WCO, no official statistics on the return rates are available.

The reverse logistics of WCO in the Region of the Médio Paraíba Fluminense (RMPF—*Região do Médio Paraíba Fluminense*) of the State of Rio de Janeiro, Brazil, has three distinct levels: (i) generation, (ii) intermediaries, which are the packaging, storage, collection and transportation locations or companies; (iii) destinations [16].

Waste cooking oil can present zero cost at its source since it is obtained through donations at collection points such as schools, which are voluntary delivery points (VDPs) in some communities in Brazil [17]. Environmental education is also essential in order for people to adopt a sustainable waste management model. The National Policy for Environmental Education, instituted by Law 9.795, of 27 April 1999, requires that environmental education is integrated into all levels and modalities of the education process [18].

The present study was justified by the need to address the relevant issue of developing appropriate WCO collection programs and correctly managing this highly polluting residue, while also including schools in the proposal. The following research question was formulated: How can a public school in the RMPF of the state of Rio de Janeiro, Brazil, act as a voluntary delivery point and support the development of a reverse supply chain for waste cooking oil?

The present article proposes an action plan for a public school to serve as a VDP in a WCO reverse supply chain. The authors provide a description of the reverse supply chain for WCO in the RMPF. A case study was carried out, using semi-structured interviews with agents who participate in the chain, including the government, waste generators, collectors, the biodiesel production industry, the community, and teachers and students of the school.

This article is divided into five sections besides the introduction. The second section presents a theoretical framework an overview of WCO reverse logistics and environmental education in schools.

The materials and methods are described in the third section. The fourth section presents the analysis of the results, followed by the fifth section, with final considerations.

2. Theoretical Framework

2.1. WCO Reverse Logistics

Green supply chain management is the integration of environmentally sound choices into the entire supply chain to improve the environmental and economic performance of individual links and of the chain as a whole; reducing the environmental impact [19]. Reverse logistics is the process of planning, implementing, and controlling the efficient and low-cost flow of raw materials, work in progress, finished products, and related information, from the point of consumption to the point of origin, with the goal of recovering value or achieving appropriate disposal [20].

Reverse logistics help reduce the loss of materials and products that would not normally be utilized. The process begins after the product is consumed, when companies must be prepared for the 4Rs: recovery, reconciliation, repair, and recycling [21]. Post-consumption products return to the production cycle through reuse, dismantling, and recycling [13]. Recycling, the last stage of the process, is the mechanism for WCO return and the focus of the present study.

Waste can be classified according to its purpose, as reverse solid waste or residue, and according to its origin, as urban, industrial, health services, rural, special, or differentiated. Waste can be disposed of (residue) in landfills or reused (reversed) through a series of physical or chemical treatments to manufacture new products [22]. Waste management is conducted through non-generation, reduction, reuse, recycling, treatment, and environmentally adequate final disposal and becomes the shared responsibility of the public and private sectors [23].

In Brazil, local governments are responsible for the adequate disposal of waste, while generators also have their own responsibilities [24]. Some Brazilian municipalities have already adopted the Municipal Plan of Integrated Solid Waste Management under the terms established by the National Solid Waste Policy [25].

Policies for the implementation of the selective collection can be carried out by cities and can be implemented in partnership with public or private recycling companies [22]. When implementing selective collection, the participation of cooperatives and other associations of collectors of recyclable and reusable material must be given priority due to its social impact [13].

However, the PNRS does not provide for the reverse logistics of WCO and the work depends on the capacity of the actors to coordinate among themselves. Intermediary agents such as cooperatives and self-employed collectors play an essential role in this chain [26,27].

Among the sustainable economic, social and environmental benefits of reusing WCO, Wildner and Hillig [28] highlighted the following:

1. Ensures income in lower-income areas, representing a permanent source of employment, and remuneration for the unskilled labor force.
2. Injects resources into local economies by creating jobs, collecting taxes, and developing the market.
3. Favors the development of environmental awareness and promotes responsible environmental behavior of companies and citizens.
4. Encourages the recycling of other materials.
5. Reduces the volume of waste generated and helps to solve the problem of the treatment of waste from consumption.

Returning WCO as a raw material involves several inter-related phases: packaging, collection, storage, and transportation to manufacturing locations [9]. Oil must be placed in 500 mL to 2 L containers in households, and in 20 L to 50 L containers in commercial establishments [9]. Households are small-quantity generators, while restaurants, hotels, snack bars, and other commercial establishments are large-quantity generators [29]. For both small and large generators, storage is necessary to reach the minimum economically feasible amount for the collection process [30].

The collection is carried out at large generators or voluntary delivery points, which receive WCO primarily from small-quantity generators [11]. The cost of having delivery points that are few and far apart is usually very high, sometimes making the return of WCO to the production cycle economically unfeasible [31]. Furthermore, route planning is essential to minimize the cost of transportation during collection [32].

Storage depends on the strategy of the collecting company. Some companies send the WCO directly to the recycling industry, while others store the oil until the right amounts are reached. In this case, it may undergo filtering to remove impurities from the foods to which the oil was exposed [9].

Using reverse logistics is sustainable when the total costs of packaging and transportation to manufacturing locations is lower than the value of the returned material, resulting in competitive advantages for recycling companies [9].

When used as material in new production chains, WCO can generate several products including soap and other biomaterials: printing ink, candles, crayons and playdough [33–36]. It can also be used to make biodiesel [37], with a WCO-to-biodiesel conversion rate of 98% via transesterification [38].

WCO-based biodiesel presents advantages in relation to other fuels. In comparison with conventional diesel, it does not emit sulfur compounds during combustion and is quickly biodegradable in soil and water. With regard to biodiesel made from other oils, it is profitable in terms of energic balance [39,40].

2.2. Environmental Education in Schools

In this context, environmental education has represented an essential awareness-raising strategy among citizens where there are no selective collection programs. Furthermore, the school environment is a formal pathway for environmental education [41].

The national policy for environmental education, through Law n. 9795/1999, Art 1, defines environmental education as the process of constructing social values, knowledge, skills, attitudes, and competencies in citizens to preserve the environment, quality of life, and sustainability [42]. As institutions that aim to shape the values and attitudes of citizens, schools can address the environmental dimension and understand its complexity and inserted cross-sectionally [41,43,44].

The policy further calls for the development and presentation of environmental education within the scope of the curricula of private and public schools. It should permeate all school relationships and activities and be developed across school subjects to reflect current issues and to encourage people to think about what type of world they want to live in. Finally, this type of education enables them to effectively carry out sustainable actions [42,45].

Among the sustainable practices that environmental education can address are an environmentally correct collection and destinations for WCO [14,46]. Whether destined for soap [28,33,47] or biodiesel production [11,48,49], WCO collection has been identified as an environmental education tool.

According to Segatto [50], environmental education has become essential in raising citizen awareness of ways to dispose of WCO and schools are promoters of this knowledge. The participation of schools in WCO collection is an instrument for fostering environmental education and sustainable actions.

Environmental education activities in and out of schools can include neighborhood campaigns, lectures on the subject, gymnastics, outdoor activities, work in student groups in schools, insertion of guidelines for sustainable practices in classes, and training of teachers on the subject [51,52]. According to Jacobi [41], the focus of these activities is seeking a holistic perspective on the relationships between people and the development of citizenship, nature, and the universe, considering that natural resources are being exhausted and that the main responsibility for their degradation lies with humans.

3. Materials and Methods

This was a qualitative, applied, and descriptive study that aimed to describe the WCO reverse supply chain in the chosen region and propose an action plan that includes public schools as a voluntary delivery point. This research used the case study strategy, considering the unit of analysis of the participation of the public school as a VDP and the subunit, the reverse supply chain for WCO in the RMPF.

The case study method permits the investigation of a contemporary phenomenon in a real environment when it is not possible to clearly differentiate the phenomenon and the context; it demands the use of different sources of evidence [53].

The Region of Médio Paraíba Fluminense in the state of Rio de Janeiro, Brazil, has 871,775 inhabitants divided into 12 municipalities: Barra do Piraí, Barra Mansa, Itatiaia, Pinheiral, Piraí, Porto Real, Quatis, Resende, Rio Claro, Rio das Flores, Valença, and Volta Redonda [54] (Figure 1).

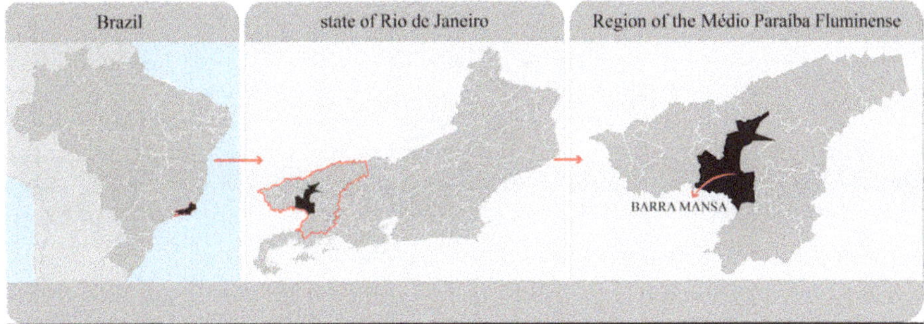

Figure 1. Maps of Brazil, the state of Rio de Janeiro, and the Region of Médio Paraíba Fluminense with emphasis on the municipality of Barra Mansa. Source: Elaborated by the authors from IBGE—Instituto Brasileiro de Geografia e Estatística (or Brazilian Institute of Geography and Statistics) [54].

The public school in question was the Washington Luiz Municipal School (CMWL—Colégio Municipal Washington Luiz), located in the municipality of Barra Mansa [55]. In March 2017, the authors conducted a document analysis by consulting the school's political-pedagogical project for 2015/2016. Internal records such as student enrollment forms and parent-teacher association (PTA) meeting protocols were also consulted to obtain information about the profile of the target audience.

The description of the reverse supply chain and development of the proposal for the school to act as a VDP in the RMPF involved the collection of primary data through on-site visits, completion of questionnaires, and interviews with the social actors involved with the issue in order to portray the current situation and propose structural and operational improvements in the operation of the system.

First, data were collected by administering a questionnaire (Appendix A) to 60 parents and guardians during PTA meetings, with the goal of assessing the community's perceptions of the topic. Under the researcher's guidance, another questionnaire (Appendix B) was administered to 50 students in the first and second stages of elementary education, between 10 and 16 years old, chosen by convenience in the classrooms. Additional data were gathered through semi-structured interviews (Appendix C) that were conducted with 8 school managers, teachers, janitorial staff, and students (Table 1).

Table 1. Education agents interviewed in the Washington Luiz Municipal School. Source: Created by the authors.

Educating Agent	Position	Level	Experience (Years)	Date
EA1	School principal	All	14	May/2017
EA2	Director of studies Responsible for the school's political-pedagogical project	All	28	June/2017
EA3	Biology teacher	Youth and adult education	32	May/2017
EA4	Ethics and sociology teacher	Technical	19	May/2017
EA5	Math teacher	Elementary (year 6 to 9)	12	June/2017
EA6	Teacher (all subjects)	Elementary (year 1 to 5)	14	June/2017
EA7	School secretary/Portuguese teacher	Elementary/Technical	26	June/2017
EA8	Janitorial staff	All	9	June/2017

Another set of semi-structured interviews (Appendixs E and F) was carried out to obtain primary data from the 17 actors who are part of the WCO reverse supply chain in the RMPF. The public sector was represented by representatives from the municipal secretary for the environment and sustainable development and the public water and sewer treatment company. In the private sector, interviews and on-site visits were conducted with commercial generators, recycling cooperatives, associations, and companies and a representative of the local biodiesel industry. Table 2 presents the profile of all those interviewed and their corresponding organizations.

Table 2. Agents (A) interviewed from the WCO reverse supply chain in RMPF. Source: Created by the authors.

Agent	Representation	Position	Date	Organization
A1	Local government	Municipal secretary	March/2017	Municipal Secretariat for Environment and Sustainable Development (SMMADS)
A2	Local government	Solid waste manager	March/2017	SMMADS
A3	Local government	Coordinator of the 'Cuidando do Óleo' and 'Ecopneu' projects	March/2017	SMMADS
A4	Municipal autonomous body	The engineer responsible for the solid waste management	May/2017	Water and Sewer Treatment Service (SAAE)
A5	Local government	Biologist responsible for the EE project	May/2017	Municipal Secretariat of Education (SME)/Environmental Park
A6	Commercial generator	Nutritionist	June/2017	Hospital restaurant
A7	Commercial generator	In charge of the snack sector	June/2017	Bakery, snack bar and restaurant
A8	Commercial generator	Owner	June/2017	Snack bar
A9	Commercial generator	President	June/2017	Association of Hotels, Restaurants, Bars, and Others
A10	VDP	Manager	May/2017	Supermarket
A11	VDP	Advertising assistant	May/2017	Sesc
A12	Collection	Supervisor responsible for oil collection	May/2017	Cooperative of collectors
A13	WCO biodiesel plant	Engineer responsible for biodiesel production	April/2017	Biodiesel company
A14	State government	Superintendent	August/2017	PROVE
A15	Collection	President	March/2017	Association of collectors
A16	Collection	Owner	March/2017	Company 1
A17	Collection	Owner	July/2017	Company 2

Notes: EE = environmental education; Sesc = Brazilian Social Service of Commerce; PROVE = Program for the Reutilization of Vegetable Oil.

Thus, this study carried out a total of 25 interviews with agents and applied a total of 110 questionnaires. At the same time, in order to achieve a better grasp of the reverse logistics of WCO in the RMPF, the authors conducted a document analysis of materials from official organizations and websites.

Finally, data analysis was conducted by triangulating the data obtained from the literature, document analysis, questionnaires, and interviews [56].

4. Analysis of Results

4.1. The Reverse Supply Chain for WCO in the RMPF

Based on the document analysis and interviews, the authors were able to describe the generation, intermediary (collection and transportation), and destination phases of the WCO reverse supply chain in the RMPF, as shown in Figure 2.

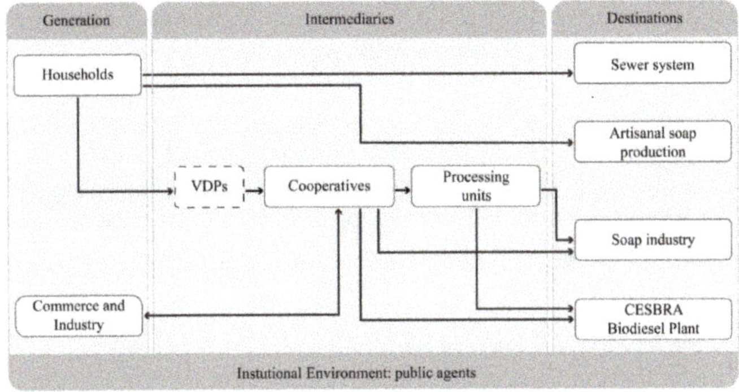

Figure 2. Waste Cooking Oil reverse supply chain in the Region of Médio Paraíba Fluminense.

4.1.1. Generation Phase

Residential generators can dispose of WCO directly in the sewer system, reutilize it in artisanal soap production, or deliver it to VDPs. Waste cooking oil delivered to VDPs is usually stored in plastic PET (Polyethylene Terephthalate) bottles. According to one of the representatives of a WCO collector (A15), households and commercial establishments should separate WCO from food material particles by filtering it before packaging, but this is not usually done.

For generators in commercial establishments and industry, WCO is collected at bars, restaurants, bakeries, snack bars, and institutions (schools, churches, local government) in barrels that are later collected by cooperatives. However, the president of the Association of Hotels, Restaurants, Bars, and Others (A9) pointed out that there is no policy or inspection regarding the correct disposal of WCO generated in the region by commercial generators. Even though the association provides guidance in their meetings about the treatment and final disposal of waste, the association itself is unaware of the correct disposal of WCO.

For example, participant A7, who delivers WCO to the collection company in exchange for cleaning supplies, said he had never bothered to learn about the WCO's final destination. The nutritionist (A6) also emphasized that the generators are not always informed of the destination for the WCO.

According to participant A8, the average amount of vegetable oil used per month varies greatly because people tend to eat more fried foods when the weather is cold. Even with the advent of recent innovations such as electric fryers, the most common technique is immersion frying in a pot or saucepan. This participant also reported that the volume of WCO waste was approximately two-thirds of the oil used for frying because the other third is absorbed into the food.

As mentioned by participant A6, the standards set by the Brazilian Health Regulatory Agency do not allow for the utilization of used oil. The reason is that if it has been used for a long period, the oil generates compounds responsible for unpleasant odors and tastes, including substances that can cause health hazards to consumers, such as gastrointestinal tract irritation and diarrhea.

The Rio de Janeiro State Plan showed that the average generation of urban solid waste in the RMPF was 0.81 kg/capita/day in 2014 [57]. There are no statistics about the amount of WCO generated in the RMPF. However, according to the experience of participants A6 and A9, the average consumption in the region is about 1.5 L/capita/month. Considering that two-thirds of this amount is disposed of and that there are 800,000 inhabitants in the region (excluding 10% of the population that does not consume fried foods), the total monthly average amount of WCO generated in the RMPF can be estimated at 800,000 L.

All of the participants in the generation phase expressed optimism regarding the growth and structuring of the WCO reverse supply chain and pointed to awareness-raising as one of the key points in this process. According to participants A2 and A3, awareness-raising actions in schools could help with WCO collection initiatives, since children and adolescents tend to influence their parents' attitudes and contribute to behavioral changes in the whole family.

4.1.2. Intermediary Phase

In the intermediary phase, represented by VDPs, companies, cooperatives and processing units, WCO is collected and transported directly from the generators or VDPs to recycling warehouses owned by companies or associations/cooperatives. Table 3 presents the profile of the organizations involved in WCO collection and the municipalities involved.

Table 3. WCO collection organizations in the Region of Médio Paraíba Fluminense. Source: Created based on data from IBGE [54] and field research.

Organization	Municipalities Where the Company Operates and an Estimated Population	Description
Óleo Local (company)	Barra do Piraí—96,261 Valença—73,154 Rio das Flores—8783 Volta Redonda—261.522 Barra Mansa—179,472 Piraí—27,311	• Founded in 2010 in Barra do Piraí • WCO collection of approximately 40,000 L/month • Supported by UNIMED (a health insurance company) as a VDP • In 2017, the company reached 60 collection points and 560 registered commercial establishments
Ecoleta (company)	Resende—123,385 Itatiaia—29,744 Quatis—13,283 Porto Real—17,663 Barra Mansa—179,472	• Founded in 2007 in Resende • WCO collection of approximately 50,000 L/month • Partnered with the Retailer's Steering Committee of Resende • Has 150 collection points and carries out awareness-raising actions in schools
Cicloóleo (company)	Volta Redonda—261,522 Pinheiral—23,488 Porto Real—17,663	• Private company located in Volta Redonda • WCO collection of approximately 45,000 L/month • All WCO destined for biodiesel production
Ecoóleo (association)	Volta Redonda—261,522	• Liquid and Solid Waste Collectors Association • Founded in 2007 in Volta Redonda • Supported by the local government • WCO collection of approximately 11,000 L/month
Coopcat (cooperative)	Barra Mansa—179,472	• Collector's cooperative of Barra Mansa • Founded in 2013 • Part of the local government's "Cuidando do Óleo" project, in partnership with the State Institute for the Environment (INEA) and the State Program for the Reutilization of Vegetable Oil (PROVE); • WCO collection of approximately 3100 L/month

The RMPF associations/cooperatives are supported by the Program for the Reutilization of Vegetable Oil (PROVE) of the State of Rio de Janeiro. According to participant A14, PROVE promotes partnerships between public agents, the third sector, companies, and representative groups. In addition to fostering the organization of collectors into cooperatives, it also provides vehicles for collection, supporting the reverse logistics of various types of waste.

Partnerships with commercial generators are trust-based. Participants A2 and A3 stated that WCO collection actions are still very informal and need to be formalized through agreements or contracts. The commercial generators registered by the cooperatives or private companies receive barrels to dispose of the WCO. When the barrels are full, the generators call the cooperatives and the full barrels are exchanged for empty ones. However, participant A7 pointed out that delays in the collection make the WCO start to release unpleasant odors in the establishments.

Participant A15 also emphasized the variability in the supply of commercial generators, which hinders collection logistics. According to him, the amount collected ranges from 100 L to 200 L per partner; in a month; this represents a 50% variation, ranging anywhere from 8000 to 12,000 L.

To encourage these partnerships, A15 suggested that commercial generators "receive cleaning materials in exchange. Chlorine is produced by cooperatives as an incentive for cooperation." In addition, enterprises/cooperatives provide green seals as a form of recognition. According to this participant, in the case of his cooperative, the seal is awarded by the Secretary of the Environment of the Volta Redonda municipality. In the case of another cooperative, according to participant A12, commercial generators are awarded a PROVE Seal for adequate WCO disposal that is valid for one year. In the case of companies, they provide their own green seals, as stated by participants A16 and A17.

However, the Secretary of the Environment and Sustainable Development of the Barra Mansa municipality does not support the exchange of WCO for cleaning products. According to A3, that is because the soap produced by the collector organizations generally does not meet technical specifications and can also harm human health and the environment.

It is also important to consider the financial incentives provided by some companies/cooperatives in exchange for WCO provided by residential and commercial generators. According to participant A1, in the near future, it will become difficult for cooperatives to obtain donations because WCO has become profitable waste. Companies and cooperatives pay between 0.80 and 1.00 Brazilian reais/L of WCO. The percentage of WCO lost and WCO sold after filtering is minimal. Sales are not just to biodiesel industries and their destination is uncertain. Nonetheless, WCO is sold for an average of 1.40 to 1.80 Brazilian reais/L. However, when soy harvests are good, and the price of vegetable oil is low, the value paid per L of WCO ranges between 1.20 and 1.30 Brazilian reais.

One of the challenges in the collection phase mentioned by the participants responsible for the collector companies and cooperatives was competition with self-employed informal collectors, who pay more per L to establish that the collection points are theirs. Another barrier to the residential collection is that even though people are aware of the possible damage caused by disposing of WCO down the sink, most do so because it is easier. Furthermore, the collector companies and cooperatives must also compete with homemade soap production.

The cooperatives and companies licensed to treat WCO generate monthly reports to government organizations that include information about the amount of WCO collected per registered commercial generator. These reports allow the local government to control the adequate disposal of this waste.

Storage and transportation require certain precautions to avoid WCO leaks; the barrels are placed on top of pallets covered with cardboard. Filtering of WCO can be performed at the collector cooperatives, or it can be sent to other processing units before being sent to biodiesel companies or the soap industry. In company or cooperative warehouses, the WCO is sieved to eliminate food particles and it decants for 24 h to separate the water from the oil. After this period, a sample is collected and tested; it should contain less than 1% of water before being approved for sale.

The intermediary links of the chain also foster awareness regarding WCO collection. Companies and cooperatives work with schools (municipal, state and private), providing students, staff, and teachers with information through lectures, posters, and pamphlets to encourage WCO collection and re-utilization in soap production and alternative sources of energy, such as biodiesel.

In some municipalities, schools already function as VDPs. According to participants A3 and A16, the "Cuidando do óleo" project, a partnership of the municipality of Barra Mansa with a cooperative, and the "Viva Óleo" project of a company in the city of Resende, held rallies at all public and private schools before they became VDPs. However, according to participants A12 and A16, these projects were canceled because of a lack of government support.

As stated by participant A17, working with households is a greater challenge. When residences are part of the collection route, they can be fit in according to market demand. He said, "It's no use saying that you're solving an environmental problem, generating 70% less polluting fuel if you're riding around in your car with no efficiency. We need to optimize and improve our relationship with generators."

One of the municipality representatives interviewed in this study (A1) emphasized the inefficiency of the residential collection strategy. According to this participant, "Housewives are not in the habit of requesting collection service by phone. They may even separate the WCO but they don't call us to collect it. So, it all gets thrown away together with the other residential trash." Participant A12 indicated that one solution to the problem of residential collection would be raising awareness through environmental education so that WCO generated in households gets delivered to VDPs.

4.1.3. Destination Phase

The most common destination of WCO is the sewer system, followed by artisanal soap production, and the soap and biodiesel industries.

According to participants A3 and A5, it is a common practice to reutilize WCO in artisanal soap production in the RMPF. Community members generate income by selling these products and churches offer soap-making courses for job and income generation. As voiced by participant A1, the greatest problem in the production of homemade soap is lack of environmental control and precautions in the manufacturing process. This participant said, "caustic soda, which is used as the reagent, can burn the person manufacturing the soap."

In Brazil, approximately 30 million L of WCO is used to produce biodiesel [58]. In the RMPF there are two biodiesel companies: CESBRA and OLFAR. However, up to the time of this study, OLFAR had not begun operations using WCO as a raw material in biodiesel production. Participant A13 indicated that the nominal production capacity of the CESBRA biodiesel plant, as authorized by the National Agency of Petroleum, Natural Gas and Biofuels, is 10,000 m^3 per bimester, for a total of 60,000 m^3 per year. Currently, the plant produces 4000 m^3 per bimester, 40% of the authorized capacity.

Biodiesel production uses degummed soybean and palm oils and several other raw materials, depending on the oilseed harvest. Regionally, WCO corresponds to 40% of the raw material used to produce biodiesel. According to participant A13, the low supply of WCO increased the value per L from 0.50 to 2.50 Brazilian reais in 2016. The amount paid for WCO is associated with the soybean harvest: "The price fluctuates according to the price of the main raw material but it is a competitive value. The last batch cost approximately 1.50 Brazilian reais/L."

Participant A13 also noted that the biodiesel industry faces some competition from the soap industry but mainly competes with some rendering and tallow companies to obtain WCO. Soap companies such as Grande Rio, Mauá, and 3Brio are not significant competitors. The JBS industry, located in Lins, São Paulo, produces approximately 20,000 m^3 of biodiesel per bimester and for participant A13, it can be considered their biggest competitor for purchasing WCO.

There are 24 specifications for biodiesel to be sold at auctions. According to participant A13, the main quality indexes are solid impurities, acidity, and amount of water in the WCO. When it is treated, WCO is heated, breaking the emulsion. The oil sinks to the bottom of the container, from which oil, flour, and all other impurities are drained. For biodiesel production, the maximum acidity is 3%. If it does not meet this target, this residue is mixed with caustic soda, a catalyst, and used to produce soap instead of biodiesel. Regarding the water content, the maximum acceptable level for biodiesel production is 0.02%. As the WCO arrives with over 1% of water, it must undergo a 12 h drying process at 150 °C in a vacuum chamber.

Regarding the yield, participant A13 estimated that 1 kg of WCO produces 990 g of biodiesel. This means that 1% of the material is lost, which is better than the 4% loss with soybeans. The biodiesel industry plans to expand its biodiesel production with WCO, based on expected growth and increasing professionalization of the WCO market.

4.1.4. Institutional Environment: Public Agents in the RMPF

The public agents interviewed in the Barra Mansa municipality represented the Municipal Secretary for the Environment and Sustainable Development (SMMADS) and the Public Water and Sewer Treatment Service (SAAE).

Participant A2 stated that the SMMADS intends to resume their environmental education activities so that municipal schools can function as VDPs; however, he indicated that no such activities have been planned yet. Participant A1 emphasized the need to resume awareness-raising and environmental education programs: "The local government does not depend on its own resources for collection programs because the Rio de Janeiro State Secretariat of the Environment provides this support. The program lacks continuity and greater adherence by the population."

It is worth noting that there was a high turnover in the persons responsible for the municipal secretariats in the RMPF during the data collection period. The same was true for the management involved in the WCO collection programs.

According to participant A4, who is with the SAAE, accumulation of oil and grease in the pipe system can cause blockages and sewer backups, and can even break the pipes that are part of the wastewater collection system. Toxic chemicals are used to remove this material and unclog pipes, creating a vicious cycle, which can even lead to significant increases in the costs of water and sewer services. However, the SAAE has no statistics available regarding the costs involved and the reasons for sewer blockages.

Although public agents are mobilizing actors to participate in WCO selective collection programs in RMPF municipalities, the application of economic instruments such as fiscal and tax incentives and the creation of more specific laws could accelerate the process and increase the volume of WCO collected to produce biodiesel.

4.2. The Washington Luiz Public School as a Voluntary Delivery Point for WCO in the RMPF

4.2.1. Educating Agents

According to participants AE1 and AE2, environmental education programs are usually proposed through partnerships between municipal government organizations in Barra Mansa that involve the Secretary of Education, the Secretary of the Environment and the SAAE. Environmental education initiatives in the school curriculum always depend on the support of external agents, i.e., partners from private institutions.

All of the participants expressed their support for an environmental education project involving WCO and the participation of schools as VDPs. For them, a permanent WCO collection program at Washington Luiz Public School (CMWL), supported by the Secretary of the Environment and other public organizations, would be of great importance in promoting environmental education and student awareness of sustainable development.

The school management expressed interest in including a recycling project and other environmental initiatives in the political-pedagogical project. However, the teachers were more skeptical, based on the discontinuity that occurred with previous programs.

According to the director of studies of the CMWL (AE2), there have previously been projects in which WCO was collected and delivered to a collector company that was a partner of the Municipal Secretary of Education. At the end of the year, the participating schools were ranked, and the schools with the highest rankings won prizes.

However, all agreed that, in addition to such actions, it is necessary to make the insertion of environmental education in the school curriculum and political-pedagogical project official. According to those interviewed, addressing the topic in specific subjects or extracurricular activities is not enough to awaken student interest and ensure their actual participation in actions directed at social-environmental issues.

4.2.2. Students

The students' views regarding environmental education were associated with preservation of the environment. All the students had had some experience with environmental education. Chart 1 presents the motivational factors regarding student participation in WCO collection programs. Thirty-six percent of the students justified their lack of interest in WCO collection activities by saying that they had not received the prizes that had been promised to them. They also said that some teachers do not release them from class to participate in these types of activities.

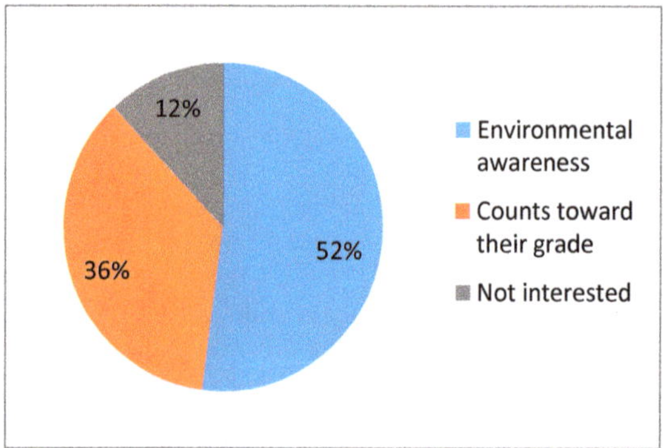

Chart 1. Motivational factors for student participation in WCO collection programs. Source: Created by the authors based on field research.

Chart 2 presents the destinations of WCO in the students' households. The most common destinations were soap production and separation for selective door-to-door collection.

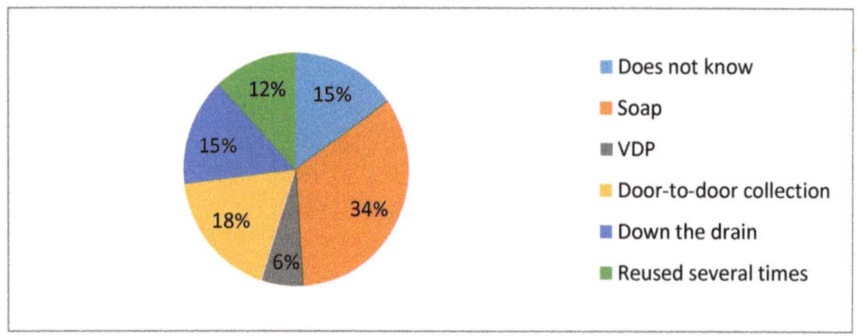

Chart 2. Destinations of WCO in students' households. Source: Created by the authors based on field research.

4.2.3. The Community

The questionnaires given to community members revealed that 23% of the participants had not finished elementary school, 34% had completed elementary school, 40% had finished high school, and another 3% had a university degree. On average, four people lived in the households of those interviewed. The monthly household income was between one and two monthly minimum wages in two-thirds of the families. Only three families received social welfare from the Bolsa Família program. Half of the participants were not aware of the meaning of environmental education. It is worth noting that in Brazil the designations of the social classes are based on the number of minimum wages (MW). In 2018, the MW was 952 Brazilian reais, and, five classes are defined: A (more than 20 MW); B (from 10 to 20 MW); C (from 4 to 10 MW); D (from 2 to 4 MW); and E (up to 2 MW).

The present study found that the mean consumption of vegetable oil was 1.4 L/month/per capita. This is below the national average of 1.6 L in 2015 [2]. All the participants reused the oil at least once before disposing of it. Regarding the destinations of the WCO, as seen in Chart 3, the main options were a donation, soap production, and pouring it down the drain. Those who poured the oil down the drain were aware of the potential for environmental damage.

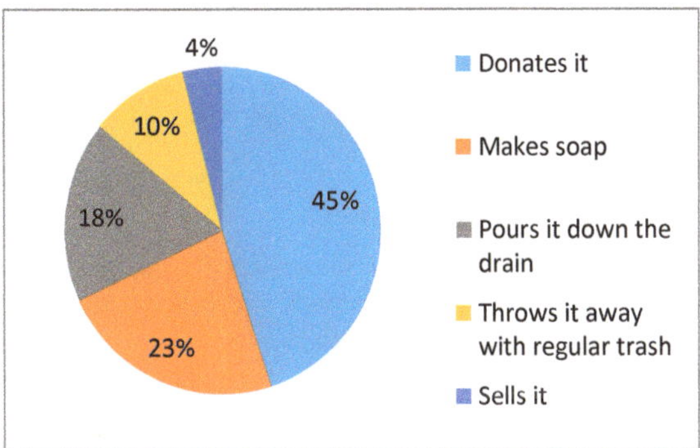

Chart 3. Destinations of WCO in the community. Source: Created by the authors based on field research.

Regarding donations, Chart 4 illustrates the competition between collection companies/ cooperatives and the soap industry to obtain residential WCO. Regarding the utility of WCO, only one-third of those interviewed knew the purposes of WCO as a raw material. Only two participants mentioned soap production in schools and churches. None of the participants knew that WCO could be used in biodiesel production.

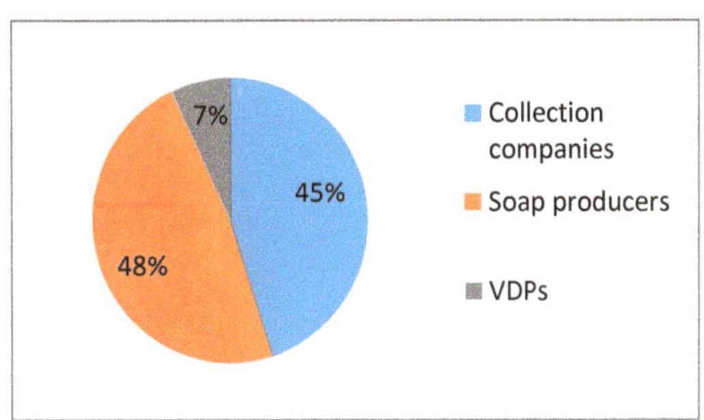

Chart 4. Destinations of community WCO donations. Source: Created by the authors based on field research.

Chart 5 shows that, among the main barriers mentioned by the community to delivering WCO to VDPs, those most cited were lack of time, distance, or being unaware of VDPs in the municipality. In general, schools represent strategic locations and can act as VDPs and facilitate access by the population.

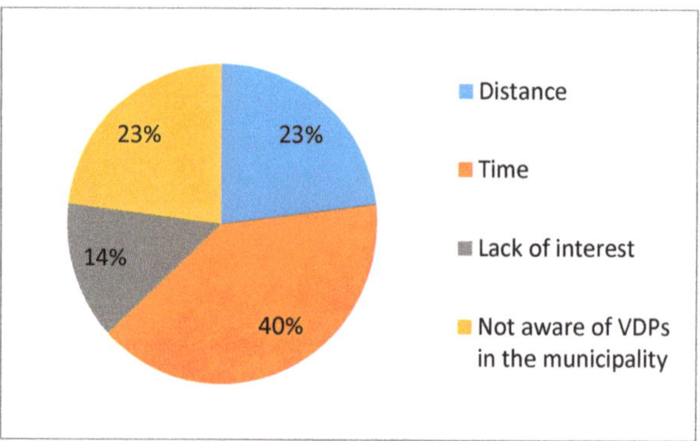

Chart 5. Difficulties cited by the community in delivering Waste Cooking Oil to Voluntary Delivery Points. Source: Created by the authors based on field research.

4.3. Action Plan for Inserting the CMWL in the WCO Reverse Supply Chain in the RMPF

Before creating an action plan for the WCO collection, it is important to involve all of the concerned actors in drafting and monitoring the school's political-pedagogical project. This is an opportunity for management, coordinators, teachers, and the community to define the profile of children's education, organizing actions to reach the proposed objectives. In this context, environmental education takes on a strategic role.

Furthermore, for the CMWL to operate as an agent of WCO reverse supply chain in the RMPF, the collaboration of other agents is essential, such as: public institutions (local governments, municipal secretariats, SAAE), commercial generators, recycling companies, associations and cooperatives, and the industries that produce biodiesel using this raw material.

Considering that a WCO reverse supply chain already exists in the RMPF, even with some bottlenecks and limitations, the action plan outlined in Table 4 describes only the internal aspects of the WCO collection operation in the school. The action plan was based on the 5W2H method, which consists of an action plan for preestablished activities that need to be as well-defined as possible. The 5W represents what (what will be done?); who (who will do it?); where (where will it be done?); when (when will it be done?); why (why will it be done?). The 2H is: how (how will it be done?); and how much (what will it cost?) [59].

Table 4. An Action plan for inserting the Washington Luiz Municipal School in the WCO reverse supply chain. Source: Created by the authors.

Who?	What?	When?	Where	Why	How	How Much? [1]
School management	Request support from the Secretariat of the Education (SME)	During the project's creation phase and later during the school year	SME	Evaluate and follow up with the school/VD Pro monitor and correct possible flaws	Memo requesting support for the project	-
School management	Request support from the Municipal Secretariat of the Environment and Sustainable Development (SMMADS)	Throughout the entire school year	SMMADS	Supply PET plastic containers	Memo requesting support for the project	400 Brazilian reais for a container of 100 liter
School management	Request support from the Water and Sewer Treatment Service (SAAE)	At the start of the project and through a request	SAAE	Awareness-raising lectures	Letter with a partnership proposal	-
School management	Include the school's WCO VDP project in its political-pedagogical project	Throughout the school year	Political-pedagogical project	Document the project	A project that includes WCO collection in the school through EE	-
School management	Creating informational material	During the awareness-raising phase	In the community	Inform the community about how to correctly manage WCO	Handing out pamphlets in the community	The estimated cost of 100 Brazilian reais
Teachers	Include EE in the annual plan for the school curriculum	Throughout the entire school year	Classrooms, after school, school premises	Raise awareness about the impacts of WCO and the need for correct collection and destinations	Multidisciplinary EE activities aimed at WCO collection involving the community and students	The estimated cost of 20 Brazilian reais per piece of educational material
Students	Systematically monitor the WCO generated at home	Monthly	Residences	Monitor the amount of WCO generated at home	A report with information about and monitoring of the actions in their households	-
PTA	Systematically monitor the WCO generated in their community and collected by the school	Monthly	Community households	Monitor adherence of generators and amount of WCO collected	Monitoring and reporting the collected data	-
Community	Effective participation in delivering WCO	Ongoing	School/VDPs	Actively participate in the project	Separate PET plastic bottles and deliver them to the school/VDP	-

[1] The amounts correspond to the quote conducted in December 2017. NOTES: EE—environmental education.

4.4. Discussion

The research question for the present study was: "How can a public school in the RMPF of the state of Rio de Janeiro, Brazil, act as a voluntary delivery point and support the development of a reverse supply chain for waste cooking oil?" In order to answer that question, we developed an action plan based on the integration of actors and efforts. The local government, through the secretaries of education and environment and the water and sewer treatment company, could provide some incentives for school management, teachers, students, the PTA, and the community to foster acceptance of the idea of WCO collection and recycling.

At the same time, as a VDP, the school needs to be integrated into the WCO reverse supply chain, connecting household generators with collectors. One important issue is the logistics of this integration. Environmental education is an important tool for teaching the community how to appropriately store WCO that is of good quality and also to convince them to take WCO to the school´s VDP.

Regarding the WCO reverse supply chain, the results show the importance of establishing contracts between commercial generators and collection cooperatives/companies, with price incentives for volume and quality, appropriate storage conditions, the frequency of generation, and avoiding exchange for cleaning products (soap) that don't meet environmental standards.

The collection cooperatives/companies should also use routing systems with enough vehicles that are managed by trained personnel, in order to minimize logistics costs. In addition, the biodiesel industry should adopt more proactive behavior, coordinating the chain and establishing a strong relationship with recycling cooperatives and companies, to ensure a continuous supply of WCO.

In addition, the local government should play a role in the WCO reverse supply chain, primarily: regulating the actors with incentives (tax) and controls; formalizing self-employed informal collectors by integrating them into existing cooperatives/associations; preventing soap production that does not follow environmental standards; maintaining the continuity of environmental education programs, along with WCO collection, in order to foster the development of more environmentally friendly citizens and reduce the cost of water and sewer treatment.

The results of the present study are in line with the literature review. Waste management related to WCO involves non-generation, reduction and recycling [23]. In order to accomplish this, the reverse supply chain needs to work well in terms of organization, cost efficiency and value generation [9], integration of small- and large-quantity generators [29], and adequate storage conditions [30]. Voluntary delivery points are strategic for collecting WCO from households (small-quantity generators) [11], without neglecting the importance of route planning to keep logistics costs down [32].

The idea of generating an action plan is also related to the National Solid Waste Policy, which demands a Municipal Plan of Integrated Solid Waste Management with shared responsibility among generators, collectors, recyclers and industry, and social inclusion of individual collectors in cooperatives and associations [13,22,24,25].

However, since there are no specific regulations for WCO reverse logistics [7], environmental education in schools is an important strategy to compensate for that gap. It is important to insert this issue across school subjects and curricula [42,45] Several other factors are also important: involvement of the community in school activities [51,52]; presenting information about the cost of collective selection programs and the convenience of VDPs [41]; highlighting biodiesel as the main product generated from WCO [37,38]. It is hoped that such programs will generate citizen awareness of ways to dispose of WCO appropriately [41,50].

5. Conclusions

The present study showed that the current structure of the WCO reverse supply chain in the RMPF is based on weak partnerships between generators and collector cooperatives and companies, and also showed lack of effective communication between the links in the chain. Other bottlenecks and requirements indicated by the participants were: lack of public awareness; a regional culture of WCO destinations being homemade soap production; the need to obtain a minimum volume for collection; competition among collectors; and quality specifications for WCO destined for biodiesel production.

In the present scenario, we evaluated the types of stakeholder participation and actions related to the selective collection, reverse logistics, and other actions regarding shared responsibility and environmental awareness. We concluded that the proper functioning of a WCO reverse supply chain depends on various factors. First, the population must be informed about the environmental damage caused by the improper disposal of this type of waste and about recycling possibilities. The reverse logistics of WCO also benefit from facilitated access to VDPs, incentives for formal contracts between cooperatives and commercial generators, optimization of collection routes, and specific incentive policies and legislation.

The development of reverse logistics is in the government's best interest. However, the actors that are part of the process are not given priority, and WCO collection programs and economic incentive programs (fiscal/tax) lack continuity. Furthermore, municipal selective collection programs established by the National Solid Waste Policy are mandatory, but local governments do not have information about or control of the destinations of all the WCO generated in their municipalities; they only control the volume processed by registered recycling cooperatives and companies.

The present analysis of the possibility of a municipal school operating as a VDP in a WCO reverse supply chain in the RMPF showed that VDP programs in schools cannot be one-offs, resulting in isolated actions such as rallies or events. These actions must be ongoing and part of the political-pedagogical program and school curriculum. Only in this way is it possible to implement the proposed action plan, as part of the environmental education program, in order to use the municipal school as a VDP.

It is important that students and the community actively participate in developing and implementing the WCO collection action plan. The study also pointed to a lack of public awareness regarding the damage caused by the improper disposal of WCO; lack of knowledge about the existence of VDPs; and criticism of discontinuity in collection and recycling initiatives.

One limitation of this study relates to the data collection phase. There was a paucity of literature about the actions of schools as VDPs and about WCO collection as a tool for environmental education. Similarly, there was a lack of a public database about WCO generation, collection and destinations. Additionally, the case study method doesn't permit generalization of results; further study is needed to support our conclusions.

Future studies should focus on the implementation of the action plan at the Washington Luiz Municipal School. It is hoped that the present investigation can guide other schools in developing similar action plans. Additionally, the authors suggest that this research is applied to other regions and that there be a more in-depth investigation about other actions that involve the participation of public schools in the management of other solid urban waste.

Funding: This research received no external funding

Conflicts of Interest: The authors declare no conflicts of interest.

Appendix A. Questionnaire Applied to the Community in General

1. What is your age? _____________
2. What is your education?
   ( ) Elementary ( ) High school ( ) Higher education
3. How many people live in your house? ________
4. What is the range of monthly family income?
   ( ) less than 1 minimum wages (MW) ( ) 1 to 2 MW ( ) 3 to 4 MW ( ) over 4 MW
   Do you receive a family grant from "Bolsa Família"? ( ) Yes ( ) No
5. Do you know what environmental education (EE) is?
   ( ) No ( ) Yes

If so, what do you think EE means?

6. What are your daily actions to help preserve the environment?
7. What do you do with the WCO in your home?
   ( ) put it down the kitchen drain ( ) throw it away with the trash ( ) Donate it ( ) Make soap with it ( ) Sell its. If you donate it, to whom?
8. What do you think most people do with this oil?
9. Do you know about any public or private initiatives for the destination of the oil? ( ) No ( ) Yes. Which one?
10. How many liters of oil does your family consume monthly?
11. How many times is the oil is used before discarding?
12. Did you know that WCO can be recycled? ( ) No ( ) Yes. And where?
13. Do you know the damage caused by the improper disposal of WCO?
14. Would you use any product made with WCO as a raw material? ( ) Yes ( ) No
15. What is the main difficulty in keeping the WCO?
16. What is the main difficulty in taking the WCO to the collection points?
    ( ) Distance ( ) Time ( ) Lack of interest ( ) Other

Appendix B. Questionnaire Applied to the Students of the Washington Luiz Municipal School

1. How old are you?
2. What is your year of schooling?
3. Do you know what environmental education is? ( ) Yes ( ) No
4. What do you think environmental education means?
5. Have you ever participated in a WCO collection program at your school? ( ) Yes ( ) No
6. What would motivate you to participate?
7. Do you know the consequences of incorrect disposal of WCO in the environment?
8. What is the destination of WCO discarded in your home?
9. Do you know which destinations the WCO could have?
10. How important are environmental issues for you, on a scale of 0 to 10, where 0 means very low and 10 means very high?

Appendix C. Interview with Education Agents (Managers, Teachers and Support Staff) of the Washington Luiz Municipal School

Interviewee's role and experience:

1. Does the school in which you work has actions in the environmental education area?
2. What is your commitment to environmental education in your school?
3. How often, per week, do you address environmental issues? ( ) Once ( ) Twice ( ) More than twice ( ) Never.
4. Do you find it difficult to work on environmental issues? ( ) Yes ( ) No
5. What do you think is missing from your school? ( ) The teachers' commitment ( ) Commitment of the management of school ( ) Commitment of the technical-pedagogical team ( ) Interest of students ( ) Commitment of teachers, management and the technical-pedagogical team ( ) Lack of technical knowledge ( ) Lack of didactic resources ( ) Other.
6. Has the school elaborated its Political Pedagogical Project (PPP)? ( ) No ( ) Yes ( ) I do not know.
7. What environmental topics are addressed/worked on? (Mark all applicable items) ( ) Public policy ( ) Pollution ( ) Climate change ( ) Biomes/Ecosystems ( ) Basic sanitation ( ) Socioenvironmental sustainability ( ) Elections ( ) Water resources ( ) Social inequality ( ) Generation of work and income ( ) Associations / cooperatives ( ) Diseases ( ) Social inclusion/Social values ( ) Conservation ( ) Food ( ) Selective collection ( ) WCO disposal ( ) Solid waste ( )

Socioenvironmental responsibility ( ) Sustainable consumption ( ) Ethnic-racial relations ( ) Biodiversity ( ) Deforestation ( ) Environmental phenomena ( ) Quality of life ( ) Health in general ( ) Other.

8. In your opinion, what are the main difficulties and obstacles currently faced by the school in the development of activities/projects?
 (Mark all applicable items) ( ) Absence of planning to guide the school ( ) Very infrequent meetings ( ) Lack of structure (rooms, physical space and equipment) ( ) Lack of teaching material ( ) Difficulty in mobilizing people and entities to collaborate with the school ( ) Members who are burdened with other professional activities ( ) Government political discontinuities ( ) Difficulty of conceptual understanding of the proposals ( ) Lack of teacher training ( ) Other.

9. How do you develop, in practice, the teaching of environmental education? (Mark all applicable items) ( ) In the curricular contents of my discipline(s) ( ) Multidisciplinary, involving more than one subject area in collective projects ( ) In individual research projects ( ) In collective research projects ( ) In actions of mobilization and social participation ( ) Other.

10. Mark the resources that are normally used to develop the actions in Question 9 above (Check all applicable items) ( ) Related texts ( ) Examples of local reality ( ) Relation of the problem with environmental, social and economic aspects ( ) Magazines and periodicals ( ) Internet, CDs, DVDs and television ( ) Playful material ( ) Political Pedagogical Project ( ) NCPs ( ) Textbook s( ) Teaching materials containing information about the environment ( ) Bibliographic or scientific material on environmental issues ( ) Specific pedagogical material son environmental issues ( ) Other.

11. Has the school developed, or is it developing, interdisciplinary project(s) around the issue of WCO? ( ) Yes ( ) No.

12. What is your level of acceptance of an action plan for environmental education with the school as a WCO collection point, on a scale of 0 to 10, where 0 means totally against and 10 strongly in favor?

Appendix D. Interview with People from the Municipal Secretariat for Environment and Sustainable Development (SMMADS)

Interviewee's role and experience:

1. What is the functioning mechanism of the secretariat? Functional organization chart?
2. Does the municipality have a specific law for the collection of WCO?
3. How does the municipality inspect the disposal of WCO in commercial organizations?
4. Is there any initiative to foster collection by the government? If not, in your opinion, why not? If yes, which one?
5. Is there technical assistance or training for collectors and cooperatives?
6. What is the local infrastructure for storing collected oil?
7. Is there any kind of association between the secretariat and the cooperatives?
8. How are the cooperatives managed?
9. Who is responsible for collection at the registered donation points?
10. How is the collected oil commercialized?
11. How are the values defined?
12. Is there any environmental education program about the consequences of incorrect oil disposal? If yes, where? If not, what is the reason?
13. Who is responsible for transporting the raw material to the industry? Are freight costs included? How are they calculated?
14. Based on your experience, what are the strengths and weaknesses of the program?
15. How are schools included in this program?

16.   What is the incentive given for environmental education?

Appendix E. Interview with WCO-Generating Agents in the Community

Interviewee's role and experience:

1.   What is the average monthly amount of vegetable oil used?
2.   What is the average monthly amount of WCO discarded?
3.   What is the process used in frying: pan, electric fryer, air-fryer?
4.   Does the frying process affect the quality of the WCO?
5.   How often is the oil reused?
6.   What is the destination of the WCO?
7.   Is there any government policy for WCO? What is it?
8.   Is there any political or social incentive given by the government?
9.   In what way is this oil collected?
10.   How is the oil stored and transported?
11.   Is this oil donated or sold?
12.   If sold, please indicate the amount?
13.   Is there any partnership with a public or private body? Which one?
14.   What are the benefits of this process?
15.   Has the consumption recently increased or decreased?
16.   To what do you attribute this variation in consumption?
17.   Does any specific factor affect consumption, such as concern about health, obesity and fitness wave?
18.   What are the obstacles to the collection or final destination?
19.   Based on your experience, is there potential for growth in the collection programs?

Appendix F. Interview with Collector Cooperatives and Associations, NGOs and Companies (Some Questions Were Used with a VDP)

Organization type:
Interviewee's role and experience:
Contact:

1.   How many families/people are involved in WCO recycling projects?
2.   Are there any political or social incentives given by the government?
3.   What is the process for registration of collection points?
4.   What kind of company can participate and what is the benefit?
5.   What is the monthly amount collected?
6.   Has collection recently has increased or decreased?
7.   To what do you attribute this variation in the collection?
8.   How is the WCO stored and transported?
9.   How is the WCO collected?
10.   Is the WCO is donated, bought or sold?
11.   If bought or sold, could you tell us how much?
12.   What is the destination of the WCO?
13.   Is there any partnership with public or private bodies? Which one (s)?
14.   Does the organization have any partnership with schools?
15.   What are the obstacles in the collection or final destination?
16.   Based on your experience, is there potential for growth in collection programs?
17.   Is there any competition in the region in terms of supply and demand?

18. Is there is any competition with other cooperatives or self-collectors? Which ones?
19. How would you rate the relationship among the collectors of WCO?
    ( )Cooperative ( ) Competitive ( ) Coopetition
20. What is the relationship between buyers and sellers?
    ( )Contract ( ) Spot market
21. What are the needs for improvement in the collection and destination process?
22. Do employees receive assistance? What kind?
23. What are the main difficulties in the reverse logistics of WCO?
24. Does the WCO have to be strained or filtered before the sale?
25. How is the processing done?
26. Are chemical reactants used?
27. Do employees use personal protective equipment?
28. How are R&D projects, if any, financed?
29. Is there adequate infrastructure for transport?
30. Is the mode of transport suitable?
31. Is the number of vehicles sufficient to meet the demand?
32. Does the frequency meet the demand?
33. Is there adequate infrastructure for storage?

Appendix G. Interview with the Biodiesel Company

Organization type:
Interviewee's role and experience:
Contact:
The installed capacity of the company:
Production capacity used:

1. Is there any program or government policy for the production of biodiesel based on WCO?
2. Is there any certification for those who purchase WCO from local cooperatives?
3. Does the company receive inspection and health monitoring services?
4. Does the company reach its production potential?
5. Is there competition in the of the region in terms of WCO demand?
6. Who are the competitors in the use of WCO in the region?
7. How do you evaluate the relationship of WCO supply and demand?
8. How do you relate with WCO suppliers?
    ( ) Contract ( ) Spot market
9. Who are your key suppliers?
10. How much do you pay for a liter of WCO?
11. Is this value competitive for biodiesel?
12. In your opinion, what are the main problems of WCO collection?
13. What could be done to increase its efficiency?
14. Are there any technological restrictions on biodiesel production from WCO?
15. What is the yield of biodiesel from WCO?
16. What are the expectations to increase biodiesel production from WCO?
17. What are the main quality indexes tracked?

References

1. Lieder, M.; Rashid, A. Towards circular economy implementation: A comprehensive review in context of manufacturing industry. *J. Clean. Prod.* **2016**, *115*, 36–51. [CrossRef]

2. Stahel, W.R. Circular economy: A new relationship with our goods and materials would save resources and energy and create local jobs, explains Walter, R. Stahel. *Nature* **2016**, *531*, 435. [CrossRef] [PubMed]

3. Hoornweg, D.; Bhada-tata, P. What a Waste: A Global Review of Solid Waste Management. In *Urban Development Series Knowledge Papers*; The World Bank: Washington, DC, USA, 2012.

4. Rathi, S. Alternative Approaches for Better Municipal Solid Waste Management in Mumbai, India. *Waste Manag.* **2005**, *26*, 1192–1200. [CrossRef] [PubMed]

5. Kalam, M.A.; Masjuki, H.H.; Jayed, M.H.; Liaquat, A.M. Emission and performance characteristics of an indirect ignition diesel engine fueled with waste cooking oil. *Energy* **2011**, *36*, 397–402. [CrossRef]

6. ABIOVE-Associação Brasileira das Indústrias de Óleos Vegetais. Available online: http://www.abiove.org.br/ (accessed on 1 July 2016). (In Portuguese)

7. Brasil. Lei 12.305, de 02 de Agosto de 2010. Dispõe Sobre a Política Nacional de Resíduos Sólidos. Diário Oficial da República Federativa do Brasil: Brasília, 03 de Agosto 2010. Available online: http://www.planalto.gov.br/ccivil_03/_ato2007-2010/2010/lei/l12305.htm (accessed on 1 July 2017). (In Portuguese)

8. Câmara Dos Deputados. Projetos de Lei n. 2.074, de 2007. Dispõe Sobre a Obrigação dos Postos de Gasolina, Hipermercados, Empresas Vendedoras ou Distribuidoras de Óleo de Cozinhas e Estabelecimentos Similares de Manter Estrutura Destinada a Coleta de óleo de Cozinha Usado e da Outras Providencias. Brasília. Available online: http://www.camara.gov.br/sileg/Prop_Detalhe.asp?id=368364&st=1 (accessed on 1 September 2016). (In Portuguese)

9. Pitta, O.S.R., Jr.; Neto, M.S.; Sacamano, J.B.; Lima, J.L.A. Reciclagem de óleo de cozinha usado: Uma contribuição para aumentar a produtividade do processo. In Proceedings of the International Workshop Advanced in Clear Production, "Key Elements for a Sustainable World: Energy, Water and Climate Change", São Paulo, Brazil, 22 May 2009. (In Portuguese)

10. Guabiroba, R.C.S.; D'Agosto, M.A.; Franca, L.S. Viability analysis of popular cooperatives in the participation of residual frying oil to biodiesel plants supply chain. *J. Transp. Lit.* **2014**, *8*, 164–186. [CrossRef]

11. Lago, S.M.S. Logística Reversa, Legislação e Sustentabilidade: Um Modelo de Coleta de óleo de Fritura Residual Como Matéria-Prima para Produção de Biodiesel. Ph.D. Thesis, Desenvolvimento Regional e Agronegócio. Universidade Estadual do Oeste do Paraná, Toledo, Paraná, 2013. (In Portuguese)

12. César, A.S.; Werderits, D.E.; Saraiva, G.L.O.; Guabiroba, R.C.S. The potential of waste cooking oil as supply for the Brazilian biodiesel chain. *Renew. Sustain. Energy Rev.* **2017**, *72*, 246–253. [CrossRef]

13. Leite, P.R. *Logística Reversa: Meio Ambiente e Competitividade*, 2nd ed.; Pearson Prentice Hall: São Paulo, Brazil, 2009; p. 240. (In Portuguese)

14. Disconzi, G.S. Coleta Seletiva do óleo Residual Doméstico: Desafios e Perspectivas para um Aproveitamento Socioambiental e Sustentável. Master's Thesis, Engenharia Ambiental-Universidade Federal de Santa Catarina, Santa Catarina, Brazil, 2014. (In Portuguese)

15. Zucatto, L.C.; Welle, I.; Silva, T.N. Cadeia reversa do óleo de cozinha: Coordenação, estrutura e aspectos relacionais. *Rev. Admin. Empres.* **2013**, *55*, 442–453. (In Portuguese) [CrossRef]

16. Ferreira, L.S. Cadeia Reversa do óleo Residual de Fritura na Região do Médio Paraíba Fluminense: Uma Proposta de Plano de Ação de Fomento a Educação Ambiental Tendo Uma Escola Municipal Como Ponto de Entrega Voluntária. Master's Thesis, Programa de Pós Graduação em Administração. Universidade Federal Fluminense, Volta Redonda, Rio de Janeiro, Brazil, 2017. (In Portuguese)

17. Araujo, V.K.W.S.A.; Hamacher, S.; Scavarda, F.F. Economic assessment of biodiesel production from waste frying oils. *Bioresour. Technol.* **2010**, *101*, 4415–4422. [CrossRef] [PubMed]

18. Brasil. Ministério da Educação. Propostas de Diretrizes da Educação Ambiental para o Ensino Formal–Resultado do II Encontro Nacional de Representantes de EA das Secretarias Estaduais e Municipais (capitais) de Educação–2001 Educação Ambiental–por um Brasil Sustentável. Available online: http://www.mma.gov.br/educacao-ambiental/politica-de-educacao-ambiental/programa-nacional-de-educacao-ambiental (accessed on 1 December 2016). (In Portuguese)

19. Srivastava, A.K. Green supply chain management: A state-of-the-art literature review. *Int. J. Manag. Rev.* **2007**, *9*, 53–80. [CrossRef]

20. Tibben-Lembke, R.S.; Rogers, D.S. Differences between forward and reverse logistics in a retail environment. Supply Chain Management. *Int. J.* **2002**, *7*, 271–282. [CrossRef]

21. Staff, L.T. The 4 R's of Reverse Logistics. *Logistics Today.* 2005. Available online: http://www.logisticstoday.com/displayStory.asp?sNO=7304 (accessed on 1 December 2015).

22. Mota, J.C.; Almeida, M.M.; Alencar, V.C.; Curi, W.F. Características e impactos ambientais causados pelos Resíduos Sólidos: Uma visão conceitual. In Proceedings of the Congresso Internacional de Meio Ambiente Subterraneo, São Paulo, Brasil, 15–18 September 2009. (In Portuguese)

23. Thode Filho, S.; Machado, C.J.S.; Vilani, R.M.; Paiva, J.L.; Marues, M.R.C. A logística reversa e a Política Nacional de Resíduos Sólidos: Desafios para a realidade brasileira. *Rev. Eletrôn. Gestão Educ. Technol. Ambient.* **2015**, *19*, 529–538. (In Portuguese)

24. Demajorovic, J.; Besen, G.; Rathsam, A.A. Os desafios da gestão compartilhada de resíduos sólidos face à lógica do mercado. In *Diálogos em Ambiente e Sociedade no Brasil*; Jacobi, P., Ferreira, L., Eds.; Annpas, Annablume: São Paulo, Brazil, 2006. (In Portuguese)

25. CEMPRE–Compromisso empresarial para a reciclagem. *Lixo Municipal: Manual de Gerenciamento Integrado*; CEMPRE: São Paulo, Brazil, 2010. (In Portuguese)

26. Machado Filho, C.A.P.; Marino, M.; Conejero, M.A. Gestão estratégica em cooperativas agroindustriais. *Caderno Pesquisas Admin.* **2004**, *11*, 61–69. (In Portuguese)

27. Santos, J. A logística reversa como ferramenta para a sustentabilidade: Um estudo sobre a importância das cooperativas de reciclagem na gestão dos resíduos sólidos urbanos. *Reuna* **2012**, *17*, 81–96. (In Portuguese)

28. Wildner, L.B.A.; Hillig, C. Reciclagem de óleo comestível e fabricação de sabão como instrumentos de educação ambiental. *Rev. Eletrôn. Gestão Educ. Technol. Ambient.* **2012**, *5*, 813–824. (In Portuguese)

29. França, C.G.B.; Guarnieri, P.; Diniz, J.D.A.S. Logística reversa de óleos e gorduras residuais (OGRs) para a produção de biodiesel. In Proceedings of the Congresso Da Sociedade Brasileira De Economia, Administracao E Sociologia Rural, Maceió, Brazil, 14–17 August 2016. (In Portuguese)

30. Gonçalves, M.F.S.; Chaves, G.L.D. Perspectiva do Óleo Residual de Cozinha (ORC) no Brasil e suas Dimensões na Logística Reversa. Rev. Espacios 2014. Available online: http://www.revistaespacios.com/a14v35n08/14350816.html (accessed on 1 October 2018). (In Portuguese)

31. Loss, M. Análise de Viabilidade Econômica do Recolhimento de Resíduos de Óleo de Fritura. In *Trabalho de Conclusão de Curso (Graduação em Engenharia Química)*; Universidade Federal do Rio Grande do Sul, Porto Alegre: Porto Alegre, Brazil, 2011. (In Portuguese)

32. Guabiroba, R.C.; Da, S.; D'agosto, M.D.A. O impacto do custo de coleta do ORF de fritura disperso em áreas urbanas no custo total de produção de biodiesel–estudo de caso. *Transportes* **2011**, *19*, 59–67. (In Portuguese) [CrossRef]

33. Franco, R.; Freire, M.; Almeida, M.F. Reaproveitamento do óleo vegetal utilizado em frituras para produção de sabão. In Proceedings of the Congresso De Pesquisa E Inovacao Da Rede Norte Nordeste De Educacao E Tecnologia, Belém, PA, USA, 2009. (In Portuguese)

34. Mello, V.M.; Oliveira, G.V.; Suarez, P.A.Z. Turning used frying oil into a new raw material to printing inks. *J. Braz. Chem. Soc.* **2013**, *24*, 314–319. [CrossRef]

35. Ocampo, D.; Aguirre, E.D.; Osorio, A.; Rios, L.A. Lacas y selladores para madera a partir de resinas alquílicas obtenidas de aceites de higuerilla, palma y usados de fritura. *Informaciontecnológica* **2014**, *25*, 157–164. [CrossRef]

36. Thode Filho, S.; Costa, A.P.S.; Rodrigues, I.; Semna, M.F.M.; Silva, E.R. Bioproducts Production from Vegetable Oil Residual: Candle, Chalkand Modelling of Mass. *Electron. J. Manag. Educ. Environ. Technol.* **2014**, *18*, 14–18.

37. Piloto-Rodriguez, R.; Melo, E.A.; Goyos-Pèrez, L.; Verhelst, S. Conversion of by-products from the vegetable oil industry into biodiesel and its use in internal combustion engines: A review. *Braz. J. Chem. Eng.* **2014**, *31*, 287–301. [CrossRef]

38. Talebian-Kiakalaieh, A.; Amin, N.A.S.; Zarei, A.; Noshadi, I. Transesterification of waste cooking oil by heteropoly acid (HPA) catalyst: Optimization and kinetic model. *Appl. Energy* **2013**, *102*, 283–292. [CrossRef]

39. Barbosa, G.N.; Pasqualetto, A. Aproveitamento do ORF de Fritura na Produção de Biodiesel. In *Trabalho Conclusão de Curso (Graduação em Engenharia Ambiental)*; Universidade Católica de Góias: Goiânia, Brazil, 2007. (In Portuguese)

40. Delatorre, A.B.; Rodrigues, P.M.; Aguiar, C.J.; Andrade, V.V.V.; Aredes, A.; Perez, V.H. Produção de Biodiesel: Considerações sobre as Diferentes Matérias Primas e Rotas Tecnológicas de Processos. Perspectivas Online-Ciências. *Biol. Saúde* **2011**, *1*, 21–47. (In Portuguese)

41. Jacobi, P. Educação ambiental, cidadania e sustentabilidade. *Cadernos de Pesquisa* **2003**, *118*, 189–205. (In Portuguese) [CrossRef]

42. Brasil. Lei 9.795, de 27 de abril de 1999. Dispõe Sobre a Educação Ambiental, Institui a Política Nacional de Educação Ambiental e dá Outras Providências. Diário Oficial da República Federativa do Brasil: Brasília, 28 de abril 1999. Available online: http://www.planalto.gov.br/ccivil_03/Leis/L9795.htm (accessed on 1 January 2017). (In Portuguese)
43. Reigota, M. *Educação Ambiental*, 6th ed.; Brasiliense: São Paulo, Brazil, 2004. (In Portuguese)
44. Tristão, M. Tecendo os fios da educação ambiental: O subjetivo e o coletivo, o pensado e o vivido. *Educ. Pesquisa* **2005**, *31*, 251–264. (In Portuguese) [CrossRef]
45. Barbieri, J.C.; Silva, D. Desenvolvimento Sustentável e Educação Ambiental: Uma trajetória comum com muitos desafios. *Rev. Admin. Mackenzie* **2011**, *12*, 51–82. (In Portuguese) [CrossRef]
46. Gonçalves, M.F.S. *Planejamento da Logística Reversa do óleo Residual de Fritura Para uma Destinação Ambientalmente Correta*; Dissertação (Mestrado em Energia); Universidade Federal do Espírito Santo: Espírito Santo, Brazil, 2016. (In Portuguese)
47. Tescarollo, I.; Thomson, J.P., Jr.; Amancio, M.S.; Alves, T.F.T. Proposta para avaliação da qualidade de sabão ecológico produzido a partir do óleo vegetal residual. *Rev. Eletrôn. Gestão Educ. Technol. Ambient.* **2015**, *19*, 871–880. (In Portuguese)
48. Benassuly, M.S.; Murta, A.L.S. Política Pública para Produção de Biodiesel a partir da Coleta Seletiva do ORF de Fritura: Estudo de Caso do Programa de Reaproveitamento de Óleo Comestível do Estado do Rio de Janeiro. *Sustain. Bus.* **2015**, *54*, 1–28. (In Portuguese)
49. Caldeira, C.; Queirós, J.; Freire, F. Biodiesel from Waste Cooking Oils in Portugal: Alternative collection systems. *Waste Biomass Valoriz.* **2015**, *6*, 771–779. [CrossRef]
50. Segatto, F.B. Conhecendo as formas de descartes do óleo saturado de cozinha para verificar a educação ambiental na escola. *Rev. Eletrôn. Gestão Educ. Technol. Ambient.* **2013**, *10*, 2122–2129. (In Portuguese) [CrossRef]
51. Trindade, N.A.D. Consciência ambiental: Coleta seletiva e reciclagem no ambiente escolar. *Encicl. Biosf.* **2011**, *7*, 1–15. (In Portuguese)
52. MMA—Ministério do Meio Ambiente. *Os Diferentes Matizes da Educação Ambiental no Brasil: 1997–2007. Secretaria de Articulação Institucional e Cidadania Ambiental*; Departamento de Educação Ambiental: Brasília, Brazil, 2008; p. 290. (In Portuguese)
53. Yin, R.K. *Case Study Research: Design and Methods*, 3rd ed.; Sage: Thousand Oaks, CA, USA, 2003.
54. IBGE Cidades Censo 2013. Available online: https://cidades.ibge.gov.br/ (accessed on 1 August 2016).
55. PPP-Projeto Político Pedagógico-Colégio Municipal Washington Luiz Washington Luiz Municipal School, 2016. (In Portuguese)
56. Abdalla, M.M.; Oliveira, L.G.L.; Azevedo, C.E.F.; Gonzalez, R.K. Qualidade em Pesquisa Qualitativa Organizacional: Tipos de triangulação como alternativa metodológica. *Admin. Ensino Pesquisa* **2018**, *19*, 66–98. (In Portuguese) [CrossRef]
57. Governo Do Estado Do Rio De Janeiro-Secretaria de Estado do Ambiente (SEA) 2014. Available online: http://www.rj.gov.br/web/sea (accessed on 1 February 2017). (In Portuguese)
58. APROBIO-Associação dos Produtores de Biodiesel do Brasil. Available online: http://aprobio.com.br/legislacao-mercado/especificacao-do-biodiesel/ (accessed on 1 August 2016).
59. Avila Neto, C.A.; Stefenon, S.F.; de Oliveira, J.R.; Coelho, A.S.; Venção, A.T.; Klaar, A.C.R. Aplicação do 5W2H para criação do manual interno de segurança do trabalho. *Rev. Espacios* **2016**, *37*, 19. (In Portuguese)

Article

A Biowaste Treatment Technology Assessment in Malawi

Wrixon Mpanang'ombe [1], Elizabeth Tilley [1,2], Imanol Zabaleta [2] and Christian Zurbrügg [2,*]

1   Department of Environmental Health, University of Malawi—The Polytechnic, Chichiri, Blantyre 3, Malawi; wrixonmpana@yahoo.com (W.M.); elizabeth.tilley@eawag.ch (E.T.)
2   Department of Sanitation, Water and Solid waste for Development (Sandec), Eawag: Swiss Federal Institute of Aquatic Science and Technology, 8600 Dübendorf, Switzerland; imanol.zabaleta@eawag.ch
*   Correspondence: christian.zurbruegg@eawag.ch; Tel.: +41-587-655-423

Received: 31 August 2018; Accepted: 21 November 2018; Published: 27 November 2018

Abstract: In the city of Blantyre, much of the generated municipal waste is biowaste, typically mixed with other waste fractions and disposed at the city's dumpsite. Energy and nutrients could be recovered; however, with many biowaste options available, choosing what technology to implement is difficult. Selecting Organic Waste Treatment Technology (SOWATT) is a tool that supports decision making for selecting a biowaste treatment option considering social, technical, and environmental aspects. SOWATT was used to evaluate options for Blantyre's Limbe Market. Anaerobic digestion, black soldier fly processing, slow pyrolysis, in-vessel composting, windrow composting, vermicomposting, and wet-biomass-briquetting were considered as options. The performance of each alternative was assessed based on five objectives by government, NGO, and market-based stakeholders in order to determine the most acceptable option for the greatest number of people: something that is rarely done, or if it is the preferences are not rigorously quantified (e.g., stakeholder workshops) and/or weighted against specific objectives. However, given the novelty of the ranking-solicitation process, some participants struggled with the variety of options presented, and further iterations of SOWATT will address this limitation. Ultimately, vermicomposting scored highest of all alternatives and could best achieve the five objectives as prioritized by the stakeholders when implemented.

Keywords: decision support system; multi criteria decision analysis (MCDA), organic waste treatment; market waste; biodegradable waste

---

1. Introduction

Appropriate management of municipal solid waste is a crucial service to uphold public health and avoid environmental pollution. With increasing urban densification, the challenge and threat of unmanaged waste becomes more acute [1]. Biowaste, the biodegradable fraction in waste, is of particular importance as it amounts to more than 50% of the total waste generated [2]. Unmanaged, it may pose considerable health and environmental risks as it attracts insects, rodents, and other disease vectors; generates leachate-polluting groundwater [3]; and emits greenhouse gases.

Biowaste management challenges are also apparent in Blantyre, Malawi's second largest city, located about 300 kilometres south of the capital, Lilongwe. As the capital of Malawi's Southern Region, Blantyre is a major commercial hub with about 1 million inhabitants [4]. The Blantyre City Council (BCC) is responsible for waste collection (formal residential areas, markets, and some institutions), transport, and disposal. All the waste collected by BCC is transported to the Mzedi dumpsite, but it is not compacted there, and the dumpsite has already exceeded its design lifespan of 20 years. More than two thirds of Blantyre's waste is organic; some materials like plastics, metals, and glass are picked up by scavengers for reselling, though the quantities are small [5].

Biowaste can be treated to recover valuable resources like energy and nutrients, thus presenting economic opportunities while reducing the negative environmental effects of open dumping and/or decomposition [6]. Biowaste management can also act as a driving force for overall waste management when, for instance, the economic value of biowaste-derived products incentivizes waste collection or new revenue opportunities enhance the financial sustainability of the waste management system [7].

Waste management-related decisions are, however, complex and must consider the many influencing factors and alternative solutions. Besides the tangible physical elements, waste management also comprises an array of "soft aspects", also referred to as governance aspects including stakeholder preferences, financial mechanisms, policies, and institutional capabilities [8,9]. Many biowaste treatment initiatives have been unsuccessful, as such issues were not sufficiently considered [6,7]. To better evaluate the advantages and disadvantages of different biowaste treatment technologies with regard to set objectives, a decision support structure can significantly help take informed decisions. A review of decision support models by Karmperis et al. [10] shows that many decision support systems in waste management rely on Life Cycle Analysis (LCA) or Cost-Benefit Analysis [11] methods, while fewer use multi-criteria decision-making approaches. Güereca et al. [12] used LCA to evaluate two biowaste management systems; however, they limited their analysis to quantifying energy and water consumption emissions to the atmosphere, and water and space requirements. Importantly, most assessment methods are used exclusively by professionals working in evaluation or planning offices making use of existing data to generate optimized decisions, but the choices rarely, if ever, include the priorities or perspectives of more than a few decision makers, and rarely the beneficiaries or end users. As such, this study used the SOWATT tool that has previously been applied in the Philippines and Colombia [13,14] to solicit and amalgamate the preferences of a cross-section of stakeholders in the selection of an appropriate biowaste treatment technology. The methodology was designed specifically for biowaste treatment considering the technical, social, environmental, and economic aspects that influence long-term sustainability, especially in the sense that end-users and future maintenance workers are involved at each step of the decision-making process [13]. This study presents the outcomes of the assessment for biowaste treatment in Blantyre, Malawi, the first of its kind for an African context.

2. Materials and Methods

2.1. SOWATT Approach

The complexity of decisions often relies on uncertainty about the future, the fact of having a variety of conflicting objectives, the existence of too many or too few alternatives, or an overwhelming number of influencing factors [15]. Decision analysis, which maximizes the benefits that could be obtained from a decision, includes tools and methods that provide a structured process and recommends a course of action. Multi-Attribute Value Theory (MAVT) is a common multi-criteria decision analysis tool (MCDA) that has been often applied in environmental management choices [16–19]. This approach decomposes complex decision problems into several components: alternatives, uncertainties, consequences of alternatives, as well as the objectives and preferences of the decision maker.

The tool used in this study, called "Selecting Organic Waste Treatment Technology" (SOWATT), is based on the MAVT methodology and was designed to facilitate the selection of a sustainable biowaste treatment technology alternative [13,14]. SOWATT considers 5 different objectives that technologies should fulfil to ensure their long term sustainability: (1) 'high technical reliability', (2) 'high social acceptance', (3) 'high environmental protection', (4) 'high hygiene and community health protection', and (5) 'high economic sustainability'. These objectives and their sub-objectives are shown in the objective-hierarchy (Figure 1). Following the SOWATT methodology, the preferences of relevant local stakeholders were assessed in order to determine the relative of importance of the objectives for the case study in Blantyre.

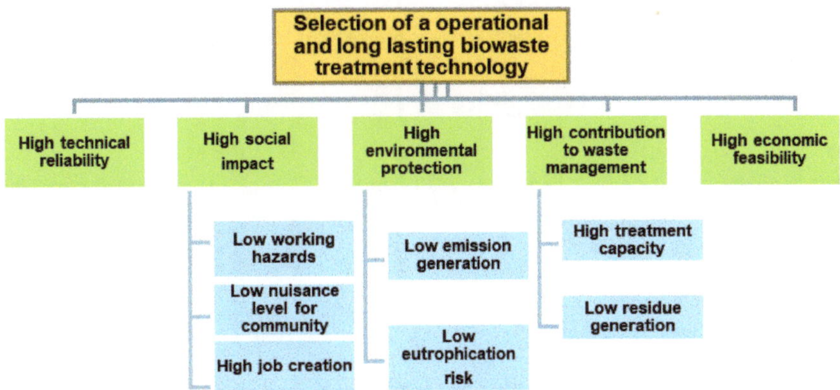

Figure 1. The default objective hierarchy defined by the SOWATT tool, adapted from [13].

2.2. Study Area

Limbe Market (LM), the largest market in Blantyre, was chosen as the focus area for the study due to the fact that biowaste was available in large, consistent quantities and was relatively pure (uncontaminated). We determined that approximately 1.1 tons of waste was generated by the market daily, of which 90% was biowaste. About 70% of the biowaste was wet fruit and vegetable waste such as banana peelings, tomatoes, leafy greens and onion leaves, while the rest was dry biowaste (15% vegetable waste and 15% paper and cardboard waste).

2.3. Biowaste Technology Options

Six technology alternatives provided by the SOWATT tool were considered in the Limbe Market study: windrow composting (WC), in-vessel composting (IC), vermicomposting (VC), anaerobic digestion (AD), slow pyrolysis (SP), and black soldier fly processing (BSF). A seventh technology, wet-biomass-briquetting (WBB), was also assessed in this case, as it is a common biowaste treatment technology in Blantyre. Of the seven technologies selected, five fall into the category of biological treatment processes, where a controlled conversion of waste is mainly driven by living organisms, either under aerobic [20,21] or anaerobic conditions [22], by bacteria and fungi or animals, i.e., worms in vermicomposting [23,24] or by insect larvae in Black Soldier Fly treatment [25]. The technology options were evaluated in terms of how they would perform if implemented at Limbe Market. The performance of the considered technologies was evaluated against 5 main objectives (Figure 1). These objectives were validated by the stakeholders during an objective validation exercise. The objectives and their attributes as provided by the SOWATT tool are presented in Table 1. The performance data (Table 2) were obtained from the SOWATT tool [13], which was established based on an extensive literature study [6,20–25], and through interviews with experts in Malawi.

Table 1. Definition and description of the evaluation objectives and their attributes (shaded objectives are main headings; unshaded objectives are sub-objectives).

Objectives	Objective Description	Attribute	Attribute Description
High technical reliability	The technology operates with as little downtime (technology breakdown or not working for whatever reason) as possible	Maximum number of consecutive days of downtime per year (days/year) *The lower this attribute, the higher is the technical reliability*	Estimated while considering a list of influencing factors that included affordability of materials for maintenance, time required to get maintenance materials from supplier, and affordability of maintenance personnel
High social acceptance	The technology is accepted by the community from a socio-cultural perspective, which is defined by four sub-objectives: (1) 'high job creation', (2) 'high working safety', (3) 'low smell impact', and (4) 'high trust in technology'		
High job creation	The technology generates employment and therefore increases social acceptance	Number of workers for each ton of biowaste treated (workers/ton)	Estimated for each technology based on similar local experiences with the technology or from literature
High working safety	The technology ensures safe working conditions, thereby increasing social acceptance	Value from 1–10. *1 is low potential of hazards (safe) and 10 is high potential of hazards (unsafe)*	Estimated considering the possible risks that the technology poses to the workers
Low smell impact	The technology does not create nuisance by smell, thereby increasing social acceptance	Number of hours per week of bad smell 20 meters away from the installation (h/week)	Estimated based on literature
High trust in technology	Past success of a technology creates a level of trust that increases social acceptance	Percentage of past experiences for each technology that are still working (%)	Estimated by dividing the number of existing installations by the total number of installations (past and current)
High environmental pollution	The technology is environmentally friendly, whereby environmental friendliness is defined by two sub-objectives: (1) 'low environmental pollution' and (2) 'high resource recovery'		

Table 1. *Cont.*

Objectives	Objective Description	Attribute	Attribute Description
Low environmental pollution	The technology generates less pollution to the atmosphere (gases) and to groundwater (leachate), which contributes to environmental protection	CO_2 equivalents emitted to the atmosphere for each ton of biowaste treated. Leachate risk (from 1 to 5) *1 being low leachate risk and 5 being high leachate risk*	Estimated based on literature
High resource recovery	The technology contributes to recovering as much phosphorus and nitrogen as possible and/or generates renewable energy from biowaste, which contributes to environmental protection	Percentage (%) of nitrogen (N) in biowaste recovered in the end-product Percentage (%) of phosphorus (P) in biowaste recovered in the end-product Energy generated, in Kilowatt hours (kWh), from each ton of waste (kWh/ton)	Estimated based on literature
High hygiene and health protection of the community	The technology contributes highly to reducing health risks and improving hygiene in the community. This objective is described by two sub-objectives: (1) 'high treatment capacity' and (2) 'low residue generation'		
High treatment capacity	The technology is able to treat a lot of the collected waste, which contributes to protection of the health of the community	Percentage (%) of the collected waste that the technology can treat	Estimated based on local experiences of the technology
Low residue generation	The technology generates less residual waste, which contributes to protection of the health of the community	Percentage (%) of the input waste that is converted into a non-marketable residue	Estimated based on local experiences and literature
High economic sustainability	The income obtained with the technology enables one to at least cover its cost and, if possible, make profit	Ratio of revenues and expenditure (dimensionless value) *The higher this ratio (value) is, the higher the economic sustainability of the technology*	Estimated by dividing the revenues and expenditures over the lifespan of the installation. Estimation based on local experiences and literature

Table 2. Estimated performance for the considered technology scenarios.

Objectives											
Sub-Objectives	Unit	AD	BSF (NT)	BSF (HT)	WBB	SP (NT)	SP (HT)	IC (NT)	IC (HT)	WC	VC
Attributes											
High technical reliability											
Downtime	days/ year	90	90	90	7	14	14	30–90	30–90	0–7	0
High social acceptance											
High job creation											
Labour productivity	workers/ton/day	1.25–2.5	2.5–5	2.5–5	3–5	3.75–7.5	3.75–7.5	1.5	1.5	2.5–5	2.5–5
High working safety											
Level of potential hazards	Scale of 1–10	7	4	4	7	9	9	3	3	4	2
Low smell impact											
Smell emissions at 20 meters distance	h/week	0	168	168	0–168	56	56	0	0	84	0–168
High trust in technology											
Percentage of projects still operational	%	20–50	0	100	25	0	100	0	100	14–57	100
High environmental protection											
Low environmental pollution											
CO_2 emission	kg CO_2 eq./ton	170–690	200–300	200–300	0–5	1600–2700	1600–2700	23–33	23–33	325–390	325–390
Leachate-risk level	Scale 1–5	4	2	2	2	1	1	1	1	5	5
High resource recovery											
Nitrogen recovered	% N	90–100	43	43	0	0	0	62.5–91	62.5–91	25–91	40–91
Phosphorus recovered	% P	95–100	67	67	0	0	0	85–99	85–99	62–99	40–99
Energy recovered	kWh/ton	600–900	0	0	500–3000	2000–3000	2000–3000	0	0	0	0
High hygiene and community health protection											
Low residue generation											
Residue output	%	0	0–20	0–20	0–5	0	0	0	0	0	0
High treatment capacity											
Applicability to biowaste collected	%	10–30	10–60	10–60	20–70	0–10	0–10	75–100	75–100	75–100	50–85
High economic sustainability											
Financial Performance	Cost-Revenue Ratio	13.31	0.17	0.17	1.69	0.04	0.04	0.94	0.94	2.86	38.2

Notes: Technology abbreviations: Anaerobic Digestion (AD), Black Soldier Fly Processing (BSF), Wet-Biomass-Briquetting (WBB), Slow Pyrolysis (SP), In-vessel Composting (IC), Windrow Composting (WC), Vermicomposting (VC). HT stands for "high trust in technology" scenario, NT for "no trust in technology" scenario.

As there were no local experiences with IC, SP, or BSF, data related to sub-objective 'high trust in technology' were not available. Hence, two scenarios were included in the analysis for each of IC, SP, and BSF, one assuming no trust (NT) and the other high trust (HT) in the technology.

2.4. Stakeholders and Preferences

The SOWATT approach depends on stakeholder inputs (preferences) in order to calculate technology scores. Potential key stakeholder clusters were identified in this study as (1) BCC officials (because BCC owns LM), (2) LM chairpersons (since they are the governing authority in the market), (3) market vendors that generate biowaste, and (4) non-governmental organizations (NGOs) that support biowaste treatment initiatives in Blantyre. From these identified stakeholder clusters, individuals were interviewed to determine their relevance for the LM case. Interviewees were asked questions that aimed at understanding how the interviewee could influence biowaste treatment practices at LM. The interviewees also suggested other potential stakeholders (who they considered to have the same influence and explained why). The interviewees that indicated that they had influence on biowaste management practices at LM were chosen as relevant stakeholders. The stakeholders were further categorised into clusters based on how similar their level of influence was (Table 3).

Table 3. Four stakeholder clusters identified.

BCC Cluster	NGO Cluster	Chair Cluster	Vendor Cluster
1. Director of Health and Social Services 2. Deputy Director of Health and Social Services 3. Blantyre Cleansing Services Officer 4. Limbe Solid Waste Management Officer	1. Centre for Community Organisation and Development (CCODE) (Representative) 2. Crown Financial Ministries (Representative) 3. Water for People (Representative) 4. Pump Aid (Representative)	1. LM Chairman 2. LM Chairlady 3. LM Waste Management Committee Chairperson	16 randomly selected vendors from the produce section of the market

In order to elicit the preferences of the stakeholders, the "swing" weighting method was used [26]. In this method, hypothetical performance scenarios of a biowaste treatment technology implemented at LM were presented, and each respondent (stakeholder) was asked to rate every scenario presented between 0 (least preferable) and 100 (most preferable). Afterwards, the "reverse swing" method was used as a consistency check. The swing questionnaire (Appendix 1) first presented a hypothetical, worst-case scenario using the worst values for all attributes; subsequent hypothetical scenarios only had one best attribute. The reverse swing questionnaire (presented after the swing questionnaire) first presented a hypothetical best-case scenario using the best desired values for all attributes, then subsequent scenarios only had one worst attribute (Appendix 2). For example, for the attribute 'levels of potential hazards' (Table 2) (under objective high social acceptance and sub-objective high working safety), hazard level 2 was selected for the best-case scenarios (no technology had a hazard level of 1), while hazard level 10 was selected for the worst-case scenarios. The best- and worst-case scenarios used in the swing and reverse swing questionnaires are presented in Figure 2.

SELECTING ORGANIC WASTE TREATMENT TECHNOLOGY FOR LIMBE MARKET

	Technical Reliability	Social Acceptance	Hygiene and Health Protection	Economic Sustainability	Environmental Protection
Best Case Scenario ☺	0 days/year downtime	2/10 Potential hazards No bad smell 20 m far from plant 8 workers/ton 100% of successful past experiences	100% of collected organic waste can be treated 0% wet waste weight as residues	39 Income expenditure ratio	0 kg CO$_2$ equivalent/ton 1/5 leachate risk 100% N recovered 100% Phosphorus recovered 3000 kWh/ton energy produced
Worst Case Scenario ☹	90 days/year downtime	10/10 Potential hazards All week bad smell 20 m far from plant 1 worker/ton 0% of successful past experiences	1% of collected organic waste can be treated 20% wet waste weight as residues	0 income- expenditure ratio	2700 kg CO$_2$ equivalent/ton 5/5 leachate risk 0% N recovered 0% Phosphorus recovered 0 kWh/ton energy produced

SOWATT Limbe   Best and Worst Case Scenarios

Figure 2. Best-case and worst-case scenarios of a hypothetical biowaste treatment technology at LM.

Each stakeholder's rankings (extracted from the questionnaires) were converted into weights between 0 (low importance) and 1 (high importance) for every considered objective. The conversion to weights was achieved using the following equations:

Equation (1): Swing method equation:

$$W_x = \frac{t_x}{\sum_i^m t_i} \tag{1}$$

Equation (2): Reverse swing method equation:

$$W_x = \frac{100 - t_x}{\sum_i^m (100 - t_i)} \tag{2}$$

in which

W_x: weight of objective or sub-objective x;

t_x: points given during the swing (in Equation (1)) or the reverse swing (in Equation (2)) method by the stakeholder to objective x; and

m: number of objectives to be considered: 5 main objectives, 4 sub-objectives for "social acceptance", 2 sub-objectives for "hygiene and health protection" and 2 sub-objectives for "environmental protection".

As a calculation example, in the swing questionnaire, the BCC Director of Health and Social Services rated 'high technical reliability' 80 points, 'high social acceptance' 50 points, 'high hygiene and health protection' 100 points, 'high economic sustainability' 40 points, and 'high environmental protection' 60 points. To calculate the Director's weight of 'high technical reliability' using Equation (1), we divided the 80 points given to this objective by the sum of all the points given to the five main objectives as follows:

$$W_{high\ technical\ reliability} = \frac{t_x}{\sum_i^m t_i} = \frac{80}{80 + 50 + 100 + 40 + 60} = 0.242$$

An average for the weights obtained from the swing method (Equation (1)) and reverse swing method (Equation (2)) was used as the stakeholder's overall weight for the objective. The calculated values were averaged to take into account the framing of the questions; asking the same question in two different ways tests for and ensures understanding and consistency. An example of the weights obtained from a stakeholder's ranking is presented in Table 4.

Table 4. BCC Director of Health and Social Services' weights and ranking of objectives.

Objectives		Swing Method		Reverse Swing		Average	Rank
		Point	Weight	Point	Weight		
Main Objectives	Technical Reliability	80	0.242	20	0.242	0.242	2
	Social Acceptance	50	0.152	50	0.152	0.151	4
	Hygiene and Health Protection	100	0.303	0	0.303	0.303	1
	Economic Sustainability	40	0.121	60	0.121	0.121	5
	Environmental Protection	60	0.182	40	0.182	0.182	3
Social Acceptance	Working Safety	100	0.333	0	0.370	0.352	1
	Smell Impact	70	0.233	50	0.185	0.209	3
	Job Creation	80	0.267	20	0.296	0.281	2
	Trust in Technology	50	0.167	60	0.148	0.157	4
Hygiene and Health Protection	Treatment Capacity	100	0.556	0	0.833	0.694	1
	Residue Generation	80	0.444	80	0.1667	0.306	2
Environmental Protection	Environmental Pollution	100	0.556	0	0.667	0.611	1
	Resource Recovery	80	0.444	50	0.333	0.389	2

Notice that for this example, the weight given by the Director for Technical Reliability (first row) is the same regardless of how the question was asked (i.e., the swing and reverse swing methods both yielded 0.242). However, there were significant differences in the weights given to Treatment Capacity: the swing format yielded a weight of 0.556, while the reverse swing format yielded a weight of 0.833. It is not expected that each respondent will assign the exact same value to each objective through each method (which is why an average is taken), but significant, consistent differences can indicate a lack of understanding and help to identify respondents that may be struggling to conceptualize the questions. In each cluster, an average for the weights obtained from every stakeholder was used as the cluster's weight (level of importance) for the respective objective (see results in Section 3.1, Figure 3).

2.5. Technology Scoring

Scores for the technology options were calculated using the weights of the objectives (stakeholder preferences) and estimated performances for each of the technology alternatives (Table 2). The values for the estimated performances were firstly normalised to obtain values between 0 and 1 for all attributes. When normalizing the values for the estimated technology performances, we assigned the normalized value 1 to the best performance values, while the normalized value 0 was assigned to the worst performance values among the technology options for the considered objective. For the objectives with the direction 'high' such as 'high economic sustainability', the value 1 was assigned to the highest performance value of that objective among the technology options. Whereas for the objectives with the direction 'low' such as 'low environmental pollution', the value 1 was assigned to the smallest performance value of that technology among the technology options. For example, (Table 2) the value 1 was assigned for 100% for the sub-objective 'high trust in technology', and the value 1 was also assigned for the sub-objective 'low leachate risk'. Where performance was estimated as a range of values, the average value was used during performance normalization. The following equations were used to normalize the estimated technology performances:

Equation (3) for "low direction" objectives:

$$N_x^y = 1 - \frac{C_x^y - m_x}{M_x - m_x} \tag{3}$$

Equation (4) for "high direction" objectives:

$$N_x^y = \frac{C_x^y - m_x}{M_x - m_x} \tag{4}$$

in which

N_x^y: normalized value of the estimated performance of technology option Y for objective X;
C_x^y: the estimated performance of technology option Y for objective X;
m_x: minimum value considered for objective X among all technology options; and
M_x: maximum value considered for objective X among all technology options.

The additive model was then used to calculate the final score of each technology. Each normalized performance value of a technology was first multiplied by the weight given to its corresponding objective. Then, the outcome scores were summed to obtain the final score for that technology. The average values for the stakeholder weights for all clusters were used to calculate the final technology scores. The additive model determined the score of a technology alternative by the following equation:

Equation (5): Score of a technology alternative:

$$v(a) = \sum_i^m w_r \cdot N_r \tag{5}$$

in which

$v(a)$: value (score) of the technology alternative A;
w_r: weight of objective r;
N_r: normalized value of the performance of technology alternative A for objective r; and
m: number of objectives.

For the objectives composed of sub-objectives, a different formula for the value of N_r was used. The objectives of 'high economic sustainability' and 'high technical reliability' do not have any sub-objectives, and therefore the value of N_r was obtained directly using Equation (4). However, for the other three objectives ('high hygiene and health protection of community', 'high social acceptance', and 'high environmental protection') the value of N_r was calculated as follows:

Equation (6): normalized performance value for objectives with sub-objectives:

$$N_r = \sum_i^m w_x \cdot n_x \tag{6}$$

in which

N_r: normalized value of the performance of alternative A for objective r;
w_x: weight of sub-objective x;
n_x: normalized value of the performance of alternative A for sub-objective x; and
m: number of sub-objectives.

3. Results

3.1. Stakeholder Preferences

The weights for the objectives determined by the BCC cluster were obtained as an average for the weights obtained from the four stakeholders in the cluster. For example, we determined the BCC Cluster's weight for the objective 'high technical reliability' as 0.122, which is an average for the weights for the same objective as obtained from the Director of Health and Social Services (0.242), the Deputy Director (0.122), the Blantyre Cleansing Services Officer (0.095), and the Limbe Solid Waste Management Officer (0.027). The same approach was used to determine all stakeholder weights. The NGO's cluster weights were determined as an average for the weights by the four NGO representatives. The Chair's cluster weights were determined as an average for the weights by the three market chairs. The Vendors' cluster weights were determined as an average of the weights from the 16 market vendors consulted.

BCC and the NGOs ranked 'high environmental protection' as their most important objective (Figure 3). Market vendors ranked 'high hygiene and health protection' as their main objective, which was not surprising considering that they are the ones affected when biowaste is poorly managed in the market. Chairpersons ranked 'high economic sustainability' as their main objective.

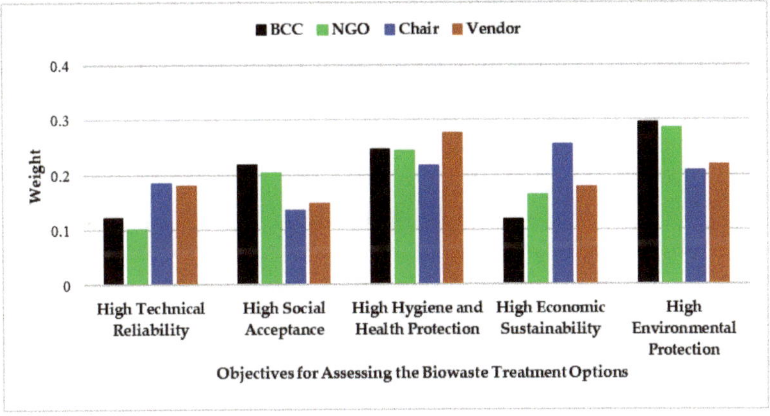

Figure 3. Weights based on stakeholder preferences for the objectives.

3.2. Technology Scores

The results of the normalised performance values multiplied by the weight of the respective objective and sum of all attribute scores for a specific technology option for the final score for that technology (Equation (5)) are shown in Figure 4. As a calculation example, for Vermicomposting (VC), the average weight (among all stakeholder clusters) for the objective 'high economic sustainability' (0.179) was multiplied by the normalised performance value for 'high economic sustainability' (0.981) to produce a 'high economic sustainability' score of 0.176 (0.179 × 0.981 = 0.176). The same approach produced scores for VC's 'high technical reliability' (0.147), 'high social acceptance' (0.130), 'high hygiene and community health protection' (0.104), and 'high environmental protection' (0.111). These objective scores added up (0.176 + 0.147 + 0.130 + 0.104 + 0.111 = 0.668) to obtain VC's overall score of 0.668.

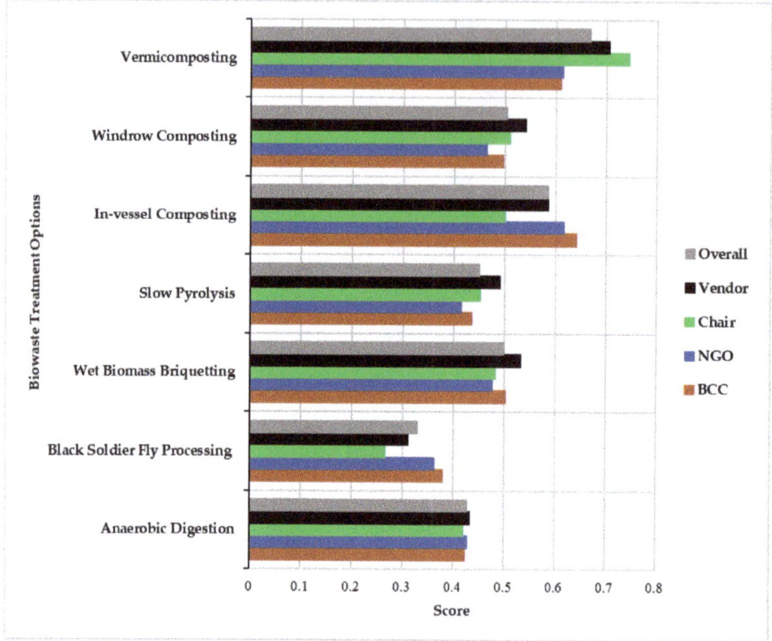

Figure 4. Ranking of the final biowaste technology scores.

VC scored the highest overall and was thus considered the most appropriate technology to implement for LM (Figure 3). In-vessel composting (IC) was ranked as the second most suitable option. Finally, BSF ranked as the least suitable technology option to implement at LM.

4. Conclusions

The SOWATT tool was successfully used for the case of Limbe market in the city of Blantyre, Malawi. The structured decision support process involved participation of different local stakeholder groups to consider seven technology alternatives for biowaste treatment: windrow composting, in-vessel composting, vermicomposting, anaerobic digestion, slow pyrolysis, black soldier fly processing, and wet-biomass-briquetting. Together with the stakeholders, the main and sub-objectives were defined, preferences for each were elicited, and technology performances were assessed. Scores for the technology alternatives were then calculated using weights and performance indicators. The results provide an evidence base for the planning and implementation of a full-scale biowaste treatment facility at LM. The results indicate (without limiting the choice) that the most appropriate technology in this context is vermicomposting. While conducting the study, certain limitations of the process became evident: the SOWATT tool requires detailed cost and performance estimations for each of the alternatives, which, for Blantyre, proved difficult, as there was limited local evidence. Estimations had to be obtained through literature from applications in similar geographic and socio-economic conditions. Although such estimations were possible, the respondents felt insecure about how well to trust this information, as there were no local implementation experiences, and therefore they could not accurately judge how such technologies might perform in Blantyre. Going through all the steps of the SOWATT procedure in a structured way proved to be quite demanding for many of the stakeholders involved, who have very seldom been confronted with such methods of evaluation. Some stakeholders, mainly vendors, found the preference elicitation method (swing and reverse swing) to be complicated. Given this experience, we therefore suggest that further studies are necessary to determine how to simplify the preference elicitation exercise for non-experts. In spite of the limitations, the Limbe Market study stimulated the stakeholders to think about different alternatives. Conducting the study also triggered their involvement and the debate on biowaste management and gave them the opportunity to reflect on the challenges and opportunities in biowaste management in Blantyre in a structured way. This assessment is also an opportunity for stakeholders to reflect on technological attributes that they might otherwise overlook when making decisions.

Supplementary Materials: The following are available online at http://www.mdpi.com/2313-4321/3/4/55/s1, Table S1: The Swing Questionnaire: points given to each scenario were used in the calculation of the stakeholder's preference weight for the objective pointed by an arrow in the respective scenario; Table S2: The Reverse Swing Questionnaire: points given to each scenario were used in the calculation of the stakeholder's preference weight for the objective pointed by an arrow in the respective scenario.

Author Contributions: The following bullets specify the individual contributions of each author: Conceptualization: I.Z. and C.Z., E.T.; Methodology: I.Z. and C.Z.; Validation: W.M., E.T. and I.Z.; Formal Analysis: W.M. and E.T.; Investigation: W.M. and E.T.; Resources: E.T. and C.Z.; Data Curation: W.M. and E.T.; Writing-Original Draft Preparation: W.M.; Writing-Review & Editing: W.M., E.T., I.Z. and C.Z.; Visualization: W.M.; Supervision: E.T. and I.Z.; Project Administration: E.T. and C.Z.; Funding Acquisition: E.T. and C.Z.

Funding: This research was funded by the Swiss Agency for Development and Cooperation (grant no. 7F-04463.03.01) and Eawag: Swiss Federal Institute of Aquatic Science and Technology.

Acknowledgments: The study was accomplished through funding from Sandec/Eawag. Special thanks to the WASHTED Centre at University of Malawi's Polytechnic for providing intellectual support and a workstation during the study. Utmost gratitude to the BCC officials, NGO representatives (CCODE, Crown, Water for People, and Pump Aid), and market chairpersons and vendors for dedicating their time by willingly participating in the study. Special recognition and thanks to the experts from Eawag, University of Malawi, Crown Financial Ministries, Malawi Industrial Research and Technology Development Centre, Green Malata, Saunders' Estate, Sustainable Development Solutions Organisation, and other individuals for their input in the technology performance estimation phase.

Conflicts of Interest: The authors declare no conflict of interest.

References

1. Guerrero, L.A.; Maas, G.; Hogland, W. Solid waste management challenges for cities in developing countries. *Waste Manag.* **2013**, *33*, 220–232. [CrossRef] [PubMed]

2. Wilson, D.C.; Rodic, L.; Scheinberg, A.; Velis, C.A.; Alabaster, G. Comparative analysis of solid waste management in 20 cities. *Waste Manag. Res.* **2012**, *30*, 237–254. [CrossRef] [PubMed]

3. Reddy, P.S.; Nandini, N. Leachate characterization and assessment of groundwater pollution near municipal solid waste landfill site. *Nat. Environ. Pollut. Technol.* **2011**, *10*, 415–418.

4. National Statistical Office. *Malawi in Figures 2016*; Government Press: Zomba, Malawi, 2016; p. 2.

5. Kasinja, C.; Tilley, E. Formalization of Informal Waste Pickers' Cooperatives in Blantyre, Malawi: A Feasibility Assessment. *Sustainability* **2018**, *10*, 1149. [CrossRef]

6. Lohri, C.R.; Diener, S.; Zabaleta, I.; Mertenat, A.; Zurbrügg, C. Treatment technologies for urban solid biowaste to create value products: A review with focus on low- and middle-income settings. *Rev. Environ. Sci. Biotechnol.* **2017**, *2017 16*, 81–130. [CrossRef]

7. Zurbrügg, C. Assessment methods for waste management decision-support in developing countries. Ph.D. Thesis, Università degli Studi di Brescia, Brescia, Italy, 2013.

8. Van de Klundert, A.; Anschütz, J. Integrated Sustainable Waste Management–The Concept. Waste Consultants: Gouda, The Netherlands, 2001; ISBN 9076639027.

9. Wilson, D.C.; Rodic, L.; Cowing, M.J.; Velis, C.A.; Whiteman, A.D.; Scheinberg, A.; Vilches, R.; Masterson, D.; Stretz, J.; Oelz, B. 'Wasteaware' benchmark indicators for integrated sustainable waste management in cities. *Waste Manag.* **2015**, *35*, 329–342. [CrossRef] [PubMed]

10. Karmperis, A.C.; Aravossis, K.; Tatsiopoulos, I.P.; Sotirchos, A. Decision support models for solid waste management: Review and game-theoretic approaches. *Waste Manag.* **2013**, *2013. 33*, 1290–1301. [CrossRef]

11. Fiorucci, P.; Minciardi, R.; Robba, M.; Sacile, R. Solid waste management in urban areas: Development and application of a decision support system. *Resour. Conserv. Recycl.* **2003**, *37*, 301–328. [CrossRef]

12. Güereca, L.P.; Gassó, S.; Baldasano, J.M.; Jiménez-Guerrero, P. Life cycle assessment of two biowaste management systems for Barcelona, Spain. *Resour. Conserv. Recycl.* **2006**, *49*, 32–48. [CrossRef]

13. Zabaleta, I.; Scholten, L.; Zurbrügg, C. Selecting Appropriate Organic Waste Treatment Options in the Philippines. *Sandec News* **2015**, *16*, 4–5.

14. Mertenat, A.; Zabaleta, I.; Zurbrügg, C. Appropriate Biowaste Treatment Technology for Aquitania, Colombia. *Sandec News* **2016**, *17*, 6–7.

15. Eisenführ, F.; Weber, M.; Langer, T. *Rational Decision Making*; Springer: Berlin, Germany, 2010.

16. Gregory, R.; Failing, L.; Harstone, M.; Long, G.; McDaniels, T.; Ohlson, D. Structured Decision Making. A Practical Guide to Environmental Management Choices. Wiley-Blackwell: Hoboken, NJ, USA, 2012; ISBN 978-1-4443-3341-1.

17. Reichert, P.; Langhans, S.D.; Lienert, J.; Schuwirth, N. The conceptual foundation of environmental decision support. *J. Environ. Manage.* **2015**, *154*, 316–332. [CrossRef] [PubMed]

18. Schuwirth, N.; Reichert, P.; Lienert, J. Methodological aspects of multi-criteria decision analysis for policy support: A case study on pharmaceutical removal from hospital wastewater. *Eur. J. Oper. Res.* **2012**, *220*, 472–483. [CrossRef]

19. Scholten, L.; Scheidegger, A.; Reichert, P.; Mauer, M.; Lienert, J. Strategic rehabilitation planning of piped water networks using multi-criteria decision analysis. *Water Res.* **2014**, *49*, 124–143. [CrossRef] [PubMed]

20. Cooperband, L. The Art and Science of Composting – A Resource for Farmers and Compost Producers. University of Wisconsin-Madison, Center for Integrated Agricultural Systems: Madison, WI, USA, 2002.

21. Rothenberger, S.; Zurbrügg, C.; Enayetullah, I.; Sinha, A. *Decentralized Composting for Cities of Low- And Middle-Income Countries—A User's Manua*; 1. Swiss Federal Institute of Aquatic Science and Technology (Eawag), Dübendorf, Switzerland and Waste Concern: Dhaka, Bangladesh, 2006.

22. Vögeli, Y.; Lohri, C.; Gallardo, A.; Diener, S.; Zurbrügg, C. *Anaerobic Digestion of Biowaste in Developing Countries—Practical Information and Case Studies*; Swiss Federal Institute of Aquatic Science and Technology (Eawag): Dübendorf, Switzerland, 2014.

23. Munroe, G. *Manual of On-Farm Vermicomposting and Vermiculture*; Organic Agriculture Centre of Canada: Nova Scotia, Canada, 2007.

24. Ali, U.; Sajid, N.; Khalid, A.; Riaz, L.; Rabbani, M.M.; Syed, J.H.; Malik, R.N. A review on vermicomposting of organic wastes. *Environ Prog Sustain Energy* **2015**, *34*, 1050–1062. [CrossRef]

25. Dortmans, B.M.A.; Diener, S.; Verstappen, B.M.; Zurbrügg, C. *Black Soldier Fly Biowaste Processing—A Step-by-Step Guide*; Eawag-Swiss Federal Institute of Aquatic Science and Technology. Department of Sanitation, Water and Solid Waste for Development (Sandec): Dübendorf, Switzerland, 2017.

26. Zheng, J.; Lienert, J. Stakeholder interviews with two MAVT preference elicitation philosophies in a Swiss water infrastructure decision: Aggregation using SWING-weighting and disaggregation using UTAGMS. *Eur. J. Oper. Res.* **2018**, *267*, 273–287. [CrossRef]

 *recycling*

Article

Recycling and Reuse Technology: Waste to Wealth Initiative in a Private Tertiary Institution, Nigeria

David O. Olukanni *, Anne O. Aipoh and Inibraniye H. Kalabo

Department of Civil Engineering, P.M.B 1023, Covenant University, Ota 121101, Ogun State, Nigeria;
anneaipoh@ymail.com (A.O.A.); ik2010@rocketmail.com (I.H.K.)
* Correspondence: david.olukanni@covenantuniversity.edu.ng; Tel.: +234-803-072-6472

Received: 5 August 2018; Accepted: 7 September 2018; Published: 8 September 2018

Abstract: The practice of collecting, treating, and managing solid waste prior to disposal has become a necessity in developing and modern societies. However, over the years, most waste has become regarded as having second-rate value and could be recovered and reused for valuable goods. However, the construction costs for conventional Material Recovery Facility(s) (MRFs) have been a major barrier for its implementation, and these technologies also require considerable technical expertise, which is not often available in developing nations for the successful operation of the MRFs. Covenant University, a private mission institution undertaking a waste-to-wealth scheme, is focused on managing and processing used materials to create reusable products. Such materials included PET bottles, paper waste, food waste from cafeterias, plastic food packs, nylon, tin cans, and others. Specific areas from the university which were chosen for the survey included the residential areas for staff and students and the two cafeterias. The waste generated was characterized so as to quantify the amount of recyclable waste generated, and also to find out which was most-occurring. The survey involved the use of structured questionnaires, on-site observations, and measurements. The study revealed that the average amount of recyclable waste generated per day in the institution were 55.56% food waste, 13.46% PET bottles, 12.64% other plastic, 9.63% nylon, 4.68% tin cans, and 4.03% paper. The study establishes that adequate waste characterization is a requirement for effective integrated solid waste management, which would boost resource recovery, reuse, and recycling.

Keywords: municipal solid waste; waste management; sustainable technology; recycling; reuse; waste to wealth

---

1. Introduction

Municipal solid waste management has emerged as one of the greatest challenges facing many developing countries. Daily human activities lead to the generation of various classes of waste, which is seen as a major environmental threat for many cities in developing nations worldwide [1,2]. The factors affecting such a high rate of change in solid waste generation includes things such as population growth, changing lifestyles, income growth, increasing use of disposable materials, excessive packaging of items, and consumer habits [3,4]. Despite the several investment opportunities that waste management offers with a very high return on investment for public and private sectors, most developing countries, including Nigeria, have solid waste management issues which are different from those found in industrialized countries in regard to composition, density, political and economic frameworks, quantity of waste, access to waste for collection, awareness, and attitude [4–6]. In developing countries, local authorities spend 77–95% of their revenue on collection and the balance on disposal [7], but are only able to collect around 50–70% of municipal solid waste [8]. In Nigeria, municipal waste densities generally range from 250–370 kg/m^3 [7,9]. Unfortunately, people in many developing countries (including Nigeria) have, until recently, regarded the issue of proper solid waste management as trivial, which may have diverted attention away from the most urgent and serious problem of achieving a fast rate of economic growth. This attitude stems from the belief that solid waste generation is an inevitable price of development [10].

According to [8], there are several factors influencing solid waste collection in Nigeria, some of which are the lack of advanced technology facilities for separation at its source, the strength of solid waste management policies and enforcement procedures, environmental education and awareness, and the economic status of individuals, among others. Mahees et al. [11] stated that better solid waste management processes should start from the solid waste generation stage. Olukanni et al. [12] and Ogwueleka [7] stated that the volume of solid waste being generated increases at a faster rate than the ability of waste management agencies to improve resources required to meet financial and technical resources needed to parallel this growth. According to Bowan and Tieroba [13], solid waste needs to be characterized by sources, generation rates, type of waste produced, and composition in order to monitor and control prevailing waste management systems while improving the existing system. A complete understanding of the composition of a waste stream as well as the activities that determine its generation is essential for effective solid waste management [14].

However, the concept of recycling is still being explored. This is the extraction and recovery of valuable materials from scraps or other discarded materials employed to supplement the production of new materials. It is essentially adding value to waste, making it economically useful [15–17]. Waste recycling has enormous economic opportunities, including job creation, poverty alleviation, and sustainable development [5]. Recyclable materials in low-, middle-, and high-income countries comprise about 17%, 43%, and 62% of the total waste stream, respectively [16]. Recyclable solid waste include textiles, construction waste, paper, plastic, ferrous and nonferrous metals, and glass. Plastic recycling industries shred plastics into pellets to manufacture other plastics and allied products. Some recycling factories process waste paper and cardboard to make tissue paper, newsprint, or bulk packaging materials. Waste glass is processed by glass or terrazzo companies, nonferrous metals are processed by aluminum smelters, and tin is recovered from aerosol cans [18]. Agunwamba [19] observed that a well-planned recycling program in Nigeria could result in savings of up to 78% in waste management costs and 79.5% in landfill avoidance costs. Aside from the economic gains of recycling, environmental benefits, such as the reduction of greenhouse gas emissions, air, and water pollution associated with production from virgin raw materials, are likely to accrue from waste recycling [16].

Literature generally reports that enormous quantities of solid waste are generated daily in the major cities of Nigeria, but exact figures are difficult to determine due to the fact that proper records of collection and disposal are not kept by the authorities responsible [20]. The project at hand used Covenant University as a case study to present an overview of the amount of municipal solid waste which is generated and studied its characterization to ascertain its economic significance. Specific areas chosen were the academic and residential areas for use by staff and students, and the two cafeterias. The aim of this study was to determine the quantity, composition, and generation rate of solid waste in the institution, with specific objectives of gathering statistical data of waste generated, presenting the current state of waste management, characterizing the solid waste generated, and quantifying waste for recycling, recovery, and reuse.

1.1. Study Area

The study site represents a typical modern community of Nigeria. The Covenant University community, within Canaan Land in Ota town, is in close proximity to the city of Lagos, Nigeria. The community hosts the world's single largest church auditorium with a capacity of 50,000, and runs five (5) worship services every Sunday. Temperatures are high throughout the year, averaging from 25 °C to 28 °C (77 °F to 82 °F). The institution has witnessed an increase in population since its inception in 2002, with a current population of above 9000 people and a daily water requirement that is estimated at 136 L/c/day. On average, one person consumes 4 bottles of water per day. Canaan Land has an expanse of 524 acres of land with an array of architectural masterpieces, which consist of the Centre for Learning Resources (university library), college buildings, a 3000-seat student chapel,

22 duplexes with 48 chalets in the Professors' Village, 64 suites at the senior staff guest house, 64 three-bedroom flats in the senior staff quarters, 100 rooms in the university guest house, two cafeterias, 96 two-bedroom apartments, and 24 one-bedroom apartments in the postgraduate halls of residence. In addition to these, there are 10 blocks of student hostels, administrative offices, lecture halls, a gymnasium, and four new engineering workshops. Figure 1 shows the master plan of the institution with selected points of interest presented in Table 1 as marked on the map.

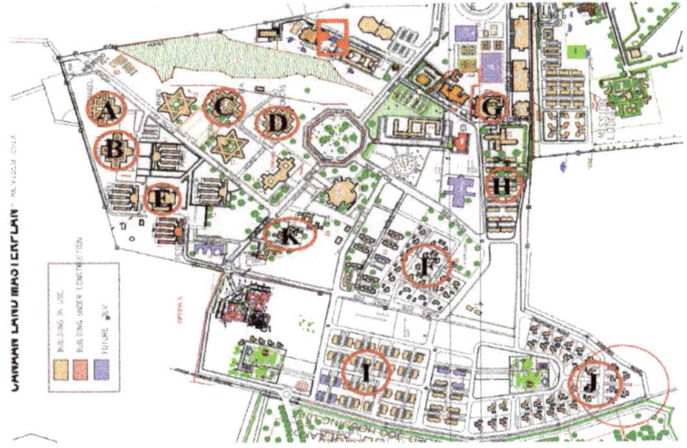

Figure 1. Master plan of the institution.

Table 1. Areas marked on the map.

Site	Building/Block	Site	Building/Block
A	Daniel Hall	F	Professor's Quarters
B	Joseph Hall	G	Cafeteria 2
C	Lydia Hall	H	University Suites
D	Dorcas Hall	I	New Estate (Block of Flats)
E	Cafeteria 1	J	New Estate (Duplexes)
		K	Post Graduate Quarters

1.2. Site-Specific Study

This study involved sampling, sorting, and weighing the individual components of the waste stream. The site-specific study required a large number of samples to be taken, ensuring that the results were not skewed or misleading. The procedures involved in municipal solid waste characterization for this project using a site-specific study were as follows.

1.3. Selection of a Representative Sample

It was very important that the samples collected were representative of the waste generation units being studied, and involved things like specifying the target population. The staff population as at 2018 was about 500 persons for academic staff and 600 persons for non-academic staff. Tables 2 and 3 show the staff and students residence populations, respectively.

Table 2. Covenant University staff residence population.

Residence	Number of Housing Units	Population
New estate	241	964
Professors village	70	280
Senior staff quarters	72	288
Suites	64	256
Post graduate quarters	120	480
Total	567	2268

Table 3. Covenant University student population as at 2018.

Female Halls of Residence	Population	Male Halls of Residence	Population
Esther Hall	773	Peter Hall	728
Mary Hall	471	Paul Hall	742
Deborah Hall	726	Joseph Hall	819
Lydia Hall	567	Daniel Hall	804
Dorcas Hall	570	John Hall	815
Total	3107	Total	3904

1.4. Sample Size

The sample size depended on the number of solid waste generation units in the sampling area. In the senior staff quarters which consisted of 72 flats (i.e., 9 blocks of 8 flats), 14 flats were sampled. In the professor's village which consisted of 22 duplexes and 48 chalets, 14 units were sampled. In the post-graduate quarters, consisting of 120 flats (i.e., 6 blocks of 20 flats, 96 two-bedroom flats and 24 one-bedroom flats), 24 flats were sampled. In the halls of residence, consisting of 10 blocks (with each block consisting of at least 8 wings), 2 blocks with 2 wings each were sampled. 13 units were sampled among the 64 suites. In the new estate, which consisted of 32 duplexes, 129 three-bedroom flats and 80 two-bedroom flats, 48 units were sampled (i.e., 6 duplexes, 26 three-bedroom flats and 16 two-bedroom flats). Both cafeterias 1 and 2 were sampled.

1.5. Sample Collection

In Covenant University, solid waste collection is carried out by the use of trucks. The trucks are usually parkers, tippers, and trucks that carry hydraulic rams to compact the waste to reduce its volume and thus be able to carry larger quantities, and this method is also known as the stationary haul collection system. The weight of the total sample was obtained before sorting, and the number of sampling units (households) included in the survey were recorded so that the average weight of waste per household per week could be determined. The solid waste in the institution was also sorted in terms of organic and inorganic materials—"organic" referring to food waste, and "inorganic" referring to PET bottles, tin cans, metal scraps, and the like. The first phase of this project dealt with the collection of waste in different bins—green bins for food waste, red for paper and disposable waste, and blue for PET bottles.

1.6. Sample Analysis

The samples were sorted into types and classes of solid waste, and the weight of each type and class was recorded. For this survey, the waste was categorized into the following classes—paper, PET bottles, nylons, tetra packs, plastic food packs, tin cans, food waste, and others. Waste was classified because we needed to get an idea of the amount of recyclable waste from the Covenant University waste stream.

2. Method of Analysis

The results were analyzed using Equations (1)–(5), respectively. Bar charts were used to express primary data collected to give the weight of characterized household waste per kg/household/day.

$$Per\ capita\ waste\ generated\ (kg/capital/day) = \frac{total\ solid\ waste\ per\ day}{total\ population\ that\ produces\ the\ waste} \qquad (1)$$

$$Average\ solid\ waste\ generated/day\ (kg/day) = \frac{total\ weight\ generated/week}{7\ days} \qquad (2)$$

$$Characterization\ of\ waste\ composition\ (\%) = \frac{weight\ of\ segregated\ waste}{weight\ of\ total\ waste} \times 100 \qquad (3)$$

$$Average\ waste\ generated\ in\ a\ household\ (kg/day,\ kg) = \frac{total\ waste\ generated\ by\ different\ households}{total\ number\ of\ households} \qquad (4)$$

$$Average\ total\ waste\ generated\ by\ population\ of\ a\ place\ (kg/day) = per\ capita\ waste \times total\ population \qquad (5)$$

3. Results

Table 4 shows the material percentage comparisons in the students' hall of residence, where waste generated from the male and female halls of residence are compared. It can be seen that the female halls of residence have a higher solid waste generation rate per day than that of the male halls of residence.

Table 4. Material comparison of waste generated in the students' hall of residence.

Sorting	Total Mass Composition of Waste in Male Halls (kg/day)	Percentage Composition of Waste in Male Halls (%)	Total Mass Composition of Waste in Female Halls (kg/day)	Percentage Composition of Waste in Female Halls (%)
PET bottles	112.50	26.95	109.13	26.00
Tetra packs	50.00	11.98	48.75	11.51
Paper	25.00	5.99	23.75	5.61
Food waste	22.50	5.39	20.00	4.72
Plastic food packs	128.75	30.84	122.50	28.92
Nylon	50.00	11.98	72.50	17.11
Tin cans	21.25	5.09	15.00	3.60
Others	7.5	1.78	12.00	2.53
Total	417.5	100	423.63	100

Total waste generated in the female halls of residence is 423.63 kg/day, while that of the male halls of residence is 417.5 kg/day. Tables 5 and 6 show a comparison between the wastes generated in the staff's residential areas of the University, while Figure 2 shows a comparison of the percentage of waste generated from the staff's residential areas.

Table 5. Material comparison of waste from residential areas for staff.

Location	PET Bottles (kg/day)	Tetra Packs (kg/day)	Paper (kg/day)	Food Waste (kg/day)	Plastic Food Packs (kg/day)	Nylon (kg/day)	Metal Cans (kg/day)
Male halls of residence	112.50	50.00	25.00	22.50	128.75	50.00	21.25
Female halls of residence	109.125	48.75	23.75	20.00	122.50	72.50	15.00
New estate	42.18	16.15	20.00	502.00	12.05	38.08	34.22
Post graduate quarters	15.60	6.90	8.70	306.30	6.12	21.00	16.80
Suites	8.00	4.48	4.48	167.04	3.20	12.80	8.00
Professors village	5.25	5.60	5.60	190.33	2.10	14.91	6.58
Total	292.66	131.88	87.53	1207.87	274.72	209.29	101.85

Table 6. Comparison of percentages of waste generated by residential areas for staff.

Sorting	Total Mass Composition (kg/day)	Total Percentage of Waste Generated
PET bottles	292.66	12.69
Tetra packs	131.88	5.72
Paper	87.53	3.79
Food waste	1207.87	52.40
Plastic	274.72	11.92
Nylon	209.29	9.07
Tin Cans	101.85	4.41
Total	2305.80	100

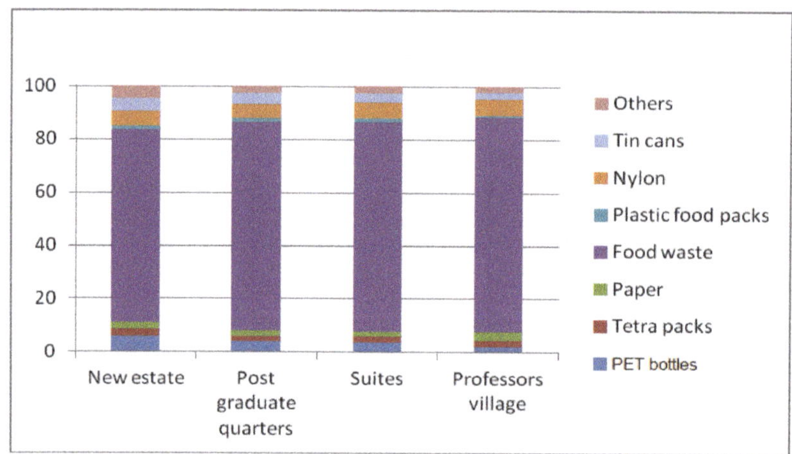

Figure 2. Comparison of percentages of waste generated by residential areas for staff.

Comparing the weight of waste generated in the various areas, it can be observed that the staff's residential areas have a larger composition of food waste. This can be attributed to the fact that staff members cook their food, unlike students who buy food which come in plastic food packs. Students' residential areas have a higher composition of plastic food packs, PET bottles, and nylons due to the frequent buying of food and drinks. It can be said that the number of housing units (i.e., the population) is a major factor affecting the rate of waste generation in the various staff residential areas. The new estate, which has the highest waste generation rate, also has the largest population. Taking an average of four (4) people per household, various estimates for each residential area can be expressed as follows:

i.   The new estate, with 141 housing units and an average population of 964 people, generates 696.71 kg/day of solid waste.

ii.   The post-graduate quarters, with 120 housing units and an average population of 480 people, generates 391 kg/day of solid waste.

iii.   The suites, with 64 housing units and an average population of 256 people, generates 212.64 kg/day of solid waste.

iv.   The professor's village, with 70 housing units and an average population of 280 people, generates 236.99 kg/day of solid waste.

Table 7 and Figure 3 compare the percentage of waste generated in the two cafeterias. It can be observed that Cafeteria 1 generated a larger amount of solid waste.

Table 7. Total waste generated in both cafeterias 1 and 2.

Sorting	Total Waste Composition (kg/day)	Total Percentage of Waste Generated (%)
PET bottles	31.85	23.24
Tetra packs	2.30	1.68
Paper	0.00	0.00
Food waste	97.68	71.26
Plastic food packs	1.00	0.77
Nylon	1.73	1.46
Tin cans	2.50	1.83
Total	137.06	100

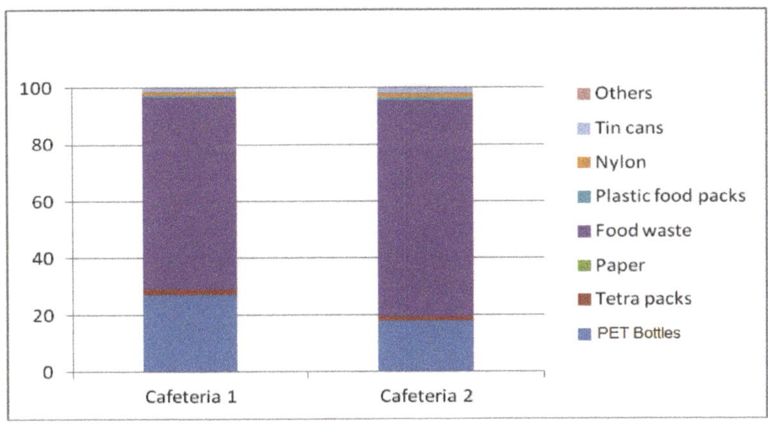

Figure 3. Percentages of waste obtained from the two cafeterias.

The disparity in results can be attributed to the difference in the number of students that visit both cafeterias—Cafeteria 1 generally has more customers, due to factors such as its proximity to the halls of residence. From Table 7 and Figure 3, it can be seen that the composition with the highest percentage in the total waste stream from the selected sites is food waste, followed by PET bottles and plastic food packs, respectively. The high composition of food waste mostly came from the staff's residential areas as they prepared their own food, thus increasing the amount of food waste generated. This corroborates with the assertions of Sridhar [21] and Ogwueleka [7], where they expressed that in Nigeria, 60 to 80 percent of waste is organic in nature. The high percentage of PET bottles and food packs is from the student residential areas, as they purchase food which comes in plastic food packs and drinks in PET bottles. Tin cans and paper were in the lowest percentile in the waste stream, and this may be attributed to the fact that people at the University rarely bought products in tin cans, and paper is also rarely used in the residential areas and cafeterias.

Table 8 and Figure 4 show that the new estate generated the largest weight of 664.68 kg/day, followed by the female halls of residence with 423.63 kg, and male halls of residence with 417.50 kg. This could be due to factors such as consumer habits, population, and others. Although the student residential areas have a higher population than that of the new estate, the heavier weight of waste in the new estate compared to that of the female and male halls of residence could be attributed to the variety of waste generated from the new estate, compared to that of the student residential areas. Because items purchased by students are limited, it reduces the weight of waste generated. Waste which comes from food preparation in the staff's residential areas weighs more than the other components in the waste stream. Table 9 shows that a significant amount of revenue could be generated from recyclable materials, and also presents the real economic values of the recyclable materials, exclusive of any processing. This is based on the sale of any of the materials in the waste stream (market price). The cost value is calculated per day, of which extrapolating to a year can be easily done. However, during the holidays, these values drop since the institution is basically residential as all students would have vacated the campus.

Table 8. Total average weight of waste generated in selected sites.

Sites	Total Average of Waste Generated (kg/day)
Male halls of residence	417.50
Female halls of residence	423.63
New estate	664.68
Post graduate quarters	381.42
Suites	208.00
Professor's Quarters	230.37
Cafeteria 1	84.475
Cafeteria 2	57.575

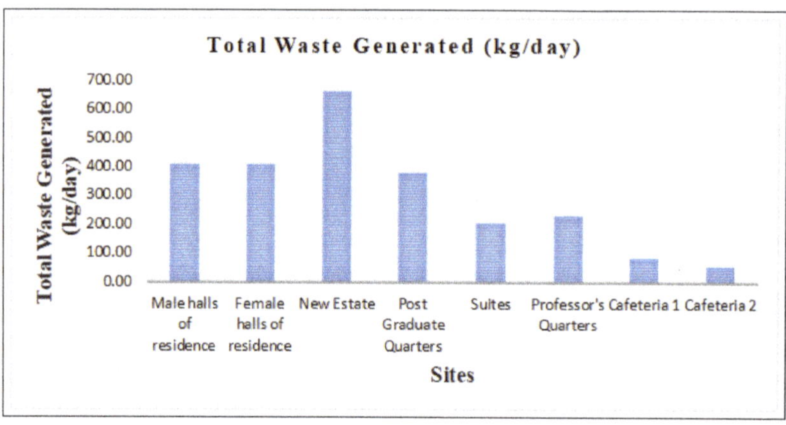

Figure 4. Comparison of total weight generated in selected areas.

Table 9. Economic value from recycling waste materials.

Recyclable Waste Materials	Average Percentage of Each Item in the Waste Stream (%)	Waste Generated per Day (kg/day)	Prices of Recyclables in Naira/kg	Total Value of Waste Generated (in Naira)
PET bottles	12.69	292.66	N55	N16,696.30
Paper	3.79	87.53	N5	N437.65
Plastic food packs	11.92	274.72	N30	N8241.60
Nylon	9.07	209.29	N30	N6278.70
Tin cans	4.41	101.85	N35	N3564.75
Tetra packs	5.72	131.88	N35	N4615.80
Food waste	52.40	1207.87	Compostable	-
Total	100.00	2305.80		N39,834.80

I. Food Waste: Compost/organic fertilizers can be obtained from food waste by composting, which is an aerobic process where micro-organisms decompose biodegradable waste to produce organic fertilizers in the presence of oxygen. In Covenant University, the main sources of food waste that can be used for composting come from the cafeterias and staff residential areas.

II. Plastic: This includes PET bottles and plastic food packs. Plastic can be recycled or reused, depending on its quality. The recycling process of plastic involves sorting, washing, drying, wet grinding, as well as extrusion, and palletizing. The final product is then packaged and sold to consumers [3]. Some of the products created from recycled plastic include office accessories, fiber for sleeping bags and duvets, polyethylene bin liners and carrier bags, and many others. The major types of recyclable plastics are polyethylene terephthalate (PET), high-density polyethylene (HDPE), low-density polyethylene (LDPE), polypropylene (PP), and so on. In Covenant University, the main sources of plastic which include PET bottles and plastic food packs are the student halls of residence and the cafeterias.

III. Nylons: Nylons can be reused and also recycled into other forms of nylon, like sachets for water and black bags used for waste disposal.

IV. Tetra Packs: Tetra Pack cartons are primarily made from paper. 75% of the Tetra Pack carton is made from paperboard, 20% from polyethylene, and 5% from aluminum. These three materials are layered together using heat and pressure to form a six-layered armor which protects the contents from light, oxygen, air, dirt, and moisture. Furthermore, Tetra Pack cartons are lightweight, easy to transport, and fully recyclable. The aseptic technology allows the product inside to stay fresh, without the need of any preservatives.

V. Tin Cans: These include drink cans, food cans, and beverage tins. They are smelted in high-temperature furnaces, and the resulting molten metal can be used to manufacture foil that is reintegrated into the manufacturing process, hence saving natural resources, energy, time, and money. Figure 5 shows the comparison of the total weight generated in the selected areas.

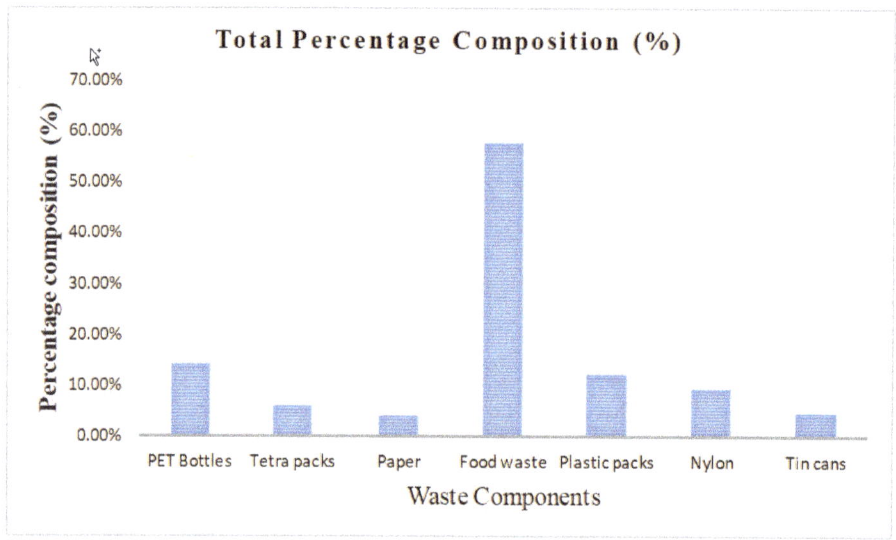

Figure 5. Comparison of total weight generated in selected areas.

Covenant University currently has some forms of recycling activities, like selling PET bottles which have been turned into pellets, and waste paper being traded in exchange for tissue use. Target marketers are companies in the states of Lagos and Ogun dealing with PET-bottle recycled products, and Chinese companies that use the materials to produce fabrics. This initiative generates income from the project for the university, and the resulting environmental sanitation and sustainability will be at its peak. Figure 6 below shows the storage site of the collected PET bottles.

Figure 6. Collection site for all used PET bottles.

4. Conclusions

Although the waste collection system at Covenant University is very efficient and is evidenced by the clean environment, however, improvements are needed in terms of its waste collection times, component separation at points of collection, and final disposal systems. Waste generated by the university are representative of municipal wastes, and the proposed methodology from this study may apply at the local, state, and federal levels in an attempt to implement and drive government policy on sustainable waste management.

Information on the characteristics of solid waste is important in evaluating systems, management programs, and plans for reuse, reduction, recycling, and final disposal activities for solid waste management. From the survey carried out in this project, a total amount of N39, 834 ($120 USD) per day can be obtained from the proper recycling of waste generated in residential areas. The highest amount can be obtained by plastic food packs with N16, 696.30 ($50 USD), and the lowest by paper is N437.50 ($1.5 USD).

From the survey carried out, it was observed that waste generation and characterization are dependent mostly on the products being supplied and sold in various areas of the University, such as the shopping mall, cafeteria, and others. In the students' halls of residence, it was observed that a sufficient amount of recyclable plastic packs and PET bottles were generated. Plastic packs from the halls of residence account for about 91.46% of the total PET bottles generated in the residential area, which is approximately 12.6% of the total waste generated in the residential areas. Food waste is the largest composition of the waste generated in the residential areas, accounting for about 52.40% of the total waste stream, with the staff residential areas contributing hugely to this at 96.48%.

The overall goal of solid waste management is to collect, treat, and dispose waste. Conclusively, it can be said that more can be done to improve solid waste management in Covenant University. Considering the amount of revenue that could be made from proper recycling, the University should make more investments in the purchase of recycling equipment for nylons, plastics, paper, and metals. The University can also partner with government agencies and private organizations and take this functional system to the wider society, starting from the immediate local government down to the state level.

Author Contributions: Conceptualization of the research work came from D.O.O.; and the Methodology adopted was carried out by all authors: D.O.O., A.O.A. and I.H.K. All the authors played a contributing role in the use of the Software, and the validation of the work was carefully checked by D.O.O. The formal Analysis and investigation was done by D.O.O. Resources and Data Curation was carried out by D.O.O., A.O.A. and I.H.K. while the Writing of Original Draft Preparation was done by D.O.O. Writing-Review Editing was done by D.O.O. Visualization, Supervision and Project Administration were done by D.O.O. It is important to note that no Funding was received for this research project.

Funding: This research received no external funding.

Acknowledgments: The management of Covenant University is well appreciated for creating enabling platform.

Conflicts of Interest: The authors declare no conflict of interest.

References

1. Klundert, A.V.; Anschiitz, J. *The Sustainability of Alliances Between Stakeholders in Waste Management-Using the Concept of Integrated Sustainable Waste Management*; UWEP/CWG Netherlands Ministry for Development Co-Operation: CW Gouda, The Netherlnds, 2000; pp. 1–22.
2. Ohaka, A.; Ozor, P.; Ohaka, C. Household Waste Disposal Practices in Owerri Municipal Council of Imo State. *Niger. J. Agric. Food Environ.* **2013**, *9*, 32–36.
3. Olukanni, D.O.; Mnenga, M.U. Municipal Solid Waste Generation and Characterization: A Case Study of Ota, Nigeria. *Int. J. Environ. Sci. Toxicol. Res.* **2015**, *3*, 1–8.
4. Olukanni, D.O.; Oresanya, O.O. Progression in Waste Management Processes in Lagos State, Nigeria. *J. Eng. Res. Afr.* **2018**, *35*, 11–23. [CrossRef]
5. Adebola, O. *Investment Opportunities in Integrated Solid Waste Management (A.K.A Waste to Wealth) for Public Private Sector*; The Abia State Environmental Summit: Umuahia, Nigeria, 2005; p. 2.
6. Olukanni, D.; Adeleke, J.; Aremu, D. A review of local factors affecting solid waste collection in Nigeria. *Pollution* **2016**, *2*, 339–356.
7. Ogwueleka, T.C. Municipal Solid Waste Characteristics and Management in Nigeria Iran. *J. Environ. Health Sci.* **2009**, *6*, 173–180.
8. Babayemi, J.O.; Dauda, K.T. Evaluation of solid waste generation, categories and disposal options in developing countries: A case study of Nigeria. *J. Appl. Sci. Environ. Manag.* **2009**, *13*, 83–88. [CrossRef]
9. Amber, I.; Kulla, D.M.; Gukop, N. Municipaal waste in Nigeria generation, characteristics and energy potential of solid. *Asian J. Eng. Sci. Technol.* **2012**, *2*, 84–88.

10. Chukwu, K.E. Recycling of used Plastic Products: Its Sanitary and Commercial Unpublication in Enugu State. In Proceedings of the Stakeholders' Workshop, Converting Waste to Wealth through Waste Recycling, Enugu State Waste Management Authority, Enugu, Nigeria, 30–31 January 2007.

11. Mahees, M.T.M.; Sivayoganathan, C.; Basnayaka, B.F.A. Consumption, Solid Waste Generation and Water Pollution in Pinga Oya Catchment area. *Trop. Agric. Res.* **2011**, *22*, 239–250. [CrossRef]

12. Olukanni, D.O.; Oladipupo, A.O.; Ede, A.N.; Akinwumi, I.I.; Ajanaku, K.O. Appraisal of Municipal Solid Waste Management. its effects and resource potential in a semi-urban city. *J. South Afr. Bus. Res.* **2014**, *2014*, 1–13. [CrossRef]

13. Bowan, P.A.; Tieroba, M.T. Characteristics and Management of Solid Waste in Ghanaian Markets-A study of WA Municipality. *Civ. Environ. Res.* **2014**, *6*, 114–119.

14. Rahman, M.D.; Atiqur, H.; Khondoker, M. Scenario of Market Waste Management and Environmental Degradation: A Case Study in Khulna City Area. In Proceedings of the Waste Safe 2013 3rd International Conference on Solid Waste Management in the Developing Countries, Khulna, Bangladesh, 10–12 February 2013.

15. Okenyi, B.E.; Ngozi-Olehi, L.C.; Njoku, B.A. Chemical Education: A Tool for Wealth Creation from Waste management. *J. Res. Natl. Dev.* **2011**, *1*, 143–147.

16. Adu, D.A.; Aremu, A.S. Impetus for Recycling Activities across the Globe: An Overview. In Proceedings of the Unilorin 4th Annual 2nd International Conference of Civil Engineering, Ilorin, Nigeria, 4–6 July 2012.

17. Awopetu, M.S.; Coker, A.O.; Awopetu, R.G.; Awopetu, S.O.; Booth, C.A.; Fullen, M.A.; Hammond, F.N.; Tannahill, K. Reduction, Reuse and Recycling of Solid Waste in the Makurdi Metropolitan Area of Nigeria. *Int. J. Educ. Res. WIT Trans. Ecol. Environ.* **2013**, *163*, 51–59.

18. Agbaeze, E.K.; Onwuka, I.O.; Agbo, C.C. Impact of Sustainable Solid Waste Management on Economic Development–Lessons from Enugu State Nigeria. *J. Econ. Sustain. Dev.* **2014**, *5*, 9.

19. Agunwamba, J.C. Analysis of scavengers' activities and recycling in some cities of Nigeria. *Environ. Manag.* **2003**, *32*, 116–127. [CrossRef]

20. Kadafa, A.A.; Latifah, A.M.; Abdullah, H.S.; Sulaiman, W.A. Current Status of Municipal Solid Waste Management Practise in FCT Abuja. *Res. J. Environ. Earth Sci.* **2013**, *5*, 295–297.

21. Sridhar, M. From urban wastes to sustainable waste management in Nigeria: A case study. In *Sustainable Environmental Management in Nigeria*; Ivbijaro, M.F.A., Akintola, F., Okechukwu, R.U., Eds.; Mattivi Productions: Ibadan, Nigeria, 2006; pp. 337–353.

Review

Public-Private Sector Involvement in Providing Efficient Solid Waste Management Services in Nigeria

David O. Olukanni * and Christiana O. Nwafor

Department of Civil Engineering, Covenant University, PMB 1022 Ota, Ogun State, Nigeria;
christiana.nwafor@stu.cu.edu.ng
* Correspondence: david.olukanni@covenantuniversity.edu.ng

Received: 21 February 2019; Accepted: 25 April 2019; Published: 27 April 2019

Abstract: This paper reviews the partnership between the public and the private sectors in providing efficient solid waste management (SWM) services. While the responsibility of providing SWM services lies with the public sector, the sector has not been able to meet the demand for efficient service delivery, especially in developing countries. In a bid to increase efficiency and lower costs incurred in rendering these services, the involvement of the private sector has been sought. With a focus on major Nigerian cities, partnerships between the local government and private operators in SWM have been analysed based on the level to which the partnership has improved the SWM services. This paper provides an understanding that the success of any public-private partnership relies on the extent to which all stakeholders perform their duties. If the public sector is slack in monitoring and supervising the activities of the private operators, the latter may focus on profit generation while neglecting efficient service delivery. Also, legislation is an important part of SWM. Without the right legislation and enforcement, waste generators will not be mandated to dispose their waste properly. The public sector as a facilitator is responsible for creating an environment for private operators to function, particularly through legislation, enforcement and public sensitization.

Keywords: municipal solid waste; waste management; environmental pollution; sustainable technology; public-private partnership

---

1. Introduction

Waste management is a complex issue in developing countries, from mega-cities to small towns and villages, and has been on the priority list of successive governments, local authorities, and international donors in recent years [1,2]. This issue creates one of the most critical health and environmental concerns confronting many administrators [3–5]. In many constitutions, it is the responsibility of the local government to provide infrastructure; including power and water supplies, public transportation, telecommunications, security, and waste management facilities. As documented in the policy guidelines of the Federal Ministry of the Environment (2005) on Solid Waste Management, urban solid waste management in Nigeria is constitutionally the responsibility of the third tier of government, that is, the local government [4]. However, this tier of government has not committed the necessary financial, material, and human resources to waste management to fulfil their responsibility [6].

The actual performance of the public sector in Nigeria has left much to be desired not only in the waste management sector but in many of the state-owned enterprises which are not responsive to the changing requirements of a growing and dynamic economy and do not seem to possess the necessary tools for effective service delivery [7]. Solid waste management institutions in Nigeria are crippled by the lack of a comprehensive legislative framework for effective solid waste management, overlapping functions, lack of funding, poor implementation of environmental policy, vested interests and corruption, and a technological deficit to match the rate at which solid waste is being generated [4,8,9]. The people also have a part to play in effective SWM, as poor waste disposal habits, corruption, refusal to pay for waste collection services and bad work attitudes hinder the efforts of the local government in providing an effective waste management service in Nigeria [4,6]. Given the failure of the public sector in effective service delivery, many have viewed the involvement of the private sector in public infrastructure provision as a way forward. This approach has gained global prominence with varying degrees of success in different countries and in different sectors of the economy [10–15].

In view of the challenges associated with providing efficient and sustainable waste management in most urban settings in Nigeria, this paper discusses the current state of public-private partnerships (PPPs) in urban solid waste management alongside management practices and strategies. The following research questions have been drawn from the aim: In what ways can and have the private sector partnered with the public sector for more efficient service delivery? How have collaborations between the public and private sectors benefited waste management in Nigeria? What lessons can be learnt from other developing countries with relative success in jointly provided waste management services? Four cities—Lagos, Ibadan, Abuja and Port Harcourt—have been chosen from three geopolitical zones in Nigeria based on their demographic and socio-economic features as a reflection of the Nigerian situation, and because of the availability of literature on these cities.

2. Private Sector Involvement in Infrastructure

Private sector involvement in infrastructure development has gained significance over the years in developed and developing countries alike [16]. The local government typically represents the public partner at the national, state, or local agency level while the private partner can be a privately-owned enterprise or consortium of businesses with a particular specialisation whose main aim is profit realization [17] defines public-private partnerships (PPPs) as a contract between a public authority and one or more private operators to transfer the control of a good or a service currently provided by the public sector, either in whole or in part, to the private operators.

2.1. Forms of Private Sector Involvement in Infrastructure

Private sector involvement (PSI) in infrastructure development has been described as a spectrum given that it can take a number of forms, depending on the degree of risk shared by the private and public sectors and the duration of the contract. In cases where the government merely outsources the provision of a basic service to the private sector, the government retains ownership of the assets involved and a larger portion, if not all, of the risk involved. Conversely, in divestitures or privatisations, the government transfers the control of an asset either partially or in full to the private sector [18,19]. PPPs are considered as alternatives to full privatisation in which there is a substantial extent of risk sharing between the public and private sectors [20,21]. Table 1 gives a brief description of the basic forms of private sector involvement practiced around the world.

Table 1. Description of types of PSI with examples from other countries.

	Type of PSI	Key Characteristics	Example with source
	Service Contract	i. Public partner owns assets and pays private contractor a service fee ii. Private partner handles service delivery	i. Waste collection and transport in Bangalore, India [15]
	Management Contracts	i. Public partner owns assets ii. Private partner manages operations partially or fully	i. Water and Wastewater service in Amman, Jordan [11]
	Leases	i. Public partner leases assets to private partner and makes capital investments ii. Private partner manages operations and maintains facility	i. Dar es Salaam Water Distribution Company, Tanzania [19]
PPPs	Build-Operate-Transfer (BOT)	i. Private partner builds new facility and owns it ii. Private partner operates facility iii. Ownership is eventually transferred to the public partner	i. MSW to Energy Plant in Wenzhou, China [22]
PPPs	Concession	i. New or Existing assets owned by private partner ii. Private partner operates the facility iii. Ownership is eventually transferred to the public partner	i. Kenya-Uganda Railways, Kenya and Uganda [19]
	Divestiture/Privatisation	i. Private partner buys asset from public partner ii. Private sector owns assets, makes capital investment and bears risk completely	i. Thames Water, UK [10]

Source: [19,23].

Table 1 above is not an exhaustive list of the forms of PSI that are practiced around the world. For example, the variations of the BOT contract (e.g. Build-Own-Operate, Build-Operate-Manage-Transfer), and the Private Finance Initiative (PFI) practiced in the UK are not included. Although BOT contracts and concessions share similar characteristics, they differ in the mode in which the private partner is paid. In many concession contracts, the private partner recovers its investment from the users of the service by imposing tolls, fees, etc., while BOT contracts usually involve the public sector paying a bulk fee to the private partner [21,23]. Also, because of the complexity of PSI projects, different means of classifying PSI exist, based on how the private partner recoups costs [18], and based on the nature of the service provided and the degree of risk sharing [21].

2.2. Success Factors of Private Sector Involvement in Infrastructure Development

Over 4000 public infrastructure projects have been implemented in collaboration with the private sector between 1990 and 2007, and a number of successes and failures have been recorded [19]. Many researchers have sought to find commonalities among PSI projects that contributed to their success or failure. One approach that has been used widely, is finding the critical success factors (CSFs) that attempt to delineate the most crucial areas that are required to ensure management success [24].

While the CSFs for PPP projects are fairly consistent around the world [25–27], the ranking of individual CSFs in developed and developing nations highlight the differences in their political and socio-economic climates. The results presented in [28] and [29] show similar CSFs for the implementation of PPP projects in Nigeria.

In Nigeria for example, the failures recorded by the state owned waste management bodies in providing non-rivalled and non-exclusive SWM service, is one of the main reasons for involving the private sector in SWM operations [4].

3. Private Sector Involvement in SWM in Nigeria

The Federal Ministry of Environment is responsible for environmental protection, natural resources conservation and sustainable development in Nigeria and effective waste management is one of its cardinal focuses. In 2005, a set of policy guidelines on solid waste management was developed by the Federal Ministry of Environment. The policy guidelines set out general objectives for solid waste management in Nigeria and defined the roles and responsibilities of the government at the federal-, state- and local-government level. The following four solid waste management options were recommended within the policy guidelines:

i.   By local government/municipal agencies
ii.   By private companies on contract with the LGA/municipality
iii.   By private companies on contract with home owners
iv.   By public-private partnership (PPP)

Like much of Nigeria's environmental laws and policies, the policy guidelines have never been effectively implemented, and there has not been any further development at the federal level of private sector involvement in solid waste management [30].

3.1. Private Sector Involvement in Solid Waste Management in Abuja, Federal Capital Territory

Abuja was created in 1976 by a decree by the then military head of state, General Murtala Mohammed. A master plan was developed for the city in 1979 that included provision of waste management infrastructure that could cater for a steady increase in population. However, due to rapid expansion, the population of Abuja now exceeds the original design capacity with a projected population of about 5.8 million people by 2026 (Federal Ministry of the Environment, 2005). Hence, waste management infrastructure has not been developed in phase with city growth and the environmental challenges of highly populated cities like Lagos are also being experienced in Abuja [31].

In Abuja, the Abuja Environmental Protection Board (AEPB) is responsible for the management of solid waste and is the principal authority for waste collection and disposal in the city. The area councils and satellite towns in the city were delineated into 22 lots for the purpose of daily waste collection from households in the councils. Collection of solid waste is contracted to 12 private operators in a concession agreement which runs for a minimum of three years. The AEPB, the Abuja Investment Company and a Ukrainian firm, entered into a joint-venture agreement for waste management activities in Abuja. Under the agreement, the AEPB and the Ukrainian technical partners shared the cost of procuring 50 compacting trucks and 12 street sweepers, with the AEPB contributing 20 percent of the funds. Figure 1 shows the distribution of waste management vehicles operating in Abuja and it can be seen that over 75% of operational waste management equipment is private-owned.

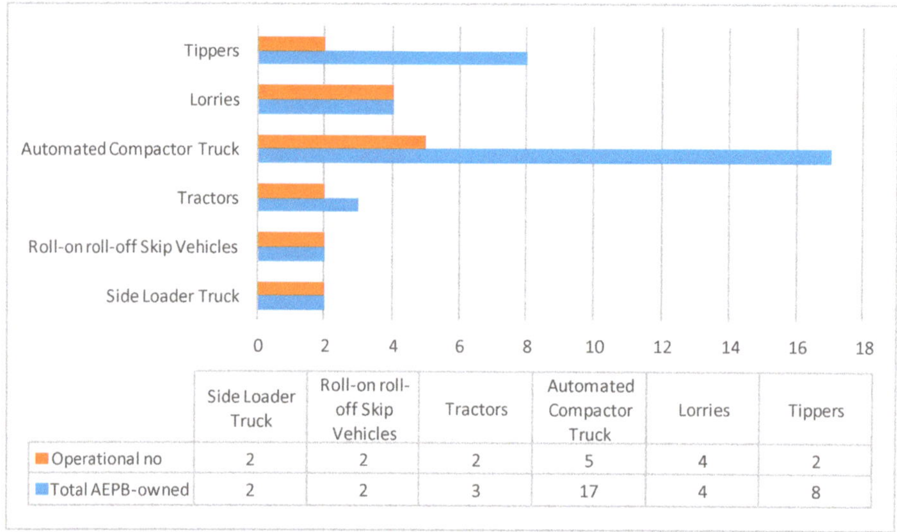

	Side Loader Truck	Roll-on roll-off Skip Vehicles	Tractors	Automated Compactor Truck	Lorries	Tippers
Operational no	2	2	2	5	4	2
Total AEPB-owned	2	2	3	17	4	8

Figure 1. Distribution of waste collection, transportation and disposal vehicles in Abuja. *Source*: [32].

It is evident from Figure 1 that the private sector owns more operational vehicles than the public sector in Abuja due to factors such as availability of capital for procurement and maintenance of these vehicles. However, efficient waste management is still a challenge in Abuja as the service provided is usually restricted to the major districts in the city. Overflow of waste into the roads due to infrequent collection of solid waste from the designated storage sites or communal bins has been reported in some satellite towns and city suburbs.

Given the current state of waste management in the city, measures should be put in place to strengthen waste governance by integrating the private public partnership and informal sector models for more effective waste collection and disposal. This approach is expected to help reduce challenges that arise as a result of conflicts across agencies, while at the same time leveraging on the financial, technical and administrative capacity of the private sector for sustainable SWM in the city [33].

3.2. Private Sector Involvement in Solid Waste Management in Lagos, Lagos State

Lagos State is the economic capital of Nigeria and the largest, most diverse single settlement in the country [33]. While MSWM operations have been organized compared to other states, the management of solid waste in Lagos State is still far from optimal. Though the Nigerian constitution charges the local governments with the responsibility of SWM, over time, the state government has taken up the responsibility.

Continuous increase in population in many cities has led to an increase in the rate of waste generation that the public agencies could not manage. The agencies lacked the financial, technological, and skilled human capital to match the rate of waste generation. This led to streets getting dirtier and exposed Nigeria to global ridicule as Lagos, the largest city in the country, became infamous for being one of the dirtiest cities in the world after the Festival of Arts and Culture in 1977 [6,34]. Even with the institution of the Lagos State Refuse Disposal Board (LSRDB) and Lagos Waste Management Authority (LAWMA) in 1977 and 1991, respectively, the state of waste management in Lagos did not improve as the medians of roads in the state were still characterized by waste heaps, drains were clogged with waste and market places and many other public places were littered with solid waste [34–36].

Prior to the incorporation of the private sector, the efforts of the government had been deterred by issues such as lack of finance for capital investments and high operational costs associated with provision of waste management service. Such problems were not peculiar to Lagos alone but were common in other states in the country. Other issues particular to the Lagos metropolis include traffic congestion which limited the number of trips collection vehicles could make per day, high rate of waste generation, unavailability of land, and rural-urban migration [37].

With 340 private sector participants (PSPs) registered in Lagos State alone, there has been an improvement in solid waste management service delivery, especially in the aspect of waste collection. The PSPs introduced waste collection at the source (that is collection from door-to-door) and residents were billed according to the value of the property and the area in the state. Residents in high income areas are billed the highest and efficiency of service delivery is reported to be highest in these areas as well [6,34]. An analysis of the state of waste management in Lagos shows that more equipment has been made available for the collection and transportation of solid waste from the source to the disposal sites.

Private sector involvement in SWM in Lagos has progressed since it was first introduced formally under the LSRDB in 1985 till 1991 when LAWMA was created. The Table 2 below outlines the history of private sector involvement in SWM in Lagos from 1997 to 2017.

Table 2. History of private sector involvement in SWM in Lagos (1997–2017).

Year	SWM Institution	Advancement in SWM	Supervising Authority
1997–1998	Lagos Waste Management Authority (LAWMA) and Private Sector Participation (PSP)	The private operators managed domestic waste while LAWMA managed industrial, commercial waste.	Ministry of the Environment and Physical Planning (MEEP)
2004	Mega PSP in waste management	PSP operators were paid monthly by the state government while tenements were billed monthly to recoup state funds.	Ministry of the Environment/Office of the Deputy Governor
2007	Mega PSP in waste management	Procurement of additional collection trucks and compactors for more efficient service delivery	LAWMA/MoE
2009	Franchised Mega PSP in waste management, PPP arrangement with an American firm	Waste to compost facility sited at Ikorodu where organic waste is converted to fertilizer	LAWMA/MoE
2015	Build-operate-manage-transfer (BOMT) contract with West Africa ENRG	Construction and operation of the first Materials Recovery Facility (MRF) in Nigeria.	LAWMA/MoE
2017	Cleaner Lagos Initiative with Visionscape Sanitation Solutions	Refurbishing, building and upgrading SWM infrastructure including engineered landfills, transfer loading stations, MRFs etc.	MoE

Source: [33].

While great strides have been made in the state [5], problems such as inadequate funding, cost recovery, unstable power supply, and traffic congestion militate against effective SWM in Lagos. Also, lack of continuity in implementing SWM policies prevents the full actualisation of the benefits of public-private partnerships: An example being the dissolution of LAWMA and the PSP scheme by the incumbent government. To replace them, the state government recently awarded contracts to Visionscape Sanitations Solutions and its strategic partners for deployment of waste management infrastructure. The transition has not been smooth, however, as the state of waste management in Lagos has since deteriorated and waste is being dumped on the streets and along the road median.

3.3. Private Sector Involvement in Solid Waste Management in Ibadan, Oyo State

Ibadan is the capital of Oyo State and the third most populous city in Nigeria with 3,088,477 residents [38]. The city is one of the foremost commercial hubs in Nigeria alongside Kano and Onitsha. The Oyo State Solid Waste Management Authority (OYOMWA) is responsible for waste management in the Ibadan metropolis. Managing the large volumes of waste generated in Ibadan has been a serious challenge for all stakeholders. Indiscriminate dumping of waste is widely practiced in the city resulting in waste heaps littering road sides and streets.

Private sector involvement in SWM in Ibadan dates as far back as 1994 when private contractors were involved in collection, transport, and disposal of solid waste [30,31]. In Oyo State, the private sector participation scheme suffered because of challenges with availability of capital for the purchase of equipment. Therefore, purchasing equipment for use such as trucks and tippers is always beyond the reach of the owners. The resultant effect of this constraint was that some firms also used their trucks for construction activities in order to maximize profit. Also due to financial constraints, the private firms are unable to purchase the necessary spare part replacements, pay staff salary and employ trained experienced manpower [39].

3.4. Private Sector Involvement in Solid Waste Management in Port Harcourt, Rivers State

Port Harcourt is the capital city of Rivers State and one of the most prominent cities in the southern region of Nigeria. Port Harcourt was initially established as a port-town during the colonial times but since the discovery of oil in Rivers State, the city has evolved into one of great economic and political importance. The city is home to 1,845,232 residents [38]; a number which is rapidly on the increase due to urbanization and rural-urban migration.

The Rivers State Waste Management Agency (RIMAWA) is responsible for solid waste management in the state under the supervision of the state Ministry of Environment. The agency engages about 88 local vendors or contractors who provide services ranging from de-silting of creeks and canals, municipal waste collection, and dumpsite management within Rivers State (RIMAWA, n.d.). The challenges of managing waste in Port Harcourt arose mainly with the uncontrolled and unplanned development of the city which resulted in overcrowding, unplanned road networks, and heaps of refuse littering various parts of the city [40]. House-to-house collection is limited to the elite areas of the city (e.g. Old GRA) while communal bins are provided in areas like Borokiri [41]. A study carried out by Stanley and Owhor [41] on 390 respondents in three neighbourhoods within the city showed that the frequency of collection also varied across the city with Ogbumnuabali enjoying more frequent collection than Borokiri and Elekahia. The above data points to the fact that waste collection is not uniform in the city even with the involvement of private contractors.

Comparing the state of solid waste management in Lagos and Port Harcourt, it is evident that the latter is only beginning to explore the benefits of collaborating with the private sector for more efficient service delivery. Waste management in Rivers State is still in its developmental stages and is mainly concerned with collection, transportation and disposal of solid waste. No attention has been given formally to waste minimization activities or the recovery of resources from the waste stream.

4. Conclusions

While there are many opportunities in private sector involvement, it should not be seen as a panacea—a supposed cure for all problems. In fact, partnerships with the private sector can worsen the state of SWM if all the stakeholders involved are not committed to fulfilling their distinct roles. The role of the public sector changes from direct involvement in construction and service provision, to ensuring that the PPP delivers value for money for the government and better services for users. In Lagos, unlike other cities in Nigeria, many of the partnerships with the private sector in SWM are with agencies with an understanding of standard SWM practices and the financial and technical capacity to deliver services efficiently.

Solid waste management in many developing countries especially Nigeria is not where it ought to be. Lapses in collection, treatment and disposal militate against the effective management of waste generated in many major cities. Relying on public-private partnerships (PPPs) without providing the necessary legal and institutional framework will not produce the desired results in effective SWM.

5. Recommendation

The World Bank recommends that cost recovery of waste management services should be ensured and the revenue collection of waste management fees from waste producers should be increased [31]. Also, a transparent procurement process for private sector participants should be established and performance-based contracts should be developed. The government should include the people in the decision making process to foster acceptance of new policies and initiatives. Finally, the act of keeping the environment safe and clean by proper disposal of wastes can be enhanced if regulatory measures are enforced on the people. In lieu of this, public health officers (PHOs) need to enforce compliance with these regulations by the residents. This will go a long way in achieving sustainable waste management because people will not want to be victims of law for fear of the penalty attached to it [28,39].

Author Contributions: Conceptualization of the research work came from D.O.O. and the methodology adopted was carried out by the authors: D.O.O. and C.O.N. All the authors played a contributing role in the use of the software, and the validation of the work was carefully checked by D.O.O. The formal analysis and investigation was done by D.O.O. Resources and data curation was carried out by D.O.O., and C.O.N. While the writing of original draft preparation was done by D.O.O. and C.O.N. Writing-review editing was done by D.O.O. Visualization, supervision and project administration was done by D.O.O.

Funding: This research received no external funding.

Acknowledgments: The authors are grateful to the management of Covenant University for providing an enabling environment in doing the report of this research work. We also appreciate the various authors of resource materials that were used which provided relevant information that led to the success of this work.

Conflicts of Interest: The authors declare no conflict of interest.

References

1. Hoornweg, D.; Bhada-Tata, P. *What a Waste: A Global Review of Solid Waste Management*; World Bank, Urban Development Series; Knowledge Papers No. 15; The World Bank eLibrary: Washington, DC, USA, 2012.
2. Olukanni, D.O. Analysis of municipal solid waste management in Ota, Ogun State, Nigeria: Potential for wealth generation. In Proceedings of the International Conference on Solid Waste Management (ICSW), Philadelphia, PA, USA, 10–13 March 2013; pp. 184–196.
3. Olukanni, D.O.; Aremu, D.O. Provisional Evaluation of Composting as Priority Option for Sustainable Waste Management in South-West Nigeria. *Pollution* **2017**, *3*, 417–428. [CrossRef]
4. Olukanni, D.O.; Adeleke, J.O.; Aremu, D.D. A Review of Local Factors Affecting Solid Waste Collection in Nigeria. *Pollution* **2016**, *2*, 339–356. [CrossRef]
5. Olukanni, D.O.; Oresanya, O.O. Progression in Waste Management Processes in Lagos State, Nigeria. *Int. J. Eng. Res. Africa* **2018**, *35*, 11–23. [CrossRef]
6. Anestina, A.I.; Adetola, A.; Odafe, I.B. Performance Assessment of Solid Waste Management following Private Partnership Operations in Lagos State, Nigeria. *J. Waste Manag.* **2014**, *2014*, 1–8. [CrossRef]
7. Asaolu, T.O. Privatization in Nigeria: Regulation, deregulation, corruption and the way forward. In *Inaugural Lecture Delivered at Oduduwa Hall, Obaremi Awolowo Univeristy, Ile-Ife*; Obafemi Awolowo University Press Limited: Ile-Ife, Osun, Nigeria, 2015; Available online: http://thenationonlineng.net/privatization-in-nigeria-regulation-deregulation-corruption-and-the-way-forward (accessed on 27 April 2019).
8. Abila, B.; Kantola, J. Municipal Solid Waste Management Problems in Nigeria: Evolving Knowledge Management Solution. *Int. J. Environ. Ecol. Geol. Min. Eng.* **2013**, *7*, 169–174.

9. Ezeah, C.; Roberts, C.L. Analysis of barriers and success factors affecting the adoption of sustainable management of municipal solid waste in Nigeria. *J. Environ. Manag.* **2012**, *103*, 9–14. [CrossRef] [PubMed]

10. Delmon, J. *Understanding Options for Private-Partnership Partnerships in Infrastructure: Sorting out the Forest from the Trees: BOT, DBFO, DCMS, Concession, Lease … ;* The World Bank eLibrary: Washington, DC, USA, 2010. [CrossRef]

11. Marin, P. *Public-Private Partnerships for Urban Water Utilities: A Review of Experiences in Developing Countries;* The World Bank eLibrary: Washington, DC, USA, 2010.

12. Spoann, V.; Fujiwara, T.; Seng, B.; Lay, C.; Yim, M. Assessment of Public–Private Partnership in Municipal Solid Waste Management in Phnom Penh, Cambodia. *Sustainability* **2019**, *11*, 1228. [CrossRef]

13. Wiafe, M. Private Sector Involvement in Solid Waste Management in the Kumasi Metropolitan Assembly. Master's Thesis, Kwame Nkrumah University of Science and Technology, Kumasi, Ashanti, Ghana, 2014. Available online: http://dspace.knust.edu.gh/bitstream/123456789/8068/1/MAXYFT%282%29final.pdf (accessed on 27 April 2019).

14. Yescombe, E.R. Public-Private Partnerships in Sub-Saharan Africa: Case Studies for Policymakers. Available online: https://www.africaportal.org/publications/public-private-partnerships-sub-saharan-africa-case-studies-policymakers-2017/ (accessed on 20 April 2018).

15. Zhu, D.; Asani, P.U.; Zurbrugg, C.; Anapolsky, S.; Mani, S. *Improving Municipal Solid Waste Management in India, A Sourcebook for Policy Makers and Practitioners;* WBI Development Studies; World Bank: Washington, DC, USA, 2008. [CrossRef]

16. Akintoye, A. PPPs for Physical Infrastructure in Developing Countries. In *Policy, Finance & Management for Public-Private Partnerships;* Wiley-Blackwell: Oxford, UK, 2009; pp. 123–144. [CrossRef]

17. Massoud, M.; El-Fadel, M. Public-private partnerships for solid waste management services. *Environ. Manag.* **2002**, *30*, 621–630. [CrossRef] [PubMed]

18. Farquharson, E.; Torres de Mästle, C.; Yescombe, E.R. *How to Engage with the Private Sector in Public-Private Partnerships in Emerging Markets;* The World Bank eLibrary: Washington, DC, USA, 2011. [CrossRef]

19. Yong, H.K. *Public–Private Partnerships Policy and Practice: A Reference Guide;* Yong, H.K., Ed.; The Commonwealth iLibrary: London, UK, 2010. [CrossRef]

20. Ahmed, S.A.; Ali, M. Partnerships for solid waste management in developing countries: Linking theories to realities. *Habitat Int.* **2004**, *28*, 467–479. [CrossRef]

21. Yescombe, E.R. *Public-Private Partnerships: Principles of Policy and Finance;* Elsevier: Amsterdam, The Netherlands, 2007.

22. Asian Development Bank. Municipal Solid Waste treatment: Case Study of Public–Private Partnerships (PPPs) in Wenzhou. 2010. Available online: https://www.adb.org/publications/municipal-solid-waste-treatment-case-study-ppps-wenzhou (accessed on 21 March 2019).

23. World Bank. *Public-Private Partnerships Reference Guide V 2.0;* The World Bank Group: Washington, DC, USA, 2014.

24. Muhammad, Z.; Sik, K.K.; Johar, F.; Sabri, S. An overview of critical success factors of publicprivate partnership in the delivery of urban infrastructure and services. *Plan. Malays. J.* **2016**, *14*, 147–162.

25. Chou, J.S.; Pramudawardhani, D. Cross-country comparisons of key drivers, critical success factors and risk allocation for public-private partnership projects. *Int. J. Proj. Manag.* **2015**, *33*, 1136–1150. [CrossRef]

26. Osei-Kyei, R.; Chan, A.P.C. Comparative Analysis of the Success Criteria for Public–Private Partnership Projects in Ghana and Hong Kong. *Proj. Manag. J.* **2017**, *48*, 80–92. [CrossRef]

27. Osei-Kyei, R.; Chan, A.P.C. Empirical comparison of critical success factors for public-private partnerships in developing and developed countries A case of Ghana and Hong Kong. *Eng. Constr. Archit. Manag.* **2017**, *24*, 1222–1245. [CrossRef]

28. Sanni, A.O. Factors determining the success of public private partnership projects in Nigeria. *Constr. Econ. Buil.* **2016**, *16*, 42–55. [CrossRef]

29. Babatunde, O.S.; Opawole, A.; Akinsiku, E.O. Critical success factors in public-private partnership (PPP) on infrastructure delivery in Nigeria. *J. Facil. Manag.* **2012**, *10*, 212–225. [CrossRef]

30. Onibokun, A.G. *Managing the Monsters: Urban Waste and Governance in Africa;* IDRC Books: Ottawa, ON, Canada, 1999.

31. World Bank. *Private Sector Participation in Solid Waste Management Activities in Ibadan, Nigeria*; The World Bank Group: Washington, DC, USA, 2017; Available online: http://documents.worldbank.org/curated/en/221251487249039986/Nigeria-Private-sector-participation-in-solid-waste-management-activities-in-Ibadan-synthesis-note (accessed on 27 April 2019).

32. Imam, A.; Mohammed, B.; Wilson, D.C.; Cheeseman, C.R. Solid waste management in Abuja, Nigeria. *Waste Manag.* **2008**, *28*, 468–472. [CrossRef]

33. Ezeah, C.; Roberts, C.L. Waste governance agenda in Nigerian cities: A comparative analysis. *Habitat Int.* **2014**, *41*, 121–128. [CrossRef]

34. Adedibu, A.A.; Okekunle, A.A. Issues in the environmental sanitation of Lagos mainland, Nigeria. *Environmentalist* **1989**, *9*, 91–100. [CrossRef]

35. Aliu, I.R.; Adeyemi, O.E.; Adebayo, A. Municipal household solid waste collection strategies in an African megacity: Analysis of public private partnership performance in Lagos. *Waste Manag. Res.* **2014**, *32* (Suppl. 9), 67–78. [CrossRef] [PubMed]

36. Adeniran, A.E.; Nubi, A.T.; Adelopo, A.O. Solid waste generation and characterization in the University of Lagos for a sustainable waste management. *Waste Manag.* **2017**, *67*, 3–10. [CrossRef]

37. Ayantoyinbo, B.; Adepoju, O. Analysis of Solid Waste Management Logistics and Its Attendant Challenges in Lagos Metropolis. *Logistics* **2018**, *2*, 11. Available online: www.africapolis.org/data (accessed on 5 February 2019). [CrossRef]

38. Alakinde, M.K. Private sector participation and sustainable solid waste management in Ibadan South West Local Government, Nigeria. *J. Emerg. Trends Econ. Manag. Sci.* **2012**, *3*, 887–892.

39. Ayotamuno, J.M.; Gobo, A.E. Municipal solid waste management in Port Harcourt, Nigeria: Obstacles and prospects. *Manag. Environ. Qual.* **2004**, *15*, 389–398. [CrossRef]

40. Ikebude, C.F. Feasibility Study on Solid Waste Management in Port Harcourt Metropolis: Causes, Effects and Possible Solutions. *Niger. J. Technol.* **2017**, *36*, 276–281. [CrossRef]

41. Stanley, H.; Owhor, A. Assessment of Solid Waste Management Practice in Port Harcourt Metropolis, Rivers State, Nigeria. *J. Geogr. Environ. Earth Sci. Int.* **2018**, *16*, 1–10. [CrossRef]

Article

Sustainability Assessment of Waste Management System for Mexico City (Mexico)—Based on Analytic Hierarchy Process

Nina Tsydenova [1,*], Alethia Vázquez Morillas [2] and Arely Areanely Cruz Salas [2]

[1] School of Computing Science, Business Administration, Economics, and Law (Faculty II),
    Carl von Ossietzky University of Oldenburg, 26129 Oldenburg, Germany
[2] Department of Energy, Area of sustainable technologies,
    Universidad Autónoma Metropolitana UAM—Azcapotzalco, 02200 Mexico City, Mexico;
    alethia@correo.azc.uam.mx (A.V.M.); areanelyc@gmail.com (A.A.C.S.)
*   Correspondence: nina.tsydenova@gmx.de; Tel.: +49-176-973-49877

Received: 28 July 2018; Accepted: 7 September 2018; Published: 12 September 2018

Abstract: Mexico City introduced the new legal waste norm Norma NADF-024-AMBT-2013 in July 2017. This report compares the proposed system with three alternatives: a baseline scenario with composting of organics, a scenario which involves anaerobic digestion of organics, and a mechanical–biological treatment scenario with no source separation. The comparison was done using an Analytic Hierarchy Process. Eleven different indicators were chosen for the evaluation: general waste performance indicators (landfill disposal and recycling rates), environmental indicators (greenhouse gas emissions, acid gas emissions, Biological Oxygen Demand (BOD), and mercury content in water and soil), economic indicators (investment and operation costs) ($ per Mg municipal solid waste (MSW)), and social indicators (jobs created and social acceptance). The scenario ranking based on pairwise comparison made by 5 experts from Mexico City showed that the most sustainable scenario, environmentally, socially, and economically, is that which corresponds to Norma NADF-024-AMBT-2013 with a ranking priority of 30.78%.

Keywords: AHP; waste management; sustainability assessment; scenario ranking

---

1. Introduction

Waste management in emerging countries has become urgent during recent years, because economic growth and rise of consumption have caused a rise in waste production. The escalation of generation of residues has resulted in a shortage of disposal sites and higher waste management costs. In Accra, Ghana, the expenditure for municipal solid waste (MSW), for example, increased from 2013 to 2015 by 8% [1]. Other important waste management issues which developing countries must face include lack of proper governance instruments, inefficient resource use, overdependence on imported equipment, improper financing methods and application of technology, inequality in service stipulation, and deficient technical expertise [2]. In many cases, especially in lower-income countries, governance issues play a more important role than technical aspects [3]. Therefore, it is vital to find implementable, knowledge-based solutions for the governments of countries with poor municipal waste management performance. For example, in Mexico City, which produces 12,920 Mg per day of MSW and experiences the problem of shortage of landfill capacity, the waste infrastructure requires an alternative scenario. However, with the variety of technologies for treatment, recovery, and energy generation from waste, it is hard for decision makers, not experts themselves, to make a correct choice. Diverse technologies have distinctive climate change effects, and different investment and operation costs. An overall assessment of a waste management system is needed in such cases.

This assessment should combine aspects such as environmental performance, economic viability, and social acceptability. Also, local conditions should always be a consideration before a system is implemented.

Extensive research has been done to determine sustainable decision-making models to assess waste management scenarios. Among the existing models, the most popular ones are life-cycle assessment (LCA), cost–benefit analysis, and multicriteria analysis. LCA calculates the environmental impact of all processes of the waste treatment from "cradle to grave"; cost–benefit analysis considers the monetary dimension, while multicriteria decision analysis (MCDA) compares social, economic, and environmental criteria [4]. There is literature available which compares different waste management systems in emerging and developing countries. Brunner and Fellner [5] conducted a study to determine appropriate waste management systems in less-developed economies. Zubruegg [6] developed an assessment tool based on a questionnaire. Elsaid and Aghezzaf [7] presented a progress-indicator-based assessment guide for integrated municipal solid waste management systems.

MCDA is often used in waste management. One of the MCDA methods widely applied in waste management is ELECTRE (Elimination and Choice Expressing the Reality). ELECTRE can incorporate many evaluation criteria for selecting an optimal alternative, coupled with the possibility of involving several decision-makers. It has been applied to the real choice process of a solid waste management system in Bosnia and Herzegovina by Vučijak et al. [8]. Another, PROMETHEE (Preference Ranking Organization Method for Enrichment Evaluations), was developed by Brans [9] and further extended by Brans and Vincke [10]. Queiruga et al. [11] used PROMETHEE, combined with a survey of experts, to rank Spanish municipalities for the installation of Waste Electrical and Electronic Equipment (WEEE) recycling plants. Demesouka et al. [12] used a combination of geographical information systems (GIS) and MCDA methods in the analysis of municipal solid waste landfill suitability.

In this study, the alternative scenarios of MSW management in Mexico City are compared based on MCDA. The purpose of the analysis is to determine the most sustainable scenario for the city. The method used was chosen because it allows the involvement of environmental, economic, and social criteria, by comparing diverse quantitative and qualitative dimensions to produce a ranking. Moreover, it allows the participation of different stakeholders with various interests. Therefore, it has been chosen as a tool to assess the sustainability of waste management in this work.

The goal of this study is to find the most sustainable waste management scenario for Mexico City based on waste composition, experts' opinions, and overall assessment of state-of-the-art technologies, by using MCDA. No research combining economic, environmental, and social dimensions has been previously conducted to determine a sustainable residues management model for Mexico City. This study is the first to visualize and compare the economic benefits, the impact of waste treatments on the environment, and social benefits of specific waste technologies for Mexico City. The outcome should help decision makers in the introduction of a successful, sustainable waste management system. The results are assumed to be transferable to other megacities in emerging countries and help decision makers in the selection of waste management scenarios with energy and resource recovery.

2. Methods

2.1. Analytical Hierarchy Process (AHP)

The present study applied the Analytical Hierarchy Process (AHP). Although AHP is one of the oldest MCDA methods, developed by Saaty [13], it is still widely used today. AHP allows the problem to be broken down into its constitutive elements, listed in relation to the main goal [14]. AHP 'is a multicriteria decision-making technique, which can concurrently consider qualitative and quantitative comparison criteria and where a lot of baseline research literature is available. Therefore, AHP is ideally suited to a project such as this, which needs to do comparative research involving many stakeholders having various interests [15]. AHP is applicable in a wide range of fields, including management, business, and policy [16,17], and is also often used to solve complex problems in environmental management. AHP application in land management is discussed by Schmoldt et al. [18].

The AHP method is widely applied in emerging countries, where waste management decisions need to be made in the absence of established sound environmental solutions. It has been used to compare solid waste treatment scenarios for cities [19] and university campuses [2] or to select the recycling strategy for specific waste categories like WEEE [20]. Taboada-González et al. [21] used AHP to select the best waste treatment with energy recovery for Ensenada, Baja California (Mexico). Araiza Aguilar et al. [22] looked for the zones suitable for the emplacement of waste management infrastructure in Mexico with the help of geographical information systems (GIS) and AHP. Martínez-Morales et al. [23] applied AHP to identify the municipalities of Mexico State with major waste management problems. Gomez Jauregui Abdo [24] discussed development of domestic water supply in Guadalajara with the help of AHP. A wide range of applications of the AHP method shows that it is a powerful decision tool for assisting decision makers in the selection of sustainable waste management strategies.

The AHP hierarchical structure allows decision makers to prioritize solutions in terms of relevant criteria. Additional criteria can be added later in the hierarchical structure after the first results are obtained. The decision procedure using the AHP is made up of four steps, as described by Saaty [25]:

(1) Define the problem and determine the kind of knowledge sought.

(2) Structure the decision hierarchy according to the goal of the decision in the following order: the objectives from a broad perspective, through the intermediate levels (criteria on which subsequent elements depend), up to the lowest level (which usually is a set of the alternatives).

(3) Construct a set of pairwise comparison matrices. Each element of the matrix in the upper level is used to compare elements in the level immediately below.

(4) Use the priorities obtained from the comparisons to weigh the priorities in the neighboring level. Do this for every element. For each element in the level below, add its weighed values and obtain its overall or global priority. Continue this process of weighing and adding until the final priorities of the alternatives in the bottom-most level are reached.

2.2. Study Area

Mexico is a diverse country with 125 million residents and abundant natural resources. It is a member of the Organization for Economic Co-operation and Development and simultaneously a developing country with a GDP per capita of 8201.3 US$ [26]. Mexico City, the capital and the most populated city, is considered in this case study. The city is located in the Valley of Mexico in the center of the country. It consists of 16 boroughs and is spread over an area of 1485 km^2. The estimated population of 9 million [27] generates 12,920 mega grams (Mg) of MSW per day with per capita production of 1.43 kg per day. However, approximately 4 million people travel for work to Mexico City from the nearby states. The composition of collected MSW is presented in Figure 1, based on the analysis made in 2 boroughs of the city, Coyoacán and Venustiano Carranza. These districts can be taken to represent a broad middle of society. Therefore, the waste composition in these areas is assumed to be representative for the whole city. However, it is to be mentioned that the drivers of waste trucks, being part of the informal recycling sector, separate the major part of cardboard, PET, and metals for further sale. The sample thus represents the composition of waste arriving at the transfer station, but not the waste coming directly from the households. The results of the waste composition analysis correspond to the outcomes of the study by the Polytechnic University in 2013 [28]. The waste composition in Mexico City is not typical for developing countries, which tend to have a higher organic fraction (more than 50%). This shows that Mexico is an emerging economy, in transition from developing to an industrialized country. However, the biggest fraction is represented by food waste (27.77%), while plastics and paper/cardboard constitute 15.79% and 10.55%, respectively. The interesting fact here is that around 6% of the MSW is represented by toilet paper, explained by cultural habits and the capacity of the sewage system.

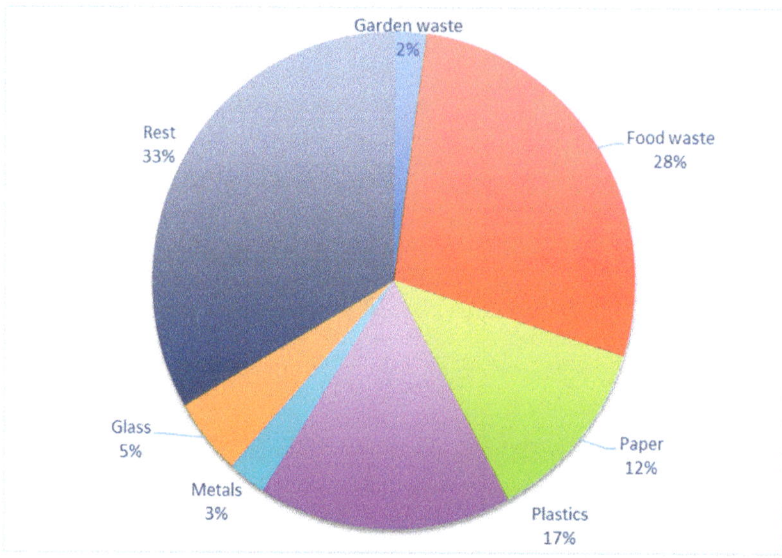

Figure 1. MSW waste composition in Mexico City 2017.

The MSW management system of the city consists of 12 transfer stations, 2 sorting facilities, and 8 composting plants. 14% of the generated waste is recycled with the help of the informal sector [29]. Almost all the collected waste is sent to the five landfills located outside of the city in the nearby state. This is a big challenge because it enormously increases transportation costs. Hence, a new waste legal norm, Norma NADF-024-AMBT-2013, was introduced in July 2017. This aims to increase the amount of collected recyclable materials and thereby decrease the quantity of landfilled material. At the same time, the city is planning to construct an incineration plant, which will treat almost half the collected residues.

2.3. Selection of Technical Alternatives

This research considers the following scenarios: (1) landfilling and composting, (2) anaerobic digestion, (3) MBT (mechanical–biological treatment) with composting as biological stage, and (4) incineration. The first scenario was chosen as a base scenario, which presents a business-as-usual scenario. It involves windrow open composting and engineered landfills. These composting piles are turned to improve porosity and oxygen content. Incineration and MBT plants were selected because these options are the most discussed options of sustainable waste management in developing countries [30–32]. The second scenario is considered to be the alternative for the others. Due to the high percentage of organics, wet anaerobic digestion can be very beneficial through energy and fertilizer supply, and is, according to Badri et al. [16], the most favorable treatment option for organic waste through energy and fertilizer supply. However, this option requires the source separation scheme of waste at households.

2.4. Waste Management Scenarios

This research assesses the sustainability of several alternative scenarios for waste management for Mexico City. Since 2003, municipal solid waste in Mexico City has been separated at source into two fractions: organics and inorganics. However, the new regulation NADF-024-AMBT-2013 mandates the new segregation of residual waste into five fractions: organic, recyclables, nonrecyclables, hazardous waste, and bulky waste [33]. This work compares the proposed system with three alternatives: a baseline scenario with composting of organics, a scenario which involves anaerobic digestion (AD) of organics, and a mechanical–biological treatment (MBT) scenario with no source separation. MBT is a collective term, mainly used in Europe, which incorporates several variations of MSW treatment, based on a combination of mechanical processing and biological treatment (in most cases aerobic or anaerobic decomposition) [34,35]. According to the new regulation Norma 024, the organics should be composted, recyclable materials should be sorted and recycled, while nonrecyclables, hazardous waste, and bulky waste would be incinerated with energy recovery. The mass flows of each scenario are presented in Figures 2 and 3. The charts were made and flows calculated with the program STAN, developed by TU Vienna [36]. The mass flows were assessed based on the following assumptions: 14% of generated MSW is recycled through the informal sector; the sorting plants receive daily 1725 Mg of recyclables from the State of Mexico. The source separation efficiency of the baseline corresponds with the official data from the Environmental Ministry of Mexico City [33]. The separation in AD and Norma 024 scenarios is determined by the waste composition from Figure 1 and corresponds to the ideal efficiency for comparison. As well, the sorting efficiency of the MBT is based on Navarotto and Llauro [37]. Detailed studies of process efficiency for MBT plants, in terms of sorting efficiencies and quality of recovered materials, are scarce in published literature. According to Cipman et al. [38], the study of Navarotto and Llauro [37] is one of the most detailed descriptions available. During their tests, the MBT Ecoparc 4 was subjected to a three-month-long campaign, and materials flows were recorded, sampled, and analyzed, including waste input, products, plant residue, and some intermediary process flows. The detailed process description of the material recovery section in the Ecoparc 4 MBT plant in Barcelona, Spain, is given in the Supplementary Material.

Hospital and hazardous waste is assumed to be gathered separately and burnt at special incineration plants, however, it is not considered in the mass flow analysis. The sources and methods used for each estimation in the mass flow diagram (MFD) are presented in Table 1. Each MFD is discussed in detail in the Supplementary Material.

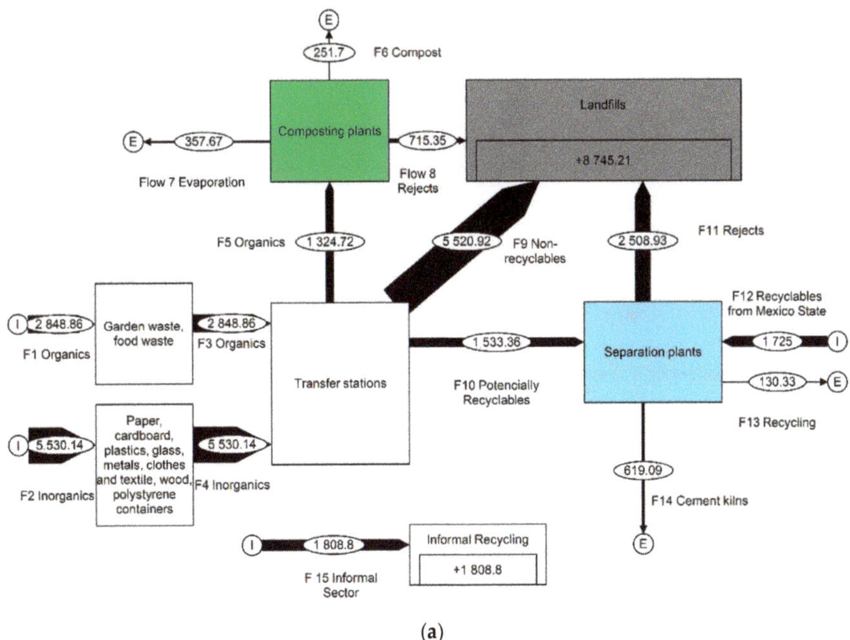

(a)

Figure 2. *Cont.*

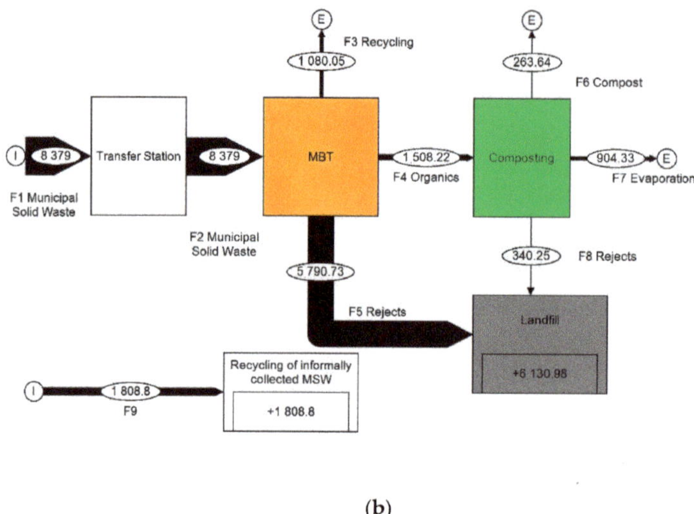

(b)

Figure 2. Mass flows of waste scenarios: (**a**) baseline scenario; (**b**) scenario with MBT.

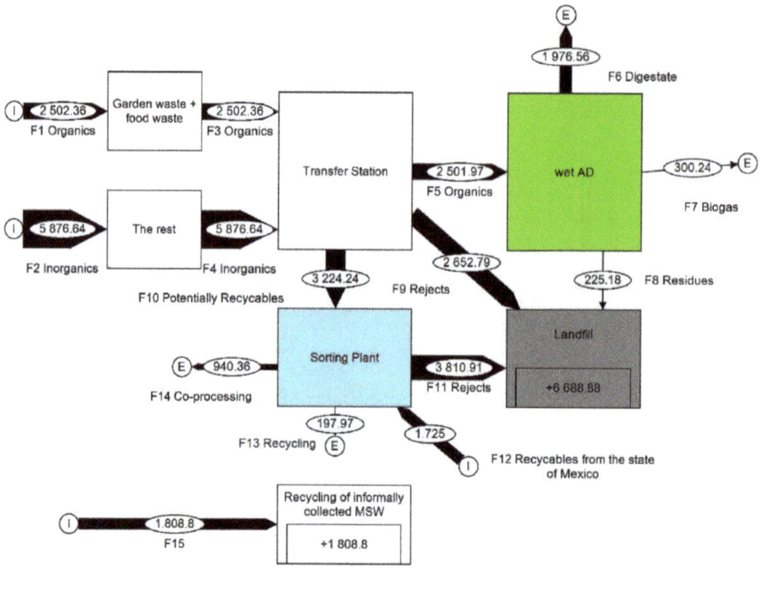

(a)

Figure 3. *Cont.*

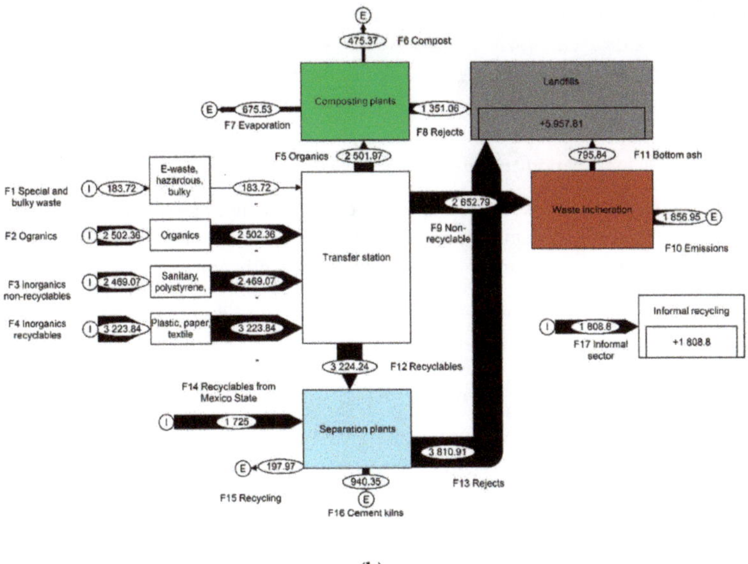

(b)

Figure 3. Mass flows of waste scenarios: (**a**) scenario with wet AD; (**b**) scenario in compliance with Norma 024.

Table 1. Description of the discussed scenarios.

Scenario	Description
Scenario a	Landfilling without landfill gas utilization 1928.33 Mg per day of waste (paper, cardboard, plastic, metal, and glass) is recycled, 1324.72 Mg per day of waste is composted, 69.09 Mg of waste is used for RDF production, 8745.21 Mg per day of waste is landfilled.
Scenario b	Mechanical–Biological treatment 2878.05 Mg per day (paper, cardboard, plastic, metal, and glass) is recycled, 1508.22 Mg per day of organic waste (food and garden waste) is composted, the rest is landfilled.
Scenario c	Anaerobic digestion with biogas utilization 2006.77 Mg per day of waste (paper, cardboard, plastic, metal, and glass) is recycled, 2501.97 Mg per day of waste is sent to anaerobic digestion plant, 940.36.98 Mg of waste is used for RDF production, the rest is landfilled.
Scenario d	Incineration with energy recovery 2006.77 Mg per day (paper, cardboard, plastic, metal, and glass) is recycled, 2652.79 Mg per day of residual waste is sent to incineration plant.

2.5. Selection of Indicators

To examine the indicators for the sustainability of a waste treatment scenario a literature review was performed. Hokkanen and Salminen [39] identified a set of 24 indicators for waste management and divided them into six groups: economic, technical, environmental, political, employment and resource recovery. Greene and Tonjes [40] applied 12 indicators: MSW recycled, MSW landfilled, MSW diverted from the landfill, diversion rate, recycling rate, curbside recycling rate, landfilling rate, recycling per capita, landfilling per capita, diversion per capita, GHG reductions, energy savings. The present study applied the sustainable indicators used in the AHP model of Milutinović et al. [41]:

- overall waste management performance: landfill disposal rate; recycling rate;
- environmental indicators: GHG emissions (CO_2 Equivalents) per Mg of MSW, acid gases emissions (nitrogen oxides) per Mg of MSW, biological oxygen demand (BOD) and mercury in soil (heavy metals in soil);
- economic indicators: investment and operational costs;
- social indicators: job creation and public acceptance.

The set of indicators was selected according to the following criteria: relevance of the indicator for local sustainability of waste management, potential measurability at the local level and power of the local authority to change the outcomes measured by the indicator [19]. No extra indicators were added during the AHP process.

2.6. Evaluation of Indicators

2.6.1. Overall Waste Management Performance

The amount of waste that remains after treatment for landfill disposal was estimated based on the mass flow modelled with the help of the STAN software. The software was used for both the representation and calculations. The baseline scenario is provided by the data presented by the Environmental Ministry of the city [33]. Information for other scenarios was based on data from the literature, since the technologies considered are not presently available in Mexico City; the data for MBT was based on Navarotto and Llauro [37], AD and incineration data on the report for the Austrian ministry of Ecology [42], and composting on Andersen et al. [43]. These sources have proven reliable when used in previous studies from Rodić and Wilson and Masood et al. [3,44]. Due to the absence of the technologies discussed in Mexico and overall in the region of Latin America, no real data could be used for the research. Therefore, the study could only consider values from the literature.

2.6.2. Environmental Indicators

Nitrogen oxides emissions to the atmosphere, BOD in water, and mercury concentration in soil were estimated within the life-cycle inventory (LCI) using EASETECH software [45] and its last available database version from July 2017. These indicators are assumed to give an overall environmental assessment, considering contamination in the atmosphere, water, and soil. The same criteria were applied in the AHP analysis in Multinovic [41].

This study has been carried out using the EASETECH model. In assessing emissions, this model calculates the emissions from the point at which a material is discarded into the waste stream to the point at which it is either converted into a useful material or finally disposed of (Kirkeby [46]). The EASEWASTE models were elaborated including waste treatment options and external processes, which can appear both upstream and downstream of a waste management system. The program also evaluates emissions associated with the fuel consumption for collection and transportation of waste. However, emissions from transportation to recycling facilities have not been included due to the lack of data on distances. Recycled materials and energy derived from the waste management system are regarded as substitutes for virgin materials or energy. Emissions into water, air, and soil alongside resource consumptions, which are avoided as a result, are subtracted from the other emissions and resource consumptions in the waste system. The model calculates emissions into water, air, and soil, along with the consumption of resources. The model applies life-cycle impact assessment (LCIA) methods for conversion of these exchanges into environmental impacts [46,47].

Climate change impact was estimated based on the International Reference Life Cycle Data System (ILCD) method. It contains the impact categories recommended by the European Commission and described by Hauschild et al. [48] (2013). The source for characterization factors for climate change at midpoint was the IPCC report for a 100-year period [49]. The calculations were made using the database of EASETECH from July 2017. All the processes were based on the premodelled technologies existing in the database, except the MBT. The MBT plant was presented as a combination of a composting and sorting plant. Among others, the emissions from the incineration process were calculated using the data from the Danish plant described by Møller et al. [50]. Emission profiles were not available for some external processes, but most impacts are covered.

2.6.3. Economic Indicators

The economic indicators include investment and operational costs for the prevailing treatment method for each scenario: AD, MBT, and incineration. The assessment was performed based on data from IADB [51], Münnich et al. [52], and Tsilemou and Panagiotakopoulos [53]. The costs of the current scenario are given in IADB [51]. The calculations were based on € Mg MSW^{-1}. For evaluation of investment costs, the expenditures for design and construction of landfills and waste treatment facilities were considered. However, taxation, amortization, and inflation were not considered.

This study does not include the collection and transportation costs, even though they may reach up to 50–80% of the entire costs of a waste management system, as in industrialized countries [54]. The expenditures for education campaigns for citizens regarding source separation are also not included due to the lack of data.

2.6.4. Social Indicators

The number of new jobs in waste treatment is calculated depending on the amount of processed waste (Mg MSW). The evaluation was done based on the data from the Environmental Ministry of Mexico City SEDEMA [33], Friends of the Earth [55], and European Commission [56]. Public acceptance is a qualitative criterion which cannot be measured, therefore, the scale established in the AHP (1-worst, 9-best) was used for the assessment of this criterion. These results were obtained during interviews with experts, where they expressed their opinion about different waste treatment options.

2.6.5. Ranking of Indicators

Indicators (not the alternatives) are compared pairwise applying a scale from 1 to 9 for making a ranking. An example questionnaire given to the experts can be found in Table 2. When "7" is chosen in the first row, the landfill elimination rate is much more important than the recycling rate for the sustainability of waste management, according to the personal opinion of the consulted experts, while "5" chosen in the second row means that the emissions of greenhouse gases are considerably more important than the rate of elimination of landfills.

Table 2. Example of ranking questionnaire for the experts.

Indicators	Ranking of Importance																	Indicators
	Extreme importance	Very strong importance	Strong importance	Moderate importance	Equal importance	Moderate importance		Strong importance		Very strong importance	Extreme importance							
Landfill disposal rate	9	8	(7)	6	5	4	3	2	1	2	3	4	5	6	7	8	9	Recycling rate
Landfill disposal rate	9	8	7	6	5	4	3	2	1	2	3	4	(5)	6	7	8	9	GHG emissions

Use the scale from 1 to 9 (where 9 is extreme importance and 1 of no importance) to indicate the relative importance of indicator in the left column to the indicator in the right column. Between 1 and 9, all situations are intermediate. Only one entry can be made in each row.

Pairwise comparisons were used to determine the relative importance of each alternative in terms of each criterion. To make a pairwise comparison and subjective priority weightings for the criteria, experts working in the waste management sector in Mexico City were consulted. This includes scientists working at the Universities UAM, UNAM, and Instituto Politécnico Nacional (the three main public universities in the City) in environmental and economic science and in the field of waste treatment, as well as experts from the government of Mexico dealing with the problem of waste management and environmental consultants. These Mexico-based scientists and experts were asked to rank the importance of the criteria with respect to the goal, selection of the most sustainable waste management scenario. The pairwise comparison was made by 5 experts. The process of filling out the questionnaire is time-consuming and the authors received only 5 filled out ones out of 14 distributed. The experts were not informed of the results of the indicators presented in Table 3. The filled-out questionnaires received by the authors are presented in Supplementary Material.

2.7. Limitations of Methodology

The methodology has some limitations. The mass flow analysis for the scenarios b, c, and d was based on Navarotto and Llauro [37], the report for the Austrian Ministry of Ecology [42], and Andersen et al. [43], due to the absence of real data in the regional context. The calculations made in the EASETECH model were based on the default data, because no chemical analysis of the waste composition was made. For the social indicators, the number of the workplaces was estimated through reports which were applied in other, previous studies. The costs of waste treatment were based on data from IADB [51], Münnich et al. [52], and Tsilemou and Panagiotakopoulos [53]. These limitations may affect the outcome of the study. The evaluation of indicators directly influences the ranking, which may lead to biased results. Nevertheless, the assumptions should to be made to implement the research.

3. Results and Discussion

3.1. Assessment of Indicators

The evaluation of the indicators (overall waste management performance, environmental, economic, and social criteria) is presented in Table 3.

Table 3. Evaluation of indicators.

Category	Criteria	Baseline	AD	MBT	Incineration	Source
Overall waste management performance	Landfill disposal rate (%)	86	64	60	56	Calculated by the authors (Mg of landfilled MSW/Mg of MSW collected)
	Recycling rate through formal sector (%)	1	2	11	2	Calculated by the authors (Mg of recycled MSW/Mg of MSW collected)
Environment	GHG (MgCO2eq Mg MSW^{-1})	0.458	0.175	0.03	−0.04	LCIA IPCC2007 Analysis made in Easetech. Available in Supplementary Material
	Nitrogen oxides (kg)	0.547	−0.14	−2.64	0.27	LCI Analysis made in Easetech. Available in Supplementary Material
	Biochemical oxygen demand (BOD) (kg)	0.27	0.21	3.66	0.39	LCI Analysis made in Easetech. Available in Supplementary Material
	Mercury in soil (kg)	0.000003	0.00009	0.000002	0.000002	LCI Analysis made in Easetech. Available in Supplementary Material
Economics	Investment costs ($ Mg^{-1})	0	13.67	415.92	94	Tsilemou and Panagiotakopoulos, 2006; Münnich, 2005
	Operation costs ($ Mg^{-1})	8.53	26.73	429.99	44.40	Tsilemou and Panagiotakopoulos, (2006), Münnich, 2005; IDB (2015)
Social	Social acceptance	9	8	6	3	Evaluations done by authors based on the interview
	Number of jobs (Persons/day)	320.76	12.83	32.77	14	Calculated by authors based on Friends of the Earth (2010), European Commission, Directorate-General Environment (2001) Employment Effects of Waste Management Policies, SEDEMA (2016)

Figure 4 shows the ranking of indicators made based on the pairwise comparison of the experts. According to the preferences of the experts, environmental criteria are most crucial for integrated sustainable waste management, while the economic indicators play the least significant role. The calculation of ranking of indicators is available in Supplementary Material.

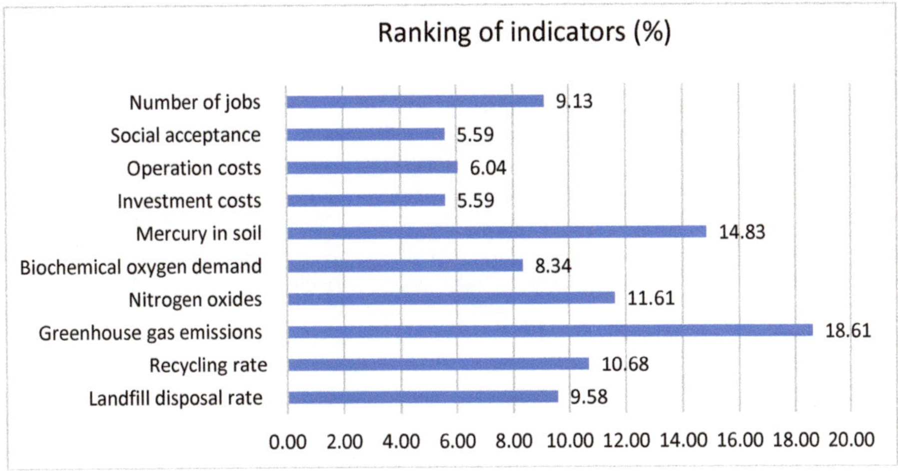

Figure 4. Ranking of indicators with respect to the goal.

3.2. Scenario Ranking

Following the pairwise criteria, the criteria weight of each scenario with respect to the goal was obtained as shown in Figure 5, presented in Supplementary Material Criteria's evaluation. The results show that Scenario 4, involving recycling of recyclable waste, composting of organic waste, and the thermal treatment of the remaining mixed waste for energy recovery (WTE), has the highest-ranking priority of 30.78%. The results show that the more separately collected categories of waste are involved in the plan, the more sustainable the scenario is considered to be by the experts.

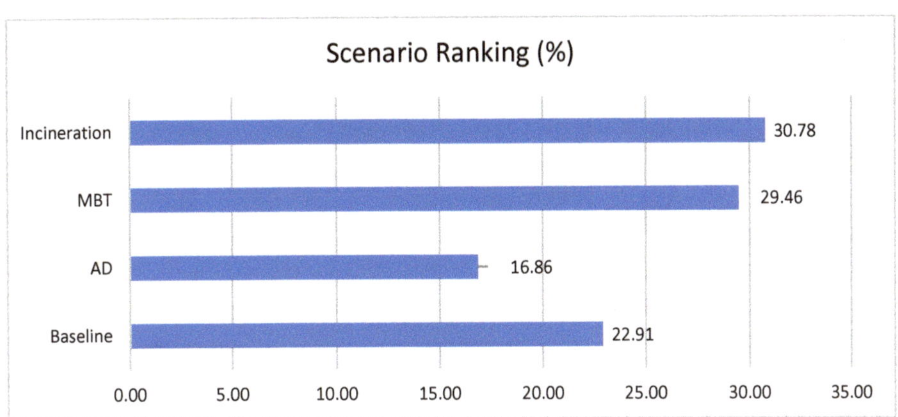

Figure 5. Ranking of scenarios.

Figure 4 indicates that in assessing the sustainability of waste treatment, the most relevant indicator is the GHG emissions. Therefore, the WTE scenario was ranked as the highest. However, it should be considered that the evaluation of indicators for this scenario was based on data from the literature, under the assumption that source separation rate is 100%. In order to achieve that result, the new incineration plant would have to comply with European standards, including emissions standards [57].

3.3. Sensitivity Analysis

The last step of the decision process using the AHP method is sensitivity analysis, where the input data of criteria weighting are modified to observe their impact on the results. If the ranking of scenarios does not change, the results are said to be robust [58]. The sensitivity analysis was performed to assess the influence of individual sustainability indicators on the proposed waste treatment scenarios. The following cases were examined following the procedure of Milutinović et al. [41]:

Case 1: All indicators had an equal weighting factor (10%). In this case, scenario ranking was changed. Baseline Scenario was top-ranked and most sustainable in terms of all indicators with a priority ranking of 29.9%. Scenario 4, which corresponds to the new waste management regulation in Mexico City and involves the incineration process and composting, was then in 2nd place.

Case 2: The group of environmental indicators was assigned a weighting factor of 100% in total (each of them had a weighting factor of 25%), while all the others had a weighting factor of 0%. In this case, Scenario 4 was ranked as the best with priority ranking of 34.62%, due to smaller CO_2 and nitrogen oxides emissions, as well as the fact that recycling greatly reduces the disposed waste volume.

Case 3: One waste indicator had a weighting factor of 100%, while all others had a weighting factor of 0%. There are 10 options, and the results showed that when economic and social waste management indicators had a weighting factor of 100%, while others had a weighting factor of 0%, Scenario 1 (baseline) got the first place. Scenario 4 (WTE) ranked best in cases where indicators of landfill disposal, CO_2 emission, and content of heavy metals in soil indicators had a weighting factor of 100%, while others had a weighting factor of 0%.

Case 4: The group of economic indicators were weighted at 100% in total (each at 50%), while all others had a weighting factor of 0%. Here, Scenario 1 is mostly preferred with a priority ranking of 54.57%. Under these conditions, the ranking changed because of the lower investment and operation costs of the existing waste management system.

Case 5: The group of social indicators was given a weighting factor of 100% in total (each 50%), while all others had a weighting factor of 0%. In this case, Scenario 3 was ranked in first place with a priority ranking of 40.68%. Scenario 4 had the lowest place in the ranking due to the small number of jobs created and public disquiet about thermal waste treatment.

All the results of the sensitivity analysis are presented in Figures 6 and 7. The calculations are presented in Supplementary Material.

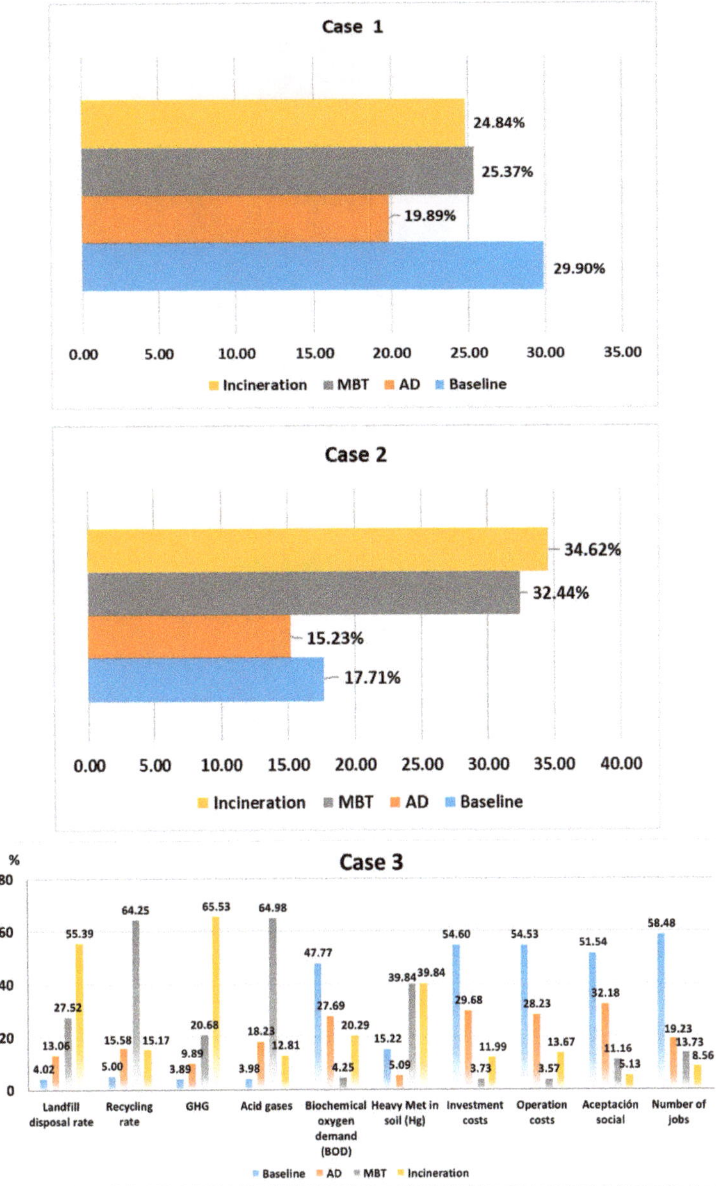

Figure 6. Results of sensitivity analysis: Case 1, Case 2, Case 3.

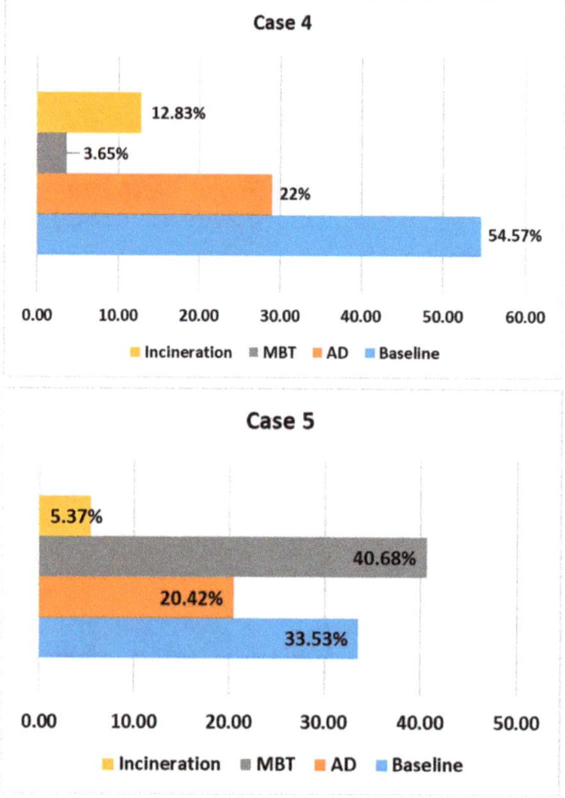

Figure 7. Results of sensitivity analysis: Case 4, Case 5.

The main conclusions from sensitivity analysis are the following:

Scenario 4 (WTE) ranked first when priority was given to the environmental indicators. However, this result is not stable due to fluctuations in the weighting of criteria. When priority was given to the economic and social impact of waste management strategies, Scenario 4 was not ranked first. In Case 4, the baseline scenario was ranked the highest, and in Case 5, the scheme involving MBT.

Scenario 2 (AD) is the most stable solution. With different indicators rankings, it is ranked third in Cases 1, 2, and 5 and never ranked first, however other scenarios change their rating so much that Scenario 2 emerges as the most stable.

4. Conclusions

AHP method was applied for sustainability assessment of waste treatment scenarios with energy and resources recovery for Mexico City. This research extends the knowledge basis of comparison of waste management in emerging countries with low municipal waste management performance. The study does not only present information on economic instruments but also ranks them.

The study was conducted to compare the new waste management plan for the city (WTE) with alternatives: "business-as-usual" scenario, AD of organic fraction, and MBT. The goal of the study was a scenario ranking using a model based on multicriteria analysis for sustainability assessment of the waste management scenarios. The pairwise comparison was done by a group of experts from Mexico City. The results of this show that the most sustainable scenario, environmentally, socially, and economically, is Scenario 4-WTE, with a ranking priority of 30.78%. These results confirm that the decision made by the government of Mexico City to introduce waste incineration is sustainable from environmental, economic, and social aspects. The analysis suggests the next steps should focus on the introduction of the new regulation, improving the infrastructure, including roads, and increasing human resources. These results can help decision makers in other megacities in developing countries in introducing a successful and sustainable waste management system. However, this is challenging for emerging economies that cannot easily afford waste-to-energy plants due to their high costs. Since the willingness to pay for waste management is not higher than 0.4% of GDP, WTE is out of the reach of countries with a per capita GDP below 3000 US$ [59].

The following constraints of this model should, however, be considered. According to the indicator ranking, environmental criteria have the highest priority, therefore, the WTE scenario was ranked first. But it should be noted that, due to the lack of data, European standards were used for the life-cycle inventory, and the efficiency rate of the source separation was assumed to be perfect. Another restriction of the present AHP analysis is that pairwise comparison was made by only 5 experts, because the process is time-consuming and the authors received only 5 filled-out questionnaires out of 14. The results of sensitivity analysis showed that the outcome of the study is not stable and, therefore, less reliable. Further research should involve a larger number of experts making the indicators ranking, to give higher validation to the results. Determination and precise evaluation of the indicators for assessing the sustainability of waste management should be focused on as well. Moreover, further study needs to be undertaken on the feasibility of WTE technology in the Mexican context, which has the typical constrains in developing countries: data acquisition, inadequate collection systems, and reduced ability to collect charges [60].

Supplementary Materials: The following are available online at http://www.mdpi.com/2313-4321/3/3/45/s1. Figure S1: Mass flow diagram of Baseline Scenario; Figure S2: Mass flow diagram of AD Scenario; Figure S3: Mass flow diagram of MBT Scenario; Figure S4: Mass flow diagram of Incineration Scenario; Table S1: Details of the flows of the MFD of Baseline Scenario; Table S2: Details of the flows of the MFD of AD Scenario; Table S3: Details of the flows of the MFD of MBT Scenario; Table S4: Details of the flows of the MFD of incineration Scenario.

Author Contributions: N.T. realized the interviews and wrote the paper. A.V.M. conceived and supervised the study. A.A.C.S. contributed in the data analysis.

Funding: The first author received financial support for the conduct of this study from DAAD (German Academic Exchange Service) within the program IPID4ALL and Heinz Neumüller Foundation.

Conflicts of Interest: The authors declare no conflict of interest.

References

1. Oduro-Appiah, K.; Scheinberg, A.; Mensah, A.; Afful, A.; Kofi Boadu, H.; de Vries, N. Assessment of the municipal solid waste management system in Accra, Ghana: A 'Wasteaware' benchmark indicator approach. *Waste Manag. Res.* **2017**, *35*, 1149–1158. [CrossRef] [PubMed]
2. Ghazvinei, P.T.; Mir, M.A.; Ariffin, J. *University Campus Solid Waste Management [Combining Life Cycle Assessment and Analytical Hierarchy Process]*; Springer: Berlin, Germany, 2017; ISBN 978-3-319-43227-4.
3. Rodić, L.; Wilson, D.C. Resolving governance issues to achieve priority sustainable development goals related to solid waste management in developing countries. *Sustainability* **2017**, *9*, 404. [CrossRef]
4. Morrissey, A.J.; Browne, J. Waste Management Models and Their Application to Sustainable Waste Management. *Waste Manag.* **2004**, *24*, 297–308. [CrossRef] [PubMed]
5. Brunner, P.H.; Fellner, J. Setting priorities for waste management strategies in developing countries. *Waste Manag. Res.* **2007**, *25*, 234–240. [CrossRef] [PubMed]
6. Assessment Methods for Waste Management Decision-Support in Developing Countries. Available online: https://www.dora.lib4ri.ch/eawag/islandora/object/eawag%3A13531/datastream/PDF/view (accessed on 30 December 2017).
7. Elsaid, S.; Aghezzaf, E.H. A progress indicator-based assessment guide for integrated municipal solid-waste management systems. *JSMCWM* **2017**, *20*, 4–11. [CrossRef]
8. Vučijak, B.; Kurtagić, S.M.; Silajdžić, I. Multicriteria decision making in selecting best solid waste management scenario: a municipal case study from Bosnia and Herzegovina. *J. Clean. Prod.* **2016**. [CrossRef]

9. Brans, J.P. Lingénierie de la décision. Elaboration dinstruments daide à la décision. Méthode PROMETHEE. In *L'aide à la décision: Nature, instruments et perspectives d'avenir*; Nadeau, R., Landry, M., Eds.; Presses de l'Université Laval: Québec city, QC, Canada, 1982; pp. 183–214.

10. Brans, J.P.; Vincke, P. A Preference Ranking Organisation Method: (The PROMETHEE Method for Multiple Criteria. Decision-Making). *Manag. Sci.* **1985**, *31*, 647–656. [CrossRef]

11. Queiruga, D.; Walther, G.; González-Benito, J.; Spengler, T. Evaluation of sites for the location of WEEE recycling plants in Spain. *Waste Manag.* **2008**, *28*, 181–190. [CrossRef] [PubMed]

12. Demesouka, O.E.; Vavatsikos, A.P.; Anagnostopoulos, K.P. GIS-based multicriteria municipal solid waste landfill suitability analysis: a review of the methodologies performed and criteria implemented. *Waste Manag. Res.* **2014**, *32*, 270–296. [CrossRef] [PubMed]

13. Saaty, T. *The Analytic Hierarchy Process, Planning, Piority Setting, Resource Allocation*; McGraw-Hill: New York, NY, USA, 1980.

14. Achillas, C.; Moussiopoulos, N.; Karagiannidis, A. The use of multi-criteria decision analysis to tackle waste management problems: a literature review. *Waste Manag. Res.* **2013**, *31*, 115–129. [CrossRef] [PubMed]

15. Kling, M.; Seyring, N.; Tzanova, P. Assessment of economic instruments for countries with low municipal waste management performance: An approach based on the analytic hierarchy process. *Waste Manag. Res.* **2016**, *34*, 912–922. [CrossRef] [PubMed]

16. Badri, M.; Al Qubaisi, A.; Mohaidat, J.; Al Dhaheri, H.; Yang, G.; Al Rashedi, A.; Greer, K. An analytic hierarchy process for school quality and inspection: Model development and application. *Int. J. Educ. Manag.* **2016**, *30*, 437–459. [CrossRef]

17. Baba, Y.; Kallas, Z.; Realini Cujó, C. A multi-criteria stated method to analyze consumers' preference and sensory evaluation towards omega-3 enriched eggs: The Analytical Hierarchy Process (AHP). In Proceedings of the 29th Conference if the IAAE, International Association of Agricultural Economists, Milan, Italy, 1–3 August 2015.

18. Schmoldt, D.; Kangas, J.; Mendoza, G.A.; Pesonen, M. *The Analytic Hierarchy Process in Natural Resource and Environmental Decision Making*; Springer: Berlin, Germany, 2013; ISBN 978-94-015-9799-9.

19. Milutinović, B.; Stefanović, G.; Dassisti, M.; Marković, D.; Vučković, G. Multi-criteria analysis as a tool for sustainability assessment of a waste management model. *Energy* **2014**, *74*, 190–201. [CrossRef]

20. Kim, M.; Jang, Y.C.; Lee, S. Application of Delphi-AHP methods to select the priorities of WEEE for recycling in a waste management decision-making tool. *J. Environ. Manag.* **2013**, *128*, 941–948. [CrossRef] [PubMed]

21. Taboada-González, P.; Aguilar-Virgen, Q.; Ojeda-Benítez, S.; Cruz-Sotelo, S. Application of analytic hierarchy process in a waste treatment technology assessment in Mexico. *Environ. Monit. Assess* **2014**, *186*, 5777–5795. [CrossRef] [PubMed]

22. Araiza Aguilar, J.A.; Nájera Aguilar, H.A.; Gutiérrez Hernandez, R.F.; Rojas Valencia, M.N. Emplacement of solid waste management infrastructure for the Frailesca Region, Chiapas, México, using GIS tools. *EJRS* **2017**. [CrossRef]

23. Martínez-Morales, I.N.; del Consuelo Mañón-Salas, M.; del Consuelo Hernández-Berriel, M.; Ojeda-Benítez, S.; Carreño-de León, M.C. Selección de Municipios del Noreste del Estado de México y Estados Aledaños Mediante el Método AHP. In Proceedings of the Memorias 9 ENERS 2018, Guadalajara, Mexico, 13 June 2018; Dra. María del Consuelo Mañón Salas: Calimaya, Estado de México, Mexico, 2018; pp. 189–199.

24. Gomez Jauregui Abdo, J.P. Sustainable Development of Domestic Water Supply in Emerging Megacities: The Case of the City of Guadalajara, Mexico. Ph.D. Thesis, Brandenburg Technical University, Cottbus, Germany, 2015.

25. Saaty, T. Decision making with the analytic hierarchy process. *Int. J. Serv. Sci.* **2008**, *1*, 83–98. [CrossRef]

26. World Bank, Data Bank, 2016. Available online: http://data.worldbank.org/country/mexico (accessed on 25 July 2018).

27. INEGI. Encuesta Intercensal. Available online: http://www.inegi.org.mx/est/contenidos/proyectos/encuestas/hogares/especiales/ei2015/doc/eic_2015_presentacion.pdf (accessed on 30 December 2017).

28. PAOT, Centro Interdisciplinario de Investigaciones y Estudios sobre Medio Ambiente y Desarrollo del IPN and Instituto Politécnico Nacional, Diagnóstico actual del flujo de residuos sólidos urbanos que se genera en el Distrito Federal, 2013. Available online: http://centro.paot.org.mx/documentos/paot/estudios/flujo_residuos_DF.pdf (accessed on 15 July 2018).

29. Gómez, F.L.; Quintero, M.J. Acuerdan GDF y Edomex canalización de basura ante cierre de Bordo Poniente. *La Jornada*. 2011. Available online: http://www.jornada.unam.mx:8810/2011/12/22/capital/036n2cap (accessed on 5 March 2018).

30. Bezama, A.; Aguayo, P.; Konrad, O.; Navia, R.; Lorber, K.E. Investigations on mechanical biological treatment of waste in South America: Towards more sustainable MSW management strategies. *Waste Manag.* **2007**, *27*, 228–237. [CrossRef] [PubMed]

31. Menikpura, N.; Sang-Arun, J.; Bengtsson, M. Mechanical biological treatment as a solution for mitigating greenhouse gas emissions from landfills in Thailand. In Proceedings of the ISWA World Solid Waste Congress, Florence, Italy, 17–19 September 2012.

32. Guidebook for the Application of Waste to Energy Technologies in Latin America and the Caribbean. Available online: http://www.seas.columbia.edu/earth/wtert/pressreleases/Guidebook_WTE_v5_July25_2013.pdf (accessed on 10 June 2016).

33. SEDEMA. El Inventario de Residuos Sólidos. 2016. Available online: http://www.sedema.cdmx.gob.mx/storage/app/media/IRS-2016.pdf (accessed on 31 December 2017).

34. Beckmann, M.; Karl, H.C.; Thome-Kozmiensky, J. Waste-derived fuels—Opportunities and problems [Ersatzbrennstoffe—Chancen und probleme]. *Aufbereit.-Tech.* **2006**, *47*, 28–37.

35. Thiel, S.; Thomé-Kozmiensky, K.J. Mechanical-Biological Pre-Treatment of Waste–Hope and Reality. In Proceedings of the ISWA World Congress 2010—Urban Development and Sustainability—A Major Challenge for Waste Management in the 21th Century, Hamburg, Germany, 15–18 November 2010.

36. Cencic, O.; Rechberger, H. Material Flow Analysis with Software STAN. *J. Environ. Eng. Manag.* **2008**, *18*, 440–447.

37. Navarotto, P.; Dominguez Llauro, R. Materials recovery from municipal solid waste: ECOPARC 4 Barcelona a case study. In *SIDISA 2012 Sustainable Technology for Environmental Protection*; Bonomo, L., Ed.; Atti del Convegno Sidisa: Milan, Italy, 2012.

38. Cimpan, C.; Maul, A.; Jansen, M.; Pretz, T.; Wenzel, H. Central sorting and recovery of MSW recyclable materials: A review of technological state-of-the-art, cases, practice and implications for materials recycling. *J. Environ. Manag.* **2015**. [CrossRef] [PubMed]

39. Hokkanen, J.; Salminen, P. Choosing a solid waste management system using multicriteria decision analysis. *Eur. J. Oper. Res.* **1997**, *98*, 19–36. [CrossRef]

40. Greene, K.L.; Tonjes, D.J. Quantitative assessments of municipal waste management systems: using different indicators to compare and rank programs in New York State. *Was Manag.* **2014**, *34*, 825–836. [CrossRef] [PubMed]

41. Milutinović, B.; Stefanović, G.; Kyoseva, V.; Yordanova, D.; Dombalov, I. Sustainability assessment and comparison of waste management systems: The Cities of Sofia and Niš case studies. *Waste Manag. Res.* **2016**, *34*, 896–904. [CrossRef] [PubMed]

42. BLFUW-Bundesministerium für Land-und Forstwirtschaft, Umwelt und Wasserwirtschaft. Benchmarking für die österreichische Abfallwirtschaft. 2015. Available online: https://publik.tuwien.ac.at/files/PubDat_247861.pdf (accessed on 31 December 2017).

43. Andersen, J.K.; Boldrin, A.; Christensen, T.H.; Scheutz, C. Mass balances and life-cycle inventory for a garden waste windrow composting plant (Aarhus, Denmark). *Waste Manag. Res.* **2012**, *30*, 1010–1020.

44. Masood, M.; Barlow, C.Y.; Wilson, D.C. An assessment of the current municipal solid waste management system in Lahore, Pakistan. *Waste Manag. Res.* **2014**, *32*, 834–847. [CrossRef] [PubMed]

45. Clavreul, J.; Baumeister, H.; Christensen, T.H.; Damgaard, A. An environmental assessment system for environmental technologies. *Environ. Model Softw.* **2014**, *60*, 18–30. [CrossRef]

46. Kirkeby, J.T.; Birgisdottir, H.; Hansen, T.L.; Christensen, T.H.; Bhander, G.S.; Hauschild, M. Environmental assessment of solid waste systems and technologies: EASEWASTE. *Waste Manag. Res* **2006**, *24*, 3–15. [CrossRef] [PubMed]

47. Kirkeby, J.T.; Birgisdottir, H.; Hansen, T.L.; Christensen, T.H.; Bhander, G.S.; Hauschild, M. Evaluation of environmental impacts from municipal solid waste management in the municipality of Aarhus, Denmark (EASEWASTE). *Waste Manag. Res.* **2006**, *24*, 793–799.

48. Hauschild, M.; Goedkoop, M.; Guinée, J.; Heijungs, R.; Huijbregts, M.; Jolliet, O.; Margni, M.; Schryver, A.D.; Humbert, S.; Laurent, A.; et al. Identifying best existing practice for characterization modeling in life cycle impact assessment. *Int. J. LCA* **2013**, *18*, 683–697. [CrossRef]

49. IPCC (2007) Climate Change: The Physical Science Basis. Contribution of Working Group I to the Fourth Assessment Report of the Intergovernmental Panel on Climate Change. 2007. Available online: http://www.ipcc.ch/publications_and_data/publications_ipcc_fourth_assessment_report_wg1_report_the_physical_science_basis.htm (accessed on 31 December 2017).

50. Møller, J.; Jensen, M.B.; Kromann, M.; Neidel, T.L.; Jakobsen, J.B. Miljø-og samfundsøkonomisk vurdering af muligheder for øget genanvendelse af papir, pap, plast, metal og organisk affald fra dagrenovation. Miljøprojekt nr. 1458. 2013. Available online: https://www2.mst.dk/Udgiv/publikationer/2013/01/978-87-92903-80-8.pdf (accessed on 31 December 2017).

51. IADB—Inter-American Development Bank (2015) Solid Waste Management in Latin America and the Caribbean. Available online: https://publications.iadb.org/handle/11319/7177#sthash.1YBgmSCS.dpuf (accessed on 31 December 2017).

52. Münnich, K.; Mahler, C.F.; Fricke, K. Pilot project of mechanical-biological treatment of waste in Brazil. *Waste Manag.* **2006**, *26*, 150–157. [CrossRef] [PubMed]

53. Tsilemou, K.; Panagiotakopoulos, D. Approximate cost functions for solid waste. *Waste Manag. Res.* **2006**, *24*, 310–322. [CrossRef] [PubMed]

54. Christensen, T. *Solid Waste Technology and Management*; Wiley: Chichester West Sussex, UK, 2011; ISBN 978-1-405-17517-3.

55. Friends of the Earth. More Jobs, Less Waste. 2010. Available online: https://www.foeeurope.org/sites/default/files/publications/foee_more_jobs_less_waste_0910.pdf (accessed on 31 December 2017).

56. European Commission, Directorate-General Environment. 2001. Effects of Waste Management Policies. Available online: http://ec.europa.eu/environment/enveco/waste/pdf/waste_management_employment.pdf (accessed on 31 December 2017).

57. Veolia (2017) Weekly Update—Invitation to Respond [Veolia—Mexico Thermo-Valorization]. Available online: https://business-humanrights.org/sites/default/files/documents/Business%20and%20Human%20Right%20Resource%20Center%20Answer_EN_veolia.pdf (accessed on 31 December 2017).

58. Ishizaka, A.; Labib, A. Analytic hierarchy process and expert choice: benefits and limitations. *OR Insight* **2009**, *22*, 201–220. [CrossRef]

59. Brunner, P.H.; Rechberger, H. Waste to energy-key element for sustainable waste management. *Waste Manag.* **2015**, *37*, 3–12. [CrossRef] [PubMed]

60. Vujic, G.; Stanisavljevic, N.; Batinic, B.; Jurakic, Z.; Ubavin, D. Barriers for implementation of "waste to energy" in developing and transition countries: a case study of Serbia. *JSMCWM* **2017**, *19*, 55–69. [CrossRef]

 *recycling*

Article

Public Policy for Solid Waste and the Organization of Waste Pickers: Potentials and Limitations to Promote Social Inclusion in Brazil

Christian Luiz da Silva * and Camille Bolson

Department of Management and Economics, Federal University of Technology, Paraná, Curitiba CEP 80230-901, Brazil; camille.bolson@gmail.com

* Correspondence: christianlsilva76@gmail.com; Tel.: +55-041-98834-2351

Received: 25 July 2018; Accepted: 30 August 2018; Published: 4 September 2018

Abstract: The management model for the National Solid Waste Policy to develop sustainable actions, proposes the social inclusion of recyclable waste pickers in the waste management system. Compliance with the law, the form of participation of the waste pickers, and incentive mechanisms are configured as a relationship open to analysis. Therefore, the aim of this work was to investigate the potentials and limitations of a recycling cooperative, in terms of social technologies and inclusion, to encourage local development. The qualitative approach was aided by structured questionnaires, semi-structured interviews, and participant observation. The resulting evidence suggests that the organization of a cooperative, enabled access to information on the legislation of the National Solid Waste Policy. It showed the need to strengthen relationships with education institutions and public authorities. Despite the low levels of education of the members of the cooperative, projects and knowledge could be developed to aid social technologies. No technological innovations were observed, nor the production of alternative artifacts for recyclable materials. This weakens the cooperative in terms of articulation among peers, most notably the integration of the Catamare cooperative in the network of Cataparaná, to support the sale of material produced for industry. It may be concluded that joining the cooperative improved the social, economic, and political conditions of the members, but there were also structural limits to the recycling production chain that were not considered in the National Solid Waste Policy; and to a certain extent this weakens the development of sustainable actions. Furthermore, the organization of the cooperative hindered the development of social technologies and the social inclusion of the waste pickers.

Keywords: recycling cooperatives; recyclable waste pickers; national solid waste policy; waste management; Brazil

---

1. Introduction

Since the 1990s, the number of studies on municipal solid waste policies [1], and discussions on cooperative organization or local impacts has grown [2]. Some studies were already concerned with the organization of the chain [3] and establishing a market for recyclable waste [4]. In this context, Curitiba, in Brazil, already stood out due to its solid waste policy [5] and raising awareness of the need for co-participation in recycling [6]. In the 2000s, studies addressed how these recycling organizations were structured, especially in developing countries [7]. Furthermore, public institutions and non-governmental organizations have prepared documents to strengthen solid waste policies [8]. Recent studies [9–11] have analyzed the optimization of the recycling chain, public policies, and the structuring of networks. The theme of lifecycles and the understanding of integrated management, are also pertinent to the theme of public recycling policies [12]. Finally, recycling cooperatives and their impact on waste management and social inclusion, are relevant themes for discussion [13,14].

The management model for solid waste, introduced by the National Solid Waste Policy in Brazil, proposes the generation of better results in terms of the development of sustainable actions in the country; and it is also an important factor regarding the social inclusion of waste pickers. The policy set fundamental goals to help eliminate landfills and dumps, through the constitution of planning instruments at the national, state, micro region, inter-municipal metropolitan, and municipal level. The policy also obliges the private sector to prepare solid waste management plans [15].

The aim of this work, was to analyze the potentials and limitations of a recycling cooperative located in Curitiba, regarding the development of social technologies and the stimulation of local development, following the sanction of the National Solid Waste Policy. The study concentrated on Curitiba, as its recycling policy is well known in Brazil and is a representative case that enabled an in-depth evaluation. Therefore, the study discussed the development of a cooperative system in a community in a situation of vulnerability and social exclusion, the Catamare Cooperative, and how it could be characterized from a theoretical framework of social technology, leading to social inclusion. This initiative was relevant because it enabled cooperation to understand the development processes of social technologies, and how the institutional context developed through these experiences.

The Catamare cooperative is representative of the empirical field of this study, as the experience of the cooperate members transcends the sanction of the National Solid Waste Policy, with the cooperative having officially existed since 2007. The cooperative is currently linked to the National Solid Waste Policy and the Eco-Citizen Project, both public initiatives that aim to promote social inclusion and generate employment and income for recyclable waste pickers.

2. Materials and Methods

To achieve the goals of this study, the methodology employed was the case study, with a qualitative and quantitative approach, the use of semi-structured questionnaires, unstructured interviews, and participant observation.

The technical procedure employed in the first stage was a document content analysis, which enabled a better understanding of the National Solid Waste Policy and waste pickers in Brazil. For four months, periodical visits were made to the cooperative for the purposes of participant observation. The visits were made periodically between July 2014 and February 2015, to enable participant observation two to six times a week. The other interviews were conducted in January 2015. At this time, interviews were conducted with the members who play strategic roles in the cooperative (the president, the sales committee, the treasurer, and the fiscal council). In these interviews, an attempt was made to seek information on: (i) How the work of the cooperative is organized; (ii) the process, methods, and techniques of the work; and (iii) the participation of the members.

In the second stage, a questionnaire was distributed to all the members of the Catamare cooperative. Of the 35 members, 26 were interviewed. The questionnaire sought information on: (i) The socio-economic condition of the members, (ii) the process and organization of the work of the members, (iii) how the members perceived the work they do, (iv) the relationships that the cooperative maintains with other organizations, (v) the importance of these relationships with other organizations to the cooperative; and (vi) how the subjects perceived the organizations with which the cooperative maintains a relationship.

The CATAMARE cooperative was representative of the empirical field of this study since: (i) The cooperative is linked to the Eco-Citizen Program, run by the Local Government of Curitiba; (ii) the experience of the cooperative members transcends the sanction of the National Solid Waste Policy; and (iii) most members were linked to the National Movement of Recyclable Waste Pickers.

Regarding how the research problem was addressed, the approach was qualitative as the focus of this study was on the process and its meaning [16]. According to Minayo [17], the intention of the qualitative approach is not to achieve the truth, but to understand the logic of what is practiced in reality, being concerned with the relationships between the world and the subject in a certain way that cannot be replicated or seen in this way, in any other place. In this sense, the study was a procedure that did not seek to generalize results, i.e., it did not strive to create models with universal goals. According to Triviños [16], a qualitative study is only intended to "obtain generalities, prominent ideas, trends that appear more defined between the people that participate in the study". In this sense, to answer the general question of this study, a qualitative case study was conducted of the Catamare cooperative. In brief, Yin [18] defines the case study as an empirical study that will seek to investigate a contemporary phenomenon in its real-life context, "when the boundaries between the context and the phenomenon are not clearly evident". Therefore, for the present work, it was necessary to use multiple sources of evidence.

3. Solid Waste Management and Social Inclusion in Brazil

The management model for solid waste, introduced by the National Solid Waste Policy in Brazil, proposes the generation of better results in terms of the development of sustainable actions in the country, ensuring a better-quality service and social inclusion for waste pickers. The innovations introduced by the policy constitute a framework to stimulate these experiences, and the possibility of inclusion for recyclable waste pickers in the waste management system. However, compliance with the law, the form of participation, and the incentive mechanisms established by the policy do not guarantee social inclusion, and constitute a relationship that is open to analysis, as will be discussed in the next sections.

3.1. National Solid Waste Policy and Social Inclusion in Brazil

Over time, there was a vacuum to be filled in Brazilian public policies regarding solid waste management due to the lack of a public environmental policy with directives and instruments, for the adequate environmental management of waste. For some time, a proposal was discussed with the participation of public agencies, representatives from the private sector, social movements, and civilian society. The central issue that hindered the process of passing the National Solid Waste Policy in Brazil, was the lack of consensus between the government, civilian society, and the private sector, regarding the model of post-consumption responsibility to be implemented, defining the responsibility of manufacturers, importers, distributors, consumers, and those in charge of public services that involved handling waste. After 21 years of deliberation, the Chamber of Deputies on 11 March 2010, followed by the Senate on 7 July 2010, passed the law. President Lula sanctioned Federal Law 12.305, of 2 August 2010, thereby enacting the National Solid Waste Policy [19].

It was in 2011, that a process was initiated to prepare the National Solid Waste Plan. Due to the complexity of the issues that it was to face, the proposal was directly related to other national plans: National Climate Change Plan, National Water Resources Plan, and the National Plan for Sustainable Production. It was also directly linked to the National Environmental Education Policy, through the proposal for the National Environmental Education Policy and the proposal of the National Basic Sanitation Plan [19].

The first action of the National Solid Waste Policy, was to set up the Inter-Ministerial Committee for the Social and Economic Inclusion of Reusable and Recyclable Waste Pickers (CIISC), created by Decree 7.405/10. The CIISC, coordinated by the General Secretariat of the President of the Republic, is made up of members of ministries of the Environment; Social Development and Combatting Hunger; Work and Employment; Social Security; Education; Health; Cities; Tourism; Mining and Energy; Economics; Science and Technology; Planning, Budget and Management; the Federal Heritage Secretariat; the General Secretariat of the President of the Republic; the Human Rights Secretariat and Legal Department of the President of the Republic; the Banco do Brasil Foundation; Caixa Econômica Federal Bank; Petrobras; National Health Foundation; and the Applied Economics Research Institute. It is implicit that other public policies are coordinated by the Committee, as waste pickers benefit from other programs of these ministries, such as the "Brazil Without Poverty" and "Family Allowance", "Literacy Brazil", "Green Brazil", "My Home, My Life", "Caring Brazil", the "Stork Network",

"Electricity for All", "ProUni", "Pronatec", "Health Has No Price", and other programs for families registered on the Single Register for Social Programs.

The creation of the CIISC, also led to the Pró-Catador (Pro Waste Picker Program), the Cataforte (Strong Waste Pickers), and to the National Center for the Defense of Human Rights of the Homeless and Recyclable Waste Pickers (CNDDH).

The Pró-Catador program has the purpose to integrate and articulate federal government actions, to support and improve the productive organization of recyclable and reusable waste pickers and improve their working conditions. It also seeks to create more opportunities for social and economic inclusion, and expand the selective collection of solid waste, encouraging reuse, and recycling through this sector.

The Cataforte Program, is the direct result of the demands from and negotiations with the National Movement of Recyclable Waste Pickers and the Federal Government. The aim of the project is to strengthen social and productive organizations, and self-managed and supportive economic ventures.

Law 12.305/2010 has 57 articles, and their essence has been maintained since the bill of law was introduced in the Chamber of Deputies, as highlighted by Grimberg [20]. The scope of the law is clearly defined, establishing directives, instruments, and responsibility for the management of solid waste.

In general, the PNRS has aimed to define strategies that add value to solid waste, increasing the competitive capacity of the productive sector, aiding social inclusion, and describing the role of the states and municipalities in solid waste management. In this context, the Federal Government, through the Ministry of the Environment, has set some parameters for the municipalities to enact the national Solid Waste Policy: reduced operational costs and values of investments; better use of technologies; the establishment of regionalized rules for the use of services; and efforts to integrate planning and shared management that include solid waste pickers.

The main guide is the Integrated Management of Solid Waste, which prioritizes a reduction in the volume of waste, more recycling, incentives to form cooperatives or other waste picker associations as a form of environmental action, and the concept of reverse logistics and shared responsibility; recognizing the need to participate in every link of the chain. This has led to the giant step of the Plano Nacional de Resíduos Sólidos (PNRS) in establishing integrated management, which is characterized by actions that seek solutions to the inherent problems of solid waste, whilst considering the political, economic, technological, environmental, cultural, and social dimension, with society assuming the role of monitoring compliance with the law. The main mechanisms of the PNRS are selective collection, reverse logistics, environmental education, scientific and technological research and accountability for the lifecycle of products, encouragement to participate and the strategic and incisive role of waste pickers, and incentives to create cooperatives and associations [19].

The general directives for the inclusion of waste pickers in the National Solid Residue Policy include: (i) Strengthening cooperatives and associations of waste pickers "seeking to raise them to a higher level of efficiency"; (ii) creating new cooperatives and associations and regulating those that already exist to "strengthen these work vehicles and socially include and formalize waste pickers that operate in isolation"; (iii) articulating networks of cooperatives and associations of waste pickers; (iv) creating mechanisms to identify and certify cooperatives; (v) strengthening initiatives to integrate and articulate federal policies, and actions to aid waste pickers; (vi) goals for the social inclusion of waste pickers, and to ensure that public policies provide alternatives for work and income for waste pickers, who cannot continue in this line of work after the extinction of landfills; (vii) goals for social inclusion, and to guarantee dignity in employment for up to 600,000 waste pickers; (viii) the participation of waste pickers in environmental education actions, and awareness in the separation of waste at the source, through adequate training and payment; (ix) bringing information systems on municipal waste and shared solid waste management up to date; (x) goals and criteria for municipalities to include waste pickers in the municipal management of solid waste; (xi) providing access for waste pickers, to urban solid waste collected selectively; and (xii) the integration of recyclable waste pickers into reverse logistics systems [21].

The role of waste pickers in local solid waste management actions can be seen in Article 19 of Federal Law 12.305/2010, which states that Municipal Solid Waste Plans should include their participation, in accordance with Article 11 of Federal Decree 7404/2010. Regarding the reduction of urban solid waste, the law includes 19 strategies for compliance. Concerning a reduction in the generation of Urban Solid Waste in landfills, the law has 22 strategies. An important strategy is to implement a selective collection, with the participation of the cooperatives as service providers, contracted by the public municipal administration, with due payment to the waste pickers to collect, sort, and direct the waste to an adequate final destination for recycling [19].

In addition to the initiatives described, the Article 80 of the Federal Decree, states that federal institutions finances should create special lines of credit for funding cooperatives and other forms of waste picker associations. These funds should be used for the acquisition of machinery and equipment, to be used in the management of solid waste, recycling, and the reuse of solid waste, and innovation and development for managing solid waste.

3.2. Recyclable Waste Picker Cooperative In Curitiba: The Eco-Citizen Project

Eco-Citizen is a social inclusion project launched in 2007, through a partnership between the city of Curitiba represented by the Municipal Environmental Secretariat, and the Social Action Foundation, the Entrepreneurial Alliance Association, and the National Waste Pickers Movement. The aim of the project was to "guide, organize and support associations and cooperatives of recyclable waste pickers" in administrative issues, infrastructure and commercial interest, and promoting the inclusion and strengthening of waste pickers in the recycling chain to help them with employment and income, to enable economic sustainability and spur local development [22].

One of the objectives of the project is to implement a mixed technology model, which will help to minimize the generation and handling of solid waste, with sorting and recovery of waste as an economic asset with social value, and the final exclusive disposal of waste material in an environmentally adequate way. With an incentive for selective collection, reverse logistics, and the sorting of materials, there will be a reduction in the amount of recyclable material in landfills [20].

Today, the Ecocidadão Project is active in 40 municipalities in Paraná State, with 47 associations of recyclable waste pickers. The Municipal Environmental Secretariat is responsible for the management of the Ecocidadão project. Since 2013, the technical side of the project has been managed by the Pró-Cidadania (Pro-Citizen) Institute (IPCC). It also helps to form the cooperatives or associations, and helps to hire the support team, acquire the necessary equipment, and other correlated actions [22].

In keeping with the National Solid Waste Plan and the State Plan for the Integrated and Associated Management of Urban Solid Waste in Paraná State, Table 1 shows the Specific Directives and the Work Directives, for the Integrated Solid Waste Plan in Curitiba.

Like the PNSR, the Integrated Solid Waste Plan of Curitiba, seeks to encourage stronger participation by cooperatives and other associations of waste pickers as providers of services, duly contracted by municipal public administrations, and developed in partnership with society. This will increase their efficiency and sustainability, mainly in the handling and sale of waste and in the methods of processing and recycling. The project also includes the creation of permanent technical and managerial training programs for waste pickers and the members of their cooperatives, according to the level of the organization through technical, teaching, research, and extension institutions from the third sector and social movements [22].

Deficiencies in waste management in the municipality, according to the Municipal Sanitation Plan [22], included: locations rife with irregular waste disposal; a need for more inspectors and technicians to analyze the Managerial Plans; a low number of cooperatives and associations of waste pickers capable of being included in the recycling chain; cooperatives and associations of waste pickers with no economic and financial self-sustainability; poor economic conditions waste handling and urban cleaning services; distances between collection areas and final disposal points; lack of new technologies for final disposal of urban solid waste; resistance from generators regarding the internalization of costs for disposal; lack of delivery points; and insufficient units for final disposal of vegetable waste and unusable wooden materials.

Table 1. Specific directive from the Integrated Solid Waste Plan of Curitiba [22].

Specific Directives	Work Directives
Environmentally correct final disposal of waste	To recover landfills, including the evaluation of environmental conditions.
Reduction of dry solid waste in landfills and inclusion of recyclable and reusable waste pickers	To promote progressive reduction of dry waste in landfills; To prepare and strengthen the organization for socio-economic inclusion of recyclable and reusable waste pickers; To seek the continued reduction of urban solid waste, taking specific locations into consideration; To adopt technologies that encourage renewable energy or other forms of using dry waste, considering technical, environmental and economic feasibility and the potential regional market.
Reduction of wet urban solid waste in landfills and treatment and landfill gas recovery	To encourage composting, using the energy from biogas produced in biodigesters or landfills or other technologies to produce energy from wet urban solid waste; To adopt technologies promoting the use of energy or other forms of using waste from through other processes, considering the technical, environmental and economic feasibility or the potential regional market.
Training in solid waste management	To strengthen urban cleaning management and the handling of urban solid waste by institutionalizing an appropriate instrument for monitoring and paying for urban cleaning and the handling of waste.
Health service waste	To strengthen the management of waste produced during health services.
Industrial waste	To strengthen the management of solid waste in industry;
Construction waste	To strengthen mechanisms for monitoring and inspecting generators. To provide voluntary delivery points for construction waste.

4. Social Technologies

In general, conventional technology can be defined based on a set of characteristics related to its effects on labor and on the environment, its large scale and production rate, the inputs used, and the type of control exercised over workers [23]. In conventional technologies, there is no emphasis on the process of technology construction or collective learning that results from it. These aspects are extremely important for social transformation because they are concerned with the engagement, participation, appropriation, and the production of knowledge of the social groups involved. The aspects of conventional technology are, in fact, efficient in maximizing private profits, but also limit their effectiveness as a strategy for social inclusion.

In this sense, as a response to the totalizing logic of capital existing in the processes of production and consumption of conventional technologies, the conception of social technology emerges as an inclusive, comprehensive, and possible approach [23]. The term social technology refers essentially to the development of technologies aimed at social inclusion, which have carried out criticism of conventional technology since its inception. Guided by a more sustainable perspective and less harmful to the environment and to the human being, social technology signals the construction of a society based on a distinct rationality, permeated by social values, such as cooperation and autonomy.

Social technologies, unlike conventional technologies, are qualified by low financial contributions, for its orientation towards the internal market, and for being able to economically make self-management ventures viable because they are exempt from a discriminatory power relationship; and above all, because they constitute a liberating potential of the direct producer. Social technologies are guided by the criteria of social inclusion, which allows the construction of more equitable socioeconomic systems in terms of income distribution, and more participative in terms of collective decision making [24].

According to Dagnino, Brandão e Novaes [23], it is possible to affirm that Conventional Technology reinforces the capitalist duality, restricting the workers to the retainers of the means of production, just as the peripheral countries are dominated by the developed countries, perpetuating and extending asymmetric powers regarding social and political relations. On the other hand, Social Technology has as its characteristic, its adaptation to small producers and consumers of low economic power, through a type of control that does not reproduce the segmentation and hierarchization of the Conventional Technology in the productive unit. Furthermore, the orientation to produce and to be able to stimulate the potential and creativity of the direct producer, making it possible to economically enable social enterprises, such as popular cooperatives, family farms, incubators, and small enterprises [25].

Based on the assumption of Social Technologies, the development and use of technologies build on the understanding that men and women should be covered in a constant process of action and reflection, so that the technical action allows expression of visions of world, social values, and political postures in face of the dominant system [26].

According to Thomas [27], with the return of the thematic of development, especially in the sustainability debates, social technologies re-entered the agenda as a fundamental discussion for the preservation of the environment, culture, and economic viability. In this sense, according to Rutkowski and Lianza [28], this model of non-predatory economic development, requires the rational use of natural resources as a strategy to improve the quality of life of those who produce and of those who consume the technologies developed. There is, for the authors, an important search for a balance between development, preservation of natural resources, and regional culture in the construction of social technologies. Moreover, beyond the productive dimension, technologies can also be understood as the result of different epistemic interactions about a process, a method or artifact, in which the integrality of the human being, its socio-historical context, and the environmental preservation are privileged [26].

5. Results and Discussion

In this section, we present the results of the research at Catamare, and discuss the potentials and limitations of a recycling cooperative, to development of social technologies and social inclusion.

5.1. Characteristics of the Case Study: The Catamare Recycled Material Cooperative

The cooperative presently has 35 members. They are organized using a hierarchical structure, as shown in Figure 1.

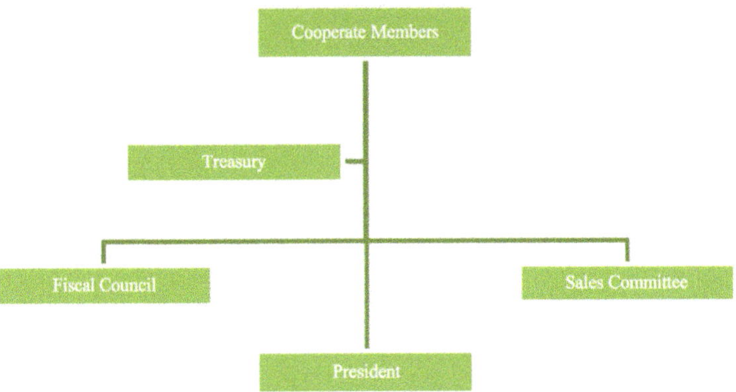

Figure 1. Hierarchical structure of Catamare.

According to the cooperative members, the group has greater decision-making power than the president and the members who hold strategic positions in the venture. The 26 members responded that all decisions of the cooperative are taken by a vote during an assembly, following a discussion.

It is at these meetings, that votes are cast to elect the members to strategic positions in the organization. Only members can attend the assembly meetings.

Every waste picker has the right to call an assembly, provided that they have the support of at least one fifth of the members. Regarding their participation at the meetings, 17 waste pickers stated that they share their opinions actively, whilst nine stated that they do not say much at the meetings, unless they are asked.

The president is legally answerable for all the activities of the cooperative. In short, he is the one who represents the cooperative at official meetings and activities and seeks solutions to the demands of the members outside the limits of the organization. The sales committee is a nucleus formed by CATAMARE members, but it is also part of a group of the Cataparaná, which studies the best options for the joint sale of the materials collected by the waste pickers, through a network to industry. The committee is responsible for participating in meetings with other associations and cooperatives in the Cataparaná. These meetings are held once a month to discuss and clarify issues, such as market prices, investments, and the costs of the network. The fiscal council is responsible for inspecting and providing guidance for the financial affairs of the cooperative. The treasury is responsible for the daily financial routine of the cooperative. It handles the cash flow, money to be received, and bills to be paid, and it applies the financial resources of the cooperative and operationalizes its financial planning.

The cooperative members agreed that each waste picker must contribute 20% of their monthly income. This money is used to pay for electricity, water, telephone bills, internet, and maintenance of machinery used at all the units of the cooperative. Some waste pickers do not agree with this arrangement. They believe that this weakens the cooperative nature of the venture and does not take the principles of the solidarity economy into consideration.

According to the members of the cooperative, those in strategic positions wish to adhere to the guidelines of the National Movement of Waste Pickers. To the members, it is only possible to build a more equal society if there are incentives for the development of solidarity ventures. Within the limits of the cooperative, this occurs through practices that combat competition and individualism through self-management and seeking mutual support between waste pickers and other workers.

The Catamare currently has two sorting units at a match factory, for safety purposes. The material is highly flammable and cannot be taken to the regular sorting units. The oldest unit is located in the Boqueirão neighborhood, and the other in Rebouças.

The cooperative has two presses, two scales, and eighteen worktables shared by all the waste pickers. In the patio of the Rebouças unit, there are electric carts (a project run by Itaipu in collaboration with the Ecocidadão Project). The cooperative has five electric carts, but the waste pickers do not use them. One of the members claimed that she does not like to use them because she thinks they are not strong enough. She said that the prototype overturns easily and cannot support the weight of the materials she collects. Therefore, she uses her own pushcart.

In the Rebouças unit, there is a conveyor belt that has yet to be used by the waste pickers. The conveyor belt has not been installed because the unit does not have the necessary electricity structure to run it, and because the waste pickers have not decided how they will use the equipment. According to the members, to sort material using the conveyor belt would mean establishing fixed working hours, the equal division of the value of the activity, and a study on the best type of material to be sorted.

For two years, the waste pickers have been divided into two spaces, rented by the local government. According to the interviewees, the cooperative members are spread over two units, because the local government was not concerned about renting one location with the necessary physical structure for all of them, after they had vacated their previous installations in the Parolim neighborhood. Several waste pickers reported this lack of concern by the local government over choosing the location of the unit because all the waste pickers that now work in the Boqueirão unit reside in Parolim or Vila Torres. It is difficult for them to go to work because of the long distance. Some waste pickers walk from Parolim to Boqueirão every day because the bus fare is too high and weighs heavily on their budget.

In general, at both units, the main stages of processing the waste are: (i) receipt of material; (ii) unloading the trucks; (iii) sorting; (iv) pressing; (v) packing; and (vi) storage of recyclable materials. Following on the spot observations, a flowchart of the production process of the cooperative was prepared, as shown in Figure 2.

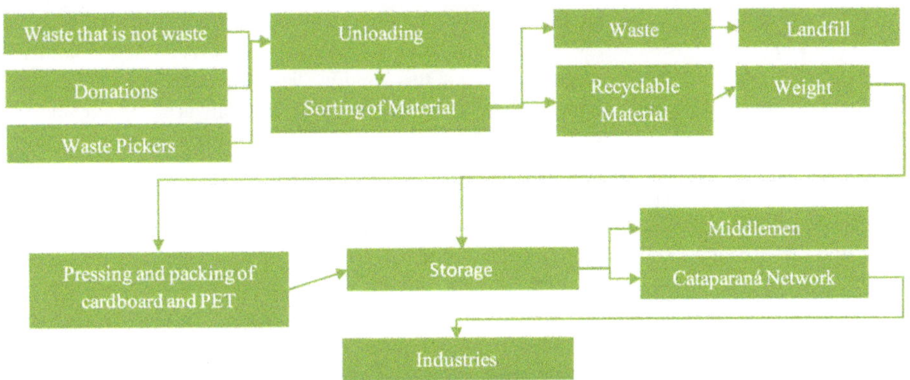

Figure 2. Flowchart of the production process at the CATAMARE Cooperative.

Every day, the cooperative receives material from the "Lixo que não é Lixo" (Waste that is not Waste) project, waste collected on the streets by the six waste pickers of the cooperative, and donations from companies and public agencies. CATAMARE has one large truck, one small truck, and a camper van for collecting donations.

According to the waste pickers on the sales committee, many donors do not act as partners of the cooperatives. The members understand that in most cases they use the cooperatives to dump what they do not want, and sell (illegally) the more valuable waste. According to the waste pickers from the Citizenship and Waste Institute, the cooperatives constantly report the illegal practices of some agencies and private companies to the Justice Department. The organizations that conduct these illegal practices are normally fined and the money is used to meet the requirements of the cooperatives in the Ecocidadão Project.

There is a certain amount of autonomy regarding the division of the materials that each waste picker sorts. At the Boqueirão unit, half of the waste pickers equally share all the materials that are donated or comes from the "Lixo que não é Lixo" program, irrespective of their origin. The other half of the waste pickers organize the separation according to the origin of the waste: Two waste pickers sort the recyclables from Hospital de Clínicas, three sort the material from the Regional Labor Tribunal, and two others sort the waste from the Polytechnic Center and the Unimed Health Insurance Company. At the Rebouças unit, all the donated materials and materials from the "Lixo que não é Lixo" are shared among all the waste pickers. Independent of this sharing, the waste is unloaded manually with the help of brooms, as neither of the trucks have unloading facilities. The materials are removed from the truck by two people and placed manually into large sacks made of raffia. The sacks are then taken to the sorting tables.

The sorting is done using a procedure where each sack of recyclable materials is opened on the table, whilst the waste pickers place several bags on the floor. Each bag is used for one type of material. During sorting, the waste pickers are exposed, as they are in direct contact with the waste, and consequently with the vectors that multiply in the unit. The members only wear gloves for handling the materials. At this stage, the presence of rats and mosquitos was observed, due to the accumulation of material, which makes the work environment unhealthy. This is because the unit does not have the necessary infrastructure for the waste pickers to stock sorted materials. The waste pickers often need to accumulate many bags of waste after separation, until they can build up the necessary volume for a package of recyclables.

The waste is separated using the following criteria: nature, color, and quality (impurities and rejects). Different types of recyclable materials are separated. The most commercialized materials are paper, cardboard, glass, soft aluminum, plastics, steel, and sections of iron. All the materials are stored in large reusable plastic bags. Once the bags are filled, they are weighed.

At the end of the day, each waste picker weighs all their material, monitored by the waste picker in charge of weighing. At the CATAMARE, the treasurer is in charge of this process at the Boqueirão unit. At the Rebouças unit, the members take turns at this activity. The numbers are noted on a control sheet that specifies all the types of materials and the weight of each. One copy is always given to the waste picker, and the other is retained by the cooperative.

Every week, the treasurer adds up the weights that each waste picker recycled and calculates their value, according to the price parameters provided by Cataparaná. This list serves two functions. It specifies the price that Cataparaná can obtain through sales and indicates the average price that the cooperative can obtain by selling the types of material that have not been sold through the network, but which can be sold to middlemen. These prices vary and are constantly updated by Cataparaná and the members of the cooperative.

After weighing, the waste pickers press and package the recyclables. All the packaged waste is transferred to storage bays. The materials are stored, following the aforementioned criteria (nature, color, and quality).

The packages are stored for a short period of time until their weekly expedition to Cataparaná, the state network for processing and selling, which is managed by waste pickers, and where they stock and process waste. If the waste is not correctly packaged, the price paid will be lower and thus the cooperative members will receive less. As the network does not yet have the infrastructure for handling and processing some types of materials, the cooperative sells them directly to a company, which takes all the material to São Paulo or some other middleman or small company that purchases material directly from the unit. The rejected material is taken to a landfill by the local government.

5.2. Analysis and Results

Although the cooperative is one of the oldest in Curitiba, the membership has varied. Approximately 46% have been members for over 6 years, but only 15% of these have been in the cooperative since it was founded. Meanwhile, 31% joined in the last year.

When asked about shared techniques, 18 members responded that they shared their knowledge with colleagues. The waste pickers indicated that most of their conversations were about the quality and types of material they handle. Other members said that they try to talk about the rights of waste pickers, with other waste pickers. Two waste pickers claimed that they like to talk about new things they hear from members of other cooperatives, or discussions at the Citizenship and Waste forum.

When asked whether the cooperative seeks ways of improving the artifacts used in the handling of materials, 23 waste pickers said that the cooperative seeks to improve the objects and machinery required for the activity. In the opinion of 14 members, this occurs mainly through projects that the cooperative develops to earn subsidies from Banco do Brasil Foundation, Funasa, Cataforte, and Ecocidadão Program.

When asked about improving the techniques and knowledge used in the processes, 21 waste pickers said that the cooperative seeks to improve knowledge of the activities necessary for collecting, sorting, and recycling materials. According to the waste pickers, this occurs through lectures and courses publicized by the Catamare, and exchanges of information with other cooperatives through Cataparaná and the Citizenship and Waste Forum. None of the waste pickers could provide concrete examples of previous experiences.

Although the waste pickers appeared interested in improving the processes at the cooperative, this was shown using an essentially instrumentalized logic. None of the waste pickers could provide alternative ways of solving the latent problems of the cooperative. According to one of the members, the lack of interest in seeking other ways of resolving the problems the cooperative faces, is one of the most worrying factors regarding the survival of the unit. Another claimed that although they all work a great deal to guarantee their income, few work with the collective of the cooperative in mind. Indeed, through observation it was possible to see the effort that the waste pickers make in their long, tiring workdays. At least three times a week the waste pickers worked over 12 hours.

When asked about the greatest difficulties in the daily life of the venture, 25 members provided a variety of answers. The recurring themes included the distance between their homes and the units, the weight of the pushcarts, and the physical strength required to carry the bags from the truck to the tables (many waste pickers were ill and felt pain caused by their work at the cooperative). They also mentioned the way people separated the materials at home (the recyclables were mixed with other waste and could not be reused), the limited space at the unit, the disorganization of the waste pickers, the physical structure of the units, and the shortage of presses, forklifts, and other machinery that could facilitate their work.

The classification of the main actors involved with the cooperative showed the importance of third sector organizations, the network is self-managed by the waste pickers and public organizations, and the general structural maintenance by the Catamare network. The organizations with a higher degree of centrality are Cataparaná and the local government. In third place, is the Citizenship and Waste Institute.

Of the 26 interviewees, 24 said that their income increased after joining the cooperative. Ten waste pickers said that they began participating more in social programs after joining the cooperative. Of the 26 members, 13 receive family allowance. Only one used the My Home, My Life program. One waste picker was a Young Apprentice at the cooperative. Five members used the Social Water Rate. Two used the Fraternal Electricity program and the Free Milk Program.

The monthly income of the waste pickers was up to one minimum wage (approximately 200 dollars) for 55% of the members, and 1.1 to 2 minimum wages for the others. Meanwhile, their family income was higher, with 27% over 1 minimum wage, 42% from 1.1 to 2 minimum wages, and 46% earning over 2 minimum wages.

Only three waste pickers claimed that their level of schooling improved after joining the cooperative, but all 26 interviewees stated that they benefited from the lectures, courses, and meetings publicized by the CATAMARE. However, 4% informed that they were illiterate, 71% said that they had up to 8 years of schooling, and the rest up to 11 years.

Of the 26 interviewees, 16 stated that their family situation had improved after joining the cooperative. The main reasons for this were higher income, and the fact that they no longer had to store waste material at home. Eight waste pickers claimed that their access to public services had improved. The main reasons for this included the exchange of information among the members, and some actions by the Citizenship and Waste Institute.

Of the 26 interviewees, 22 claimed that their working conditions at the cooperative were better than their previous situation, but only 14 said they had more free time for other activities. Furthermore, 24 claimed that their relationship networks increased positively after they became members. However, only 14 waste pickers felt that their work became more socially acceptable after they joined the cooperative.

5.3. Discussion on the Limits and Potentials of Recycling Cooperatives

This study was done to analyze the limits and potential of the cooperative as an alternative for the development of social technologies, and to promote social inclusion. The relevance of this initiative lies in the possibility of cooperating to understand the processes of development of social technologies, as well as the institutional context that develops from these experiences.

Firstly, there are serious structural and historical restrictions on the recycling industry, which continue to hinder the social inclusion of waste pickers and the development of social technologies, even when cooperatives are transformed into a solidarity venture. When analyzing the historical relationship between recycling and waste pickers, it should be understood that it is not plausible to determine the composition of the workforce of waste pickers through the existence of cutting-edge technologies available for recycling waste materials [23].

Brazilian recycling only became possible on a large scale when the collection and sorting of waste were shown to be feasible at a low cost by workers whose pay compensated for investments in costly technologies, which led to the emergence of the recycling sector. Even today, despite the political gains for these workers, the cooperatives function as a kind of outsourced labor for big recycling companies, which exploit the precarious work of the members of these cooperatives. Irrespective of the organization of this type of labor, the profit rate has to compete with prices set by the world market, responsible for petroleum derivatives and aluminum and cellulose production. In other words, the work of waste pickers must be submitted to the variables of production and technological standards imposed by the recycling industry.

There is clearly an impasse between the conventional technologies demanded by recycling companies, and the reality of production on the part of waste pickers—who do not need conventional technologies—but end up depending on them because it is difficult to develop alternative ways of doing their work. Furthermore, it was observed that a minority of the waste pickers believed that the introduction of technology used by private companies at the cooperative would guarantee the economic progress of the members. Public policies also reproduced this logic.

In this sense, another important factor is the political and social awareness of the waste pickers. When asked about the meaning that they attribute to social inclusion, half of the waste pickers had a fairly wide-ranging view of the concept. In line with the definition presented in the theoretical development of the study, several waste pickers spoke of gaining rights. The most frequently mentioned were the right to education, healthcare, housing, work, and access to public spaces. Many waste pickers mentioned racial issues, and issues regarding respect for sexual orientation for social inclusion to be achieved. Others included in their understanding of the term, the right to take part in decisions that affect their lives.

However, during the interviews, the study identified that they work in a permanent condition of coercion from multiple social actors. These actors include some public authorities, the communities where the units are located, and the businesspeople that work with recycling or should donate material to the cooperative.

The relationship between the cooperative and the public authorities is highly contradictory in nature. While some public agencies comply with the law and donate recyclable materials to the cooperatives, there is no inspection or control over the diversion of resources by some institutions, and the public authorities do not pay for the service provided by waste pickers. If on the one hand, there are federal, state, and municipal projects that allow the funding of equipment and artifacts to improve the working conditions of the waste pickers, then on the other hand, this occurs within a totally determinist rationality [29]. According to the waste pickers, there are restrictions on the types of materials that can be obtained using federal funding. Normally, it is only possible to acquire certain kinds of machinery, and it is not possible to invest in technological research. If, indeed, the local government helps Catamare waste pickers by paying the rent on their units, it does not take the needs of the workers into consideration when it comes to the space they have to conduct their work. The structure is deficient and there is no maintenance. The units are small and cannot accommodate all the workers. There is a shortage of equipment for handling the material. In other words, while there is potential in the relationship between the cooperative and the authorities, this relationship imposes limits on improving the social inclusion of the waste pickers, and the development of alternative technologies to meet their demands.

The fluctuating prices of recyclable materials is another factor that compromises not only Catamare, but all waste picker cooperatives. According to the waste pickers, there is a seasonal decline in the values of recyclables and this fall in prices is related to the international market, with economic crises and currency exchange policies, which directly affect the price of the materials. The oligopolistic nature of the recycling industry, according to the waste pickers, also affects the prices paid for materials. This almost entirely limits their sales margin, and consequently, the amount they are paid for their work. Furthermore, there are situations in which the material passes through the hands of more than one buyer, which possibly reduces the prices paid to waste pickers; in order to maintain the profit margin on recyclables acquired by companies [30].

In this situation, it is perfectly admissible to declare that the organization of the work, the working hours, and the income of the waste pickers are completely determined by the price of recyclable materials. A point in question are the materials that continue to be sent to landfills because there are no buyers in the region.

Therefore, the political organization of waste pickers is a potential for empowering these workers. The main forms of organization are the National Movement of Recyclable Waste Pickers and the Cataparaná network. The Movement seeks to ensure the independence of these workers in relation to political parties, governments, and entrepreneurs. At the same time, fighting for the integrated management of solid waste, with the active participation of organized waste pickers throughout the collection, sorting, and processing of recyclable materials. The movement also believes that it is strategic to develop feasible technologies, which guarantee control of the production chain of the waste pickers.

The waste pickers collect and sort waste to earn their income. They play an important social and environmental role, as they help to minimize environmental impacts by collecting recyclable material around the city, and generate income by sorting and processing it, so that it can be reused by industry instead of being placed in landfills. Therefore, it can be inferred that waste pickers work, but are excluded, because of the type of work they do by the working class they represent. This is compounded by discrimination in terms of gender and race, as most of them are black women and single mothers. In this sense, it could be said that the inclusion of these waste pickers, even when they are members of associations and cooperatives, is somewhat perverse.

The problems of the cooperative are a challenge to the usual distinction between the technical and the social. They are, indeed, both social and technical. The solutions employed to combat the problems should be understood as being conditioned by social and technical problems. In this sense, to avoid social exclusion through partial exclusion through work, it is necessary for the authorities and support institutions, to consider the specific demands of waste pickers. They must invest in research and development to find adequate solutions, for the technological route of popular recycling, and to articulate different policies for education, healthcare, housing, the eradication of child labor, and social work for waste pickers and other requirements that must be heard and observed in the experience of these workers.

6. Conclusions

It can be proved that waste pickers when organized into cooperatives can gain some economic and social benefits that improve their quality of life. Moreover, as they become involved in social movements that are concerned with their work, they begin to perceive the important role they play in preserving the environment and discover that they have the right to citizenship. On the other hand, the interviews with the waste pickers showed that they work in a permanent condition of coercion by multiple social subjects, in addition to the veiled competition between the waste pickers themselves. The study also found that the political organization of waste pickers and awareness of this group, together with other relationships that the cooperative has built with the public authorities, enable the development of social technologies, although this is still in its early stages. This study reaffirmed that isolated programs do not totally alter the perverse social exclusion of waste pickers via inclusion through precarious work. In this sense, there must be guarantees of the universalization and integration of public policies in all spheres of public administration.

The issues raised in this study indicated that the organization of the work of the waste pickers, serves the interests of the capital involved in the chain of purchasing, recycling, and selling all the collected waste. In this context, the theme is understood as being highly relevant to scientific research, managers of public administration, and cooperatives of waste pickers, who are partners in the process of improving social inclusion and gaining citizenship, for these most important environmental actors in large urban centers.

Among the suggestions for future studies, is the need to produce researches that have the cooperative network as a study object. It could also be interesting to understand the gender implications that permeate the activity, since most of the collectors were women and most of them suffered or had already suffered domestic violence. It could be seen that this influences the construction of the identity of the activity, and there are direct relationships between the work and the autonomy of women. Further works on the recycling industry are needed, with emphasis on the cooperatives experience after the National Solid Waste Policy has been sanctioned. Research needs to be done on other less costly and more efficient processes and recycling methods, which consider the collectors activity. It is also necessary to analyze public projects, such as the Cataforte, and the methodologies applied by the institutions in the development of these projects.

Author Contributions: Conceptualization, C.L.d.S. and C.B.; Methodology, C.L.d.S. and C.B.; Formal Analysis, C.L.d.S. and C.B.; Investigation, C.B.; Resources, C.L.d.S.; Writing-Original Draft Preparation, C.L.d.S. and C.B.; Writing-Review & Editing, C.L.d.S.; Visualization, C.L.d.S.; Supervision, C.L.d.S.; Project Administration, C.L.d.S.; Funding Acquisition, C.L.d.S.

Funding: Please add: This research was funded by CNPq (Conselho Nacional de Desenvolvimento Científico e Tecnológico) grant number 308634/2013-6.

Conflicts of Interest: The authors declare no conflict of interest.

References

1. Callan, S.; Thomas, M. The impact of state and local policies on the recycling effort. *East. Econ. J.* **1997**, *23*, 411–423.
2. Folz, D.; Hazlett, J. Public participation and recycling performance: Explaining program success. *Public Adm. Rev.* **1991**, *51*, 526–532. [CrossRef]
3. Podolsky, M.; Spiegel, M. Municipal waste disposal: Unit pricing and recycling opportunities. *Public Works Manag. Policy* **1998**, *3*, 27–39. [CrossRef]
4. Sundin, E.; Björkman, M.; Eklund, M.; Eklund, J.; Engkvist, I.L. Improving the layout of recycling centres by use of lean production principles. *Waste Manag.* **2011**, *31*, 1121–1132. [CrossRef] [PubMed]
5. Silva, C.; Bollmann, H. Avaliação das Relações Sociais em Redes de Políticas Públicas para Consolidação de Programas de Gestão de Resíduos Sólidos Urbanos: Um estudo aplicado sobre o Programa Lixo que Não é Lixo de Curitiba. *Rev. Bras. Ciênc. Ambient.* **2011**, *21*, 31–47. (In Portuguese)
6. Oliveira, M. A trajetória do discurso ambiental em Curitiba (1960–2000). *Rev. Sociol. Polit.* **2001**, *16*, 97–106. [CrossRef]
7. Wilson, D.C.; Whiteman, A.; Tormin, A. *Strategic Planning Guide for Municipal Solid Waste Management*; World Bank: Washington, DC, USA, 2001.
8. Kanat, G. Municipal solid-waste management in Istanbul. *Waste Manag.* **2010**, *30*, 1737–1745. [CrossRef] [PubMed]
9. United States, Environmental Protection Agency—Office of Solid Waste and Emergency Response. *Solid Waste Management: A Local Challenge with Global Impacts*; Indiana State Library: Washington, DC, USA, 2002; pp. 1–22.
10. Wan, C.; Shen, Q.; Yu, A. The role of perceived effectiveness of policy measures in predicting recycling behaviour in Hong Kong. *Resour. Conserv. Recycl.* **2014**, *83*, 141–151. [CrossRef]
11. Moh, Y.; Manaf, L. Overview of household solid waste recycling policy status and challenges in Malaysia. *Resour. Conserv. Recycl.* **2014**, *82*, 50–61. [CrossRef]

12. Lakhan, C. Exploring the relationship between municipal promotion and education investments and recycling rate performance in Ontario, Canada. *Resour. Conserv. Recycl.* **2014**, *92*, 222–229. [CrossRef]

13. De Feo, G.; De Gisi, S. Public opinion and awareness towards MSW and separate collection programmes: A sociological procedure for selecting areas and citizens with a low level of knowledge. *Waste Manag.* **2010**, *30*, 958–976. [CrossRef] [PubMed]

14. Gouveia, N. Resíduos sólidos urbanos: Impactos socioambientais e perspectiva de manejo sustentável com inclusão social. *Cien. Saude Colet.* **2012**, *17*, 1503–1510. [CrossRef]

15. Da Silva, C.L. Proposal of a dynamic model to evaluate public policies for the circular economy: Scenarios applied to the municipality of Curitiba. *Waste Manag.* **2018**, *78*, 456–466. [CrossRef]

16. Triviños, A. *Introdução à Pesquisa em Ciências Sociais: A Pesquisa Qualitativa em Educação*; Atlas: São Paulo, Brazil, 1987; p. 83.

17. Minayo, M. *Pesquisa Social: Teoria, Método e Criatividade*, 22th ed.; Vozes: Rio de Janeiro, Brazil, 2003.

18. Yin, R.K.; Thousand, S. *Case Study Research: Design and Methods*; Blackwell Science Ltd.: London, UK, 1984; p. 23.

19. Brasil. Lei No 12.305, de 2 de Agosto de 2010. Institui a Política Nacional de Resíduos Sólidos; Altera a Lei No 9.605, de 12 de Fevereiro de 1998; e dá Outras Providências. Available online: http://www.planalto.gov.br/ccivil_03/_ato2007-2010/2010/lei/l12305.htm (accessed on 10 July 2014).

20. Política Nacional de Resíduos Sólidos: O Desafio Continua. Available online: http://www.polis.org.br/uploads/571/571.pdf (accessed on 14 January 2017).

21. Plano Nacional de Resíduos Sólidos. Available online: http://www.sinir.gov.br/documents/10180/12308/PNRS_Revisao_Decreto_280812.pdf/e183f0e7-5255-4544-b9fd-15fc779a3657 (accessed on 30 June 2014).

22. Prefeitura Municipal de Curitiba. Plano de Gestão Integrada de Resíduos Sólidos. Available online: http://multimidia.curitiba.pr.gov.br/2017/00211737.pdf (accessed on 17 February 2018).

23. Dagnino, R.; Brandão, F.; Novaes, H. Sobre o marco analítico-conceitual da tecnologia social. In *Tecnologia Social: Uma Estratégia Para o Desenvolvimento*; Lassance, J.R., Ed.; Fundação Banco do Brasil: Rio de Janeiro, Brasil, 2004.

24. Thomas, H.; Fressoli, M. En búsqueda de una metodología para investigar Tecnologías Sociales. In *Tecnología Social: Ferramenta Para Construir Outra Sociedade Campinas*; Dagnino, R., Ed.; Unicamp: Campinas, Brasil, 2009.

25. Dagnino, R. *Tecnologia Social: Ferramenta Para Construir Outra Sociedade*; Unicamp: Campinas, Brazil, 2010.

26. Jesus, V.M.B.D. Elementos transformadores e obstáculos para superação da resistência sociotécnica em experiências de tecnologia social. *Ciênc. Tecnol. Soc.* **2013**, *1*, 54–75.

27. Thomas, H. Estructuras cerradas vs. Procesos dinámicos: Trayectorias y estilos de innovación y cambio tecnológico. In *Actos, Actores y Artefactos: Sociología de la Tecnología*; Thomas, H., Alfonso Buch, A., Eds.; Universidad Nacional de Quilmes: Bernal, Argentina, 2008; pp. 217–262.

28. Rutkowski, J.; Lianza, S. Sustentabilidade de empreendimentos solidários: Que papel espera-se da tecnologia? In *Tecnologia Social: Uma Estratégia Para o Desenvolvimento*; Lassance, J.R., Ed.; Fundação Banco do Brasil: Rio de Janeiro, Brazil, 2004.

29. Bosi, A. A organização capitalista do trabalho "informal": O caso dos catadores de recicláveis. *Rev. Bras. Cienc. Soc.* **2008**, *23*, 101–116.

30. Forlin, F.; Faria, J. Reciclagem de embalagens plásticas. *Polímeros* **2002**, *12*, 1–10. [CrossRef]

 *recycling*

Article

Applying the Theory of Planned Behavior to Recycling Behavior in South Africa

Wilma F. Strydom

Natural Resources and the Environment, Council for Scientific and Industrial Research,
P.O. Box 395, Pretoria 0001, South Africa; wstrydom@csir.co.za

Received: 30 July 2018; Accepted: 5 September 2018; Published: 8 September 2018

Abstract: This paper reports on an application of the Theory of Planned Behavior to understand the relationships between the determinants (latent variables) comprising the Theory of Planned Behavior and, based on these findings, to guide decision-making related to household recycling in South Africa. Data from a representative sample of respondents in large urban areas ($n = 2004$) was analyzed using Structural Equation Modeling (SEM). The results of the SEM analysis showed a good fit of the survey data to the Theory of Planned Behavior theoretical model. The Theory of Planned Behavior explains 26.4% of the variance in recycling behavior and 46.4% of the variance in intention to recycle. Only 3.3% of South Africans in large urban areas show dedicated recycling behavior, considering the recycling of five materials: paper, plastic, glass, metal, and compostable organic waste. The recycling frequency item in the recycling behavior construct is the most likely to be over-reported. South Africans lack sufficient knowledge, positive attitudes, social pressure, and perceived control that would encourage recycling behavior. Awareness drives containing moral values (injunctive norms) and information about available recycling schemes, combined with the provision of a curbside collection service for recyclables, have the greatest chance to positively influence recycling behavior amongst South Africa's city dwellers.

Keywords: recycling behavior; Theory of Planned Behavior (TPB); Structural Equation Modeling (SEM); South Africa

---

1. Introduction

The waste management challenge is not unique to South Africa. The urban areas in "most low- and middle-income countries" receive unreliable and inadequate municipal solid waste management services (Bartone 2004:3). Similar to other developing countries [1–5], many municipalities in South Africa struggle to supply adequate waste management services [6,7]. Waste collection coverage in Africa varies between 25% and 70% [8]. In 2012 about 68 million tons of the estimated 125 million tons of waste generated in Africa, was collected [9]. It is projected that by the year 2025 urban waste generation in Africa will reach 441,840 tons per day [10]. In addition to population growth that adds to the volumes of waste generated, increased consumption rates, excessive packaging, and throw-away attitudes aggravate the waste problem and puts pressure on the environment and on limited resources [11–13].

In South Africa, the implementation of the waste management hierarchy, as envisaged in national legislation [14,15], provides the required enabling regulatory environment to support a circular economy, i.e., a closed loop system where secondary resources are reintroduced back into the economy. One of the many benefits of moving waste up the hierarchy is that recycling and energy recovery from waste displace the use of virgin resources, which in turn reduce the costs (financial, social, and environmental) associated with virgin resource extraction [16–18]. In 2011, the annual resource value of waste in South Africa was estimated at R25.2b—about 0.86% of South Africa's gross domestic product (GDP) [18]. Ambitious targets for diverting these recyclables from landfill add to the waste management challenge [19].

In an attempt to divert recyclables from landfill, the South African recycling sector has mostly been active to recover recyclables from preconsumer waste, i.e., the recovery of recyclable materials from commercial and industrial processes without a consumer being involved as the end-user. The important role of the informal sector in postconsumer recycling is acknowledged [20], but postconsumer recycling should receive more attention in order to increase recycling rates on a national level, especially if the targets for diversion is to be reached [21,22].

This paper reports on the findings from a baseline study in South Africa to ascertain recycling behavior at household level at a given point in time before the National Environmental Management: Waste Act (NEMWA) (Act No. 59 of 2008) [14] was widely implemented. As a theoretical framework, the Theory of Planned Behavior (TPB) is applied to show the relationships between attitude, social pressures, and perceived control over the act of recycling, as determinants of intention to recycle as well as recycling behavior. It is envisaged that the results from this study would inform waste management decision-making and highlight areas of possible intervention that would have the most impact at household level to positively change behavior towards increased recycling. Knowing which variables have a greater effect on recycling behavior can guide government and industry initiatives and interventions towards implementation of the NEMWA and reaching separation at source targets. Decision-making related to household recycling includes the structure, functioning, and placement of recycling programs, the infrastructure requirements to support behavioral expectations, and the focus of communication strategies and awareness programs.

Following on this short introduction, the rationale for selecting the TPB and a brief introduction to the theoretical framework is provided in the literature section. The third section describes the methodological approach, including a description of the questionnaire content, followed by the statistical method in the fourth section. Section 5 presents the results and discussion and the sixth section summarizes the main conclusions.

2. Literature

For decades researchers have been searching for variables that influence behavior and to identify the variables affecting behavior the most. Published in the mid-1970s, some of the first studies on recycling behavior include the effect of information and incentives on paper recycling amongst resident university students [23], the effect of attitude and personality on recycling [24], the ascription of recycling behavior to ideological and demographic variables [25], and the willingness to recycle glass and paper [26]. Over the years, conflicting results were published, for example, Weigel (1977) ascribed recycling to, amongst others, demographic variables [25], but later studies showed no direct effect of demographic variables on behavior [27–30].

In the waste management domain popular theories that have been applied in an attempt to explain recycling behavior include, amongst others, the Theory of Reasoned Action (TRA), the TPB, and Schwartz's (1977) Theory of Normative Conduct [31], to name a few. Over the years the TPB has been used in many research fields to understand behavior [32]. Examples include human health (refer to the 1996 Godin and Kok review of studies using the TPB in behavior related to health [33]), water conservation [34], and waste recycling [35–38]. Thus, a wealth of empirical data supports and contributes to the popularity of the TPB [39].

Critique against the TPB is also documented. One of the main criticisms of the TPB is that the model finds it difficult to predict behavior that is not out of choice or preference, or that requires resources and skills [40]. In addition, acknowledging that beliefs and attitudes are important; Boldero (1995) concluded that the TPB is inadequate to explain recycling behavior [41]. Ajzen and Madden (1986) noted that although a person can decide to act a certain way, the successful execution of the behavior relies on external factors such as the availability of resources [42]. If barriers that prevent the action are removed, the action is also more likely to happen [43]. Despite being criticized [41,44], the TPB is one of the most widely-used and most-supported theories to explain recycling behavior [45–48].

2.1. Theoretical Framework

Based on the concepts explained in Dulany's Theory of Propositional Control, Ajzen and Fishbein developed the TRA, the forerunner of the TPB, with the addition of predictions of specific intentions and behavior [49]. The Theory of Propositional Control explains that behavior is not necessarily strengthened by reward or discouraged by punishment [50,51]. Rather, "people form a conscious intention [a] behavioral intention" to act a certain way [50,51] (p. 440). The behavioral intention (BI) is determined by the beliefs that a specific behavior will have a desired outcome, the value attributed to the outcome, the perception that a specific behavior will contribute to the outcome, the perception of the correctness of the behavior, and perceptions of the degree to which the specific behavior is expected [50]. In other words, people's intentions depend on their motivation to comply with what they believe is a desired action, what they feel is expected of them and "what they think they are supposed to do" [50] (p. 440). Dulany's theory of BI does not nullify the value of positive reinforcement, but emphasizes the role of people's beliefs [50].

According to the TRA, an intention to act is a precursor to the behavior related to the same act [49]. In turn, attitude towards the behavior and social pressures (subjective norm) are determining factors of the intention to perform a certain action [49]. Attitude is a personal factor which refers to a person's evaluation of the behavior. Subjective norm is a social factor which refers to the "perceived social pressure" to comply with a certain behavior, where social pressure is defined as the perceptions, beliefs, and judgments of other household members and community members related to recycling. Both attitude and subjective norm are grounded in the belief systems of a person [42] (p. 454). While subjective norms are the construct where the "influence of relevant others" are expressed [52], attitude is a more personal construct—an expression of the "self" [31].

The TPB expand on the TRA by including perceived behavioral control (PBC), which is a reflection of people's beliefs or "confidence in their ability to perform" a certain action [42] (p. 457), as well as an indication of the "available resources and opportunities" [42] (p. 459) (Figure 1). PBC has an additional effect on people's intention to act, which is independent of either attitude or subjective norm [42]. Apart from the intention to act, the executing of a behavior is also dependent on a person's perception of the ability to perform the specific behavior—how easy or difficult it is to perform the specific action. Thus, PBC exerts pressure on the intention to behave, but also independently on the behavior itself [53].

To allow comparison with similar studies conducted in developed and developing countries, and with guidance for phrasing of the construct statements of the TPB model [36,54,55], the TPB is also used in this study to explain and help understand recycling behavior in urban households in South Africa.

2.2. Research Question

This study addresses the following research question. Within the context of the status of postconsumer recycling in South Africa, which interventions at household level would be the most successful to encourage and maximize postconsumer recycling behavior in South Africa?

The subquestions interrogated in this paper are listed below.

- Which variables have the greater effect on recycling behavior (within the framework of the TPB)?
- Are the findings from the South African baseline study similar to the findings of international studies of domestic waste recycling behavior to allow generalization and thus co-learning across boundaries?

2.3. Hypotheses

Figure 1 shows the TPB model as described in the theoretical framework (Section 2.1). The hypotheses are defined as follows:

H_1 A positive attitude towards recycling has a positive and direct effect on the intention to recycle.

H_2 Subjective norm, i.e., the social pressures to recycle (the beliefs and the judgments) has a positive and direct effect on the intention to recycle.

H_3 Perceived behavioral control has a positive and direct effect on the intention to recycle.

H_4 Intention to recycle has a positive and direct effect on recycling behavior.

H_5 Perceived behavioral control has a positive and direct effect on recycling behavior.

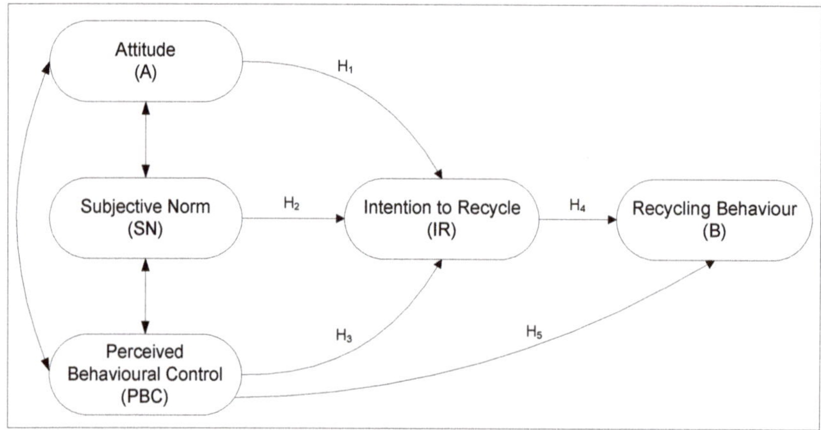

Figure 1. The Theory of Planned Behavior (adapted from Azjen & Madden, 1986 [42]).

3. Method

3.1. Research Design

A quantitative research method was followed to apply the TPB to waste recycling behavior in South Africa. A fixed form survey which allowed the selection of options was used to gather data from a relatively large sample (n = 2004) within a short period of time [56,57]. A structured questionnaire standardized the interview process and ensured that the same questions were posed in the same way [57].

3.2. Sampling

A random probability sampling method was followed to obtain a representative sample of the South African population in the large urban areas [56], i.e., metropolitan areas and cities with a population size of 250,000 or more and included Cape Town, Johannesburg, Durban, East Rand, Pretoria, Port Elizabeth, the Vaal area, East London, Pietermaritzburg, Bloemfontein, and Welkom (Figure 2). For the standard random selection procedure starting points were selected at random and a Kish-grid used to select individual respondents [58]. A large sample size of 2000 was chosen to reduce the effect of sampling errors [56,59,60].

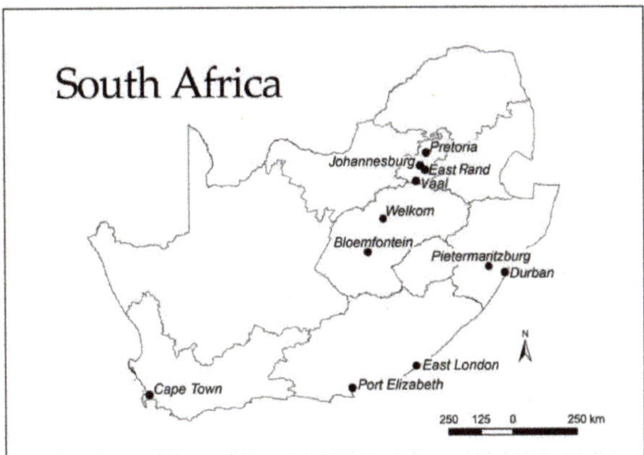

Figure 2. Metropolitan areas and cities included in the study.

3.3. Questionnaire Design

Being part of a larger survey, one of the questionnaire sections made up the constructs of the TPB. The TPB constructs contain statements which measured respondents' behavior, intentions to behave, attitudes, subjective norms, and perceived behavioral control. The sequence of questions can influence survey results [60]. Therefore, the behavior questions preceded the attitude questions to prevent alignment of the easier factual accounts of behavior to the aspirations acquired with answering the attitude and other questions [60]. The starting point of items within a set, e.g., the attitude items, was also rotated.

Where applicable, and similar to many other TPB studies [54], 7-point Likert-type rating scales [61] and 7-point semantic differential scales were used. The statements were anchored on either side with strongly agree/strongly disagree or, for example, with bad/good.

3.3.1. Theory of Planned Behavior Latent Variables

A summary of the items that made up each of the constructs of the TPB is provided below.

- Recycling behavior—seven items in total: recycling frequency (one item with seven options: never, almost never, seldom, sometimes, often, almost always, always); taking responsibility for recycling in the household (one item with four options: no one, not me/someone else, I am, I am and sometimes someone else is); and, a qualitative measurement of recycling quantity (five items with seven options: nothing, very little, some things, about half, most of everything, almost all, everything) for five streams of recyclables with space to specify additional recyclables.
- Intention to recycle—seven items, four items testing the probability of recycling under various circumstances and three items testing the probability of executing various recycling activities: sorting; putting out recyclables for collection; and, taking recyclables to collection points (refer to Table 3 for set of items).
- Attitude towards recycling—seven items starting with "For your household to recycle is" anchored on either side with, e.g., bad/good (refer to Table 4 for set of items).
- Subjective norm—five items: two statements about motivation to comply and three items about the perceptions of others (refer to Table 5 for set of items).
- Perceived behavioral control—six items: three items measure opportunity to recycle and three items ability to recycle (refer to Table 6 for set of items).

3.3.2. Testing of the Questionnaire

The draft questionnaire was tested in a pilot study to be able to sharpen the measuring instruments, to identify ambiguities and questions which might cause uneasiness, and to ensure workability of the data [56,60]. During the course of one day, tenants and shoppers in a shopping center were approached and the purpose of the study explained. The volunteers were offered a paid-for hot drink at a pre-arranged coffee shop where they could sit and relax while completing the questionnaire.

Fifty-nine individuals completed the pilot questionnaire on an anonymous basis. Each construct (measuring tool) was tested statistically for consistency and reliability. After completion of the pilot study the questionnaire was shortened to only include those questions that would add the most value.

3.3.3. Data Collection

Due to the nature, sample size, time frames, and geographical distribution of the proposed sample, a professional survey company that was able to meet the survey requirements, was contracted to include the questions in their annual survey. The interviews were conducted face-to-face at respondents' homes. If a respondent refused to participate, the respondent was substituted using the same procedure in another household. At no stage was or can the identity of the individual respondent be linked back to the data.

The CSIR provided the wording of an accompanying briefing pamphlet which stated the purpose of the study, how the information will be used, and on whose behalf the specific part of the survey is conducted. The relevant contact details of the principle researcher and the CSIR Research Ethics Committee member were provided. The first statement on the questionnaire was a tick box with an acknowledgement from the participant that he/she received the pamphlet and was sufficiently informed before answering the questions. Participation was on a voluntary basis without receiving any rewards or incentives for taking part in the survey.

4. Statistical Method

4.1. Analysis

The TPB was tested by fitting a structural equation model (SEM) to the latent variables. First, each of the constructs was examined for reliability (i.e., if similar results are likely to be obtained with a retest) and unidimensionality (whether the items in a construct work together to measure one latent trait). The Guttman's lambda 6 (λ_6) [62], as well as the criticized [63], but still widely used Cronbach alpha (α) [64] and McDonald's omega (ω) report reliability, and Revelle's beta (β) report internal consistency [65].

Thereafter, the mean score of the items within each construct was calculated in order to obtain a summary score for each construct (measuring instrument). The next step was to test the effect(s) of the constructs on recycling behavior. The objective of the statistical analysis was to test whether the theoretical TPB model could be supported with a statistical significant model based on the collected data. Structural equation modeling (SEM), using partial least squares path modeling (PLSPM) was used. As a final check, classification and regression trees and random forests (results not shown) were also used to confirm the predictive ability of the various constructs on recycling behavior. The package R was used for the statistical analysis [66,67].

MS Excel was used for descriptive statistics (graphs, frequency tables, etc.), to describe averages and for determining measures for variability and relationships between variables (correlation and regression analyses).

4.1.1. Treatment of Inappropriate Answers

A small number of respondents gave "not applicable" or "do not know" answers which do not form part of the 7-point scales. Of the 33 respondents that indicated they do not know how often their households recycle, 29 respondents selected the "nothing" option in the statements that tested the quantity households recycle and their "do not know" option was subsequently changed to "never". The remaining four respondents' "do not know" option for frequency of recycling was also scored to match their recycling quantity scores. The 33 respondents represent 1.65% of the total sample group. Given the large sample size ($n = 2004$), these adjustments should have no significant effect, even if the modified scores are incorrect by more than 2 units. The occurrence of inappropriate answers to other items was low and was thus treated as missing. The statistical methods make provision for such instance by dropping those observations with missing values in the specific analyses affected by the missing information, but retaining information of non-missing values in other analyses.

4.1.2. Assumption

It was assumed that if the respondent reports no recycling behavior in his/her household, then the respondent do not recycle at home. However, some over-reporting of individual recycling behavior is possible where the respondent do not recycle but someone else in the household does. This possibility is captured in the responsibility item of the recycling behavior construct.

5. Results and Discussion

5.1. Sample Profile (Demographic Composition)

This study targeted a representative sample of households in the larger urban areas in South Africa. A total of 2004 households in 11 large urban areas, including all metropolitan municipalities, were interviewed. The demographic composition of the sample is provided in Appendix A, Table A1. Table 1 shows the descriptive statistics for each of the constructs (the latent variables), i.e., the means, as well as the values per percentiles. Tables 2–6 include descriptive statistics for each of the items per construct. The results from each construct are discussed in detail in the sections to follow, concluding with the TPB structural equation model (Tables 7–10).

Table 1. Descriptive statistics for the latent variables, mean scores, and percentiles ($n = 2004$).

Construct (Latent Variables)	$\bar{x}$	Percentiles						
		0.05	0.10	0.25	0.50	0.75	0.90	0.95
B (Recycling behavior)	1.44	1.00	1.00	1.00	1.00	1.14	2.86	3.57
IR (Intention to recycle)	3.76	1.00	1.43	2.57	3.86	4.86	5.71	6.29
A (Attitude)	3.86	1.45	2.00	2.96	3.86	4.71	5.67	6.00
SN (Subjective norm)	3.37	1.40	1.60	2.40	3.40	4.20	5.00	5.60
PBC (Perceived behavioral control)	3.30	1.17	1.67	2.33	3.33	4.00	5.00	5.50

Where 1 = the lowest possible score, and 7 = highest possible score.

5.2. Recycling Behavior

Of the total sample group ($n = 2004$), 540 of the respondents' households (26.9%) reported recycling behavior. The rest of the respondents ($n = 1464$; 73.1%) with a recycling behavior score of 1 (B = 1), reported no recycling activity in their households. The 26.9% recycling households includes those who, for example, reported very little recycling of one type of recyclable material only. Eighteen percent (18.5%) show very little recycling activity, which is indicated with a recycling score of 2–3. Only 3.3% of the respondents reported that their households often recycle about half or more of all their recyclables (recycling scores greater than 4). The low mean recycling behavior score (x-bar =1.44) confirms the low percentage of households in which recycling behavior is reported (Table 2).

Table 2. Descriptive statistics for the recycling behavior measurement and the separate items that make up the construct.

Item	Recycling Behavior Average and per Item		Main Components	
	$\bar{x}$	SD	$\bar{x}$	SD
Recycling behavior (Average of all items below)	1.44	0.94		
Recycling frequency	1.76	1.55	1.76	1.55
Taking responsibility for recycling in your household	1.44	0.89	1.44	0.89
Recycling quantities *: (Average of items below)	-	-	1.35	1.11
Paper	1.48	1.32	-	-
Plastic	1.44	1.24	-	-
Glass	1.41	1.17	-	-
Metal	1.27	0.95	-	-
Compost	1.27	1.06	-	-

Where 1 represents no recycling activity and 7 represents the best possible mean value for recycling behavior.
* Qualitative measurement of recycling quantities (refer to Section 3.3.1).

The recycling behavior construct consists of three components: a "recycling frequency"-item; a "taking responsibility for recycling in the household"-item; and a "recycling quantity"-item (Table 2). Recycling quantity is measured by the average of the "quantities" reported to be recycled of each of five recyclables (paper, plastic, glass, metal, and compost) reported as being recycled. A comparison of the three main components that make up the recycling behavior construct shows that the average for recycling frequency (x-bar = 1.76) is higher than for taking responsibility (x-bar = 1.44) and for quantities recycled (x-bar = 1.35) (Table 2, main components). In addition, some respondents (3.4%) indicated that their households sometimes recycle, but failed to indicate recycling quantities of any recyclable materials. These 68 respondents form part of the 18.5% respondents that reported very little recycling activity. The data thus suggest some over-reporting of recycling behavior which mainly originates from the frequency item. Being one of seven items in the behavior construct, the effect of the over-reporting of recycling frequency on recycling behavior is smaller than it would have been if recycling frequency was the only item measured for recycling behavior.

The overall higher self-reporting for recycling frequency compared to recycling quantities is most probably due to two reasons: firstly, it is easier to over-report recycling behavior on a "soft" frequency measurement than on actual physical quantities of items recycled; and secondly, the frequency question was asked first and thus before the reality check of the actual quantities. A possible third reason is that there is no correlation between recycling frequency and recycling quantities. However, the data show that the correlations are significant ($p < 0.001$) and of medium to high strength, depending on the recyclable material [68], i.e., correlation factors between recycling frequency and the "quantity" of paper, glass, metal, plastic, and compostables are 0.723, 0.648, 0.496, 0.671, and 0.467, respectively (Appendix A, Table A2). It is noteworthy that paper recycling is probably the most-established in South Africa, with the Ronnie bag collection system operating in many areas for more than 30 years.

Due to the random probability-sampling method the results for recycling behavior can be extrapolated to the South African population in the larger urban areas of the country. Given that the self-reported recycling behavior is expected to be higher than what it would be if measured [30,45], the reported results may reflect an optimistic view of the domestic recycling situation in South Africa. However, it should be noted that the purpose of the study was not to gather actual recycling-rate data. Although self-reported, the recycling behavior results provide valuable insight into recycling tendencies in South Africa at a given point in time—after the NEWWA came into effect but before wide implementaton of separation of waste.

The domestic recycling reality, as indicated by the results of this South African study, is that, at the time of the waste recycling survey (November 2010), the majority of South African households (73.1%) in large urban areas did not recycle. Only a small fraction of urban households (3.3%) recycled most of their household waste on a fairly frequent basis.

5.3. Intention to Recycle

The majority of respondents either expressed no intention to recycle or low levels of intention to recycle (x-bar = 3.76) (Table 3). The results suggest that respondents are more likely to recycle if their recyclables are collected at curbside (x-bar = 4.21) than when they have to take recyclables to collection points (x-bar = 3.42). The likelihood that respondents will recycle also decrease the further the collection points are from their homes. The item that shows the lowest mean score (x-bar = 3.33) is the one implying travelling to a collection point the furthest away.

Table 3. Descriptive statistics for the intention of respondents to recycle.

Item	$\bar{x}$	SD
Intention to recycle (Average of items below)	3.76	1.54
The willingness or likelihood to recycle . . . If curbside collection for recyclables in area	4.21	2.01
If to put recyclables out separately for curbside collection	3.92	1.91
If to sort your recyclable waste from rest of household waste	3.86	0.98
If have to take recyclables to a collection point in area close to house	3.81	1.91
If have to take recyclables to support a charity initiative	3.75	1.86
If to take recyclables to collection points	3.42	1.77
If have to take recyclables to a collection point further away from house	3.33	1.73

Where 1 = the option very unlikely or not willing at all (no intention), and 7 = the option very likely or very willing (high level of intention).

It should be noted that the items are phrased to capture perceived distances, because a 2 km distance is just around the corner for someone who can drive there, but for someone who has to carry a bag of recyclables, it is a long distance. The role of the convenience factor in intention to recycle is emphasized by the difference in mean scores of two of the items, namely " . . . if curbside collection for recyclables in area" (x-bar = 4.21) and "if to put recyclables out separately for curbside collection" (x-bar = 3.92). The latter, through the use of the word "separately", implies multiple sorting of recyclables, which is not so explicitly expressed in the first item. The results thus suggest that people would be more willing to recycle should they be serviced with a 2-bag system which is collected at curbside, compared to multiseparation of recyclables.

The results suggest that the intention to recycle is overruled by the practical reality of being able to recycle. The curbside collection item is the only item with a positive score (x-bar > 4.00) in the IR construct. The majority of respondents feel negative about taking their recyclables to drop-off points (x-bar < 4.00). The willingness to take recyclables to collection points decrease significantly the further the perceived distance to the collection point is. Since the majority of the respondents reported that the household does not have a motor vehicle in the household (Appendix A, Table A1), longer distances to recycling points are problematic for household recycling behavior.

5.4. Attitude Towards Recycling

With the mean score for attitude towards recycling (x-bar = 3.86) being less than the neutral point of 4.0 (Table 4), the majority of respondents has a negative attitude towards recycling. Only four respondents in total chose the "do not know"-option and only on a single item, which suggest that the respondents do have an attitude and that this attitude leans towards the negative, rather than not having formed an attitude, yet. Due to the random sampling method the results can thus be extrapolated to suggest the existence of an overall negative attitude towards recycling among South African city dwellers.

Table 4. Descriptive statistics for the attitude of respondents towards recycling.

Item	$\bar{x}$	SD
Attitude (Average of items below)	3.86	1.34
For your household to recycle is . . .		
Bad/good	4.12	1.60
Useless/useful	3.99	1.59
Unimportant/important	3.95	1.65
Awakens negative emotions/positive emotions	3.89	1.44
A waste of time/useful	3.82	1.64
Undesirable/desirable	3.76	1.67
A hassle/easy	3.48	1.58

Where 1 implicates a most negative attitude towards recycling, and 7 implicates a most positive attitude towards recycling.

Within the attitude construct, the moral component as represented by the item "for your household to recycle is bad/good", shows the highest score (x-bar = 4.12) of all the attitude items. The "for your household to recycle is a hassle/easy" item shows the lowest mean score (x-bar = 3.48), and could be an indication of the influence of perceived convenience of recycling on householders' attitude towards recycling.

5.5. Social Pressure to Recycle (Subjective Norm)

The majority of respondents reported that they experience a lack of social pressure to recycle (x-bar = 3.37) (Table 5). The two items, "most of the people important to you want you to recycle" and "it is expected of you to recycle" show the lowest mean scores (x-bar = 2.96 and 3.27, respectively), and could indeed be a true reflection of the situation in South Africa, given the small percentage (3.3%) of respondents that reported that they engage in meaningful recycling (B > 4). Thus, extrapolated to the South African population, the individuals among family, friends, neighbors, and other significant people that would expect of others or exert pressure to recycle on others is part of a small group of South Africans. Nonrecyclers do not recycle and therefore would not be able to either be a recycling role-model or exert pressure to recycle in the manner that recyclers would be able to. In fact, someone who recycles would fall outside the norm of this-is-how-things-are-done-around-here, which suggests that the descriptive norm would not be pro-recycling. This is in line with the conclusion of Cialdini and coworkers [69] (p. 231) that an intervention which focuses on the descriptive social norm will only be successful in cases where the majority of people already conform to the desired behavior. If the majority do not recycle, the person who recycles would rather be considered and feel the odd one out. Minato (2012) also warns that the pressure through descriptive norms decline due to degrading social networks [70].

Table 5. Descriptive statistics for the subjective norm variable.

Item	$\bar{x}$	SD
Social pressure to recycle (Average of items below)	3.37	1.30
Your household does not want to recycle/wants to recycle	3.70	1.62
For your household to recycle is difficult/easy	3.55	1.55
Your municipality thinks it is important for your household to recycle: strongly disagree/strongly agree	3.37	1.75
It is expected of your household to recycle: strongly disagree/strongly agree	3.27	1.79
Most of the people important to you want you to recycle: strongly disagree/strongly agree	2.96	1.67

Where 1 implicates a most negative attitude towards recycling, and 7 implicates a most positive attitude towards recycling.

Another angle from which to interpret the results is that one would suspect that the respondents with high recycling behavior scores are represented by the high subjective norm scores, but with a correlation of 0.49 this only suggest a relationship of medium strength (Appendix A, Table A3). A large percentage of those respondents from reportedly recycling households also do not experience any social pressure to recycle. Thus, it can be argued that the recyclers tap their motivation to recycle from a source independent of what others expect of them. It can be speculated that injunctive norms (moral values) could be a driver for recycling behavior, but because the questionnaire is weak on injunctive norm items, this possibility should be further researched.

5.6. Perceived Control over the Act of Recycling

The average (x-bar = 3.30) of the perceived behavioral control measurement is less than the neutral point (x-bar = 4.00). This result suggests that respondents do not feel that they have control over their ability to recycle (Table 6). All items making up this construct scored less than the neutral point. Albeit negative (x-bar < 4.00), the item "you know how to recycle" which represents a knowledge component, has the highest average (x-bar = 3.81) of all the perceived behavioral control items. The item "To recycle is difficult/easy" addressing the perceived difficulty to recycle and, also suggesting an underlying knowledge component, has the second highest score (x-bar = 3.59). The data thus suggests that, although the knowledge component of perceived behavioral control is not the main hurdle to overcome to change people's perceptions of their control over their ability to recycle, there is still a lack of sufficient knowledge among the majority of South Africans.

Table 6. Descriptive statistics for respondents' perceived behavioral control over recycling behavior.

Item	$\bar{x}$	SD
Perceived control over the act of recycling (Average of items below)	3.30	1.25
You know how to recycle false/true	3.81	1.65
To recycle is difficult/easy	3.59	1.58
The opportunities for you to recycle are none/plenty	3.49	1.56
There are recycling schemes in your area: strongly disagree/strongly agree	3.01	1.74
The necessary resources and facilities are available that allow you to recycle: strongly disagree/strongly agree	3.01	1.68
You are aware of recycling schemes in your area: strongly disagree/ strongly agree	2.94	1.74

Where 1 implicates a most negative attitude towards recycling, and 7 implicates a most positive attitude towards recycling.

The respondents perceived the opportunity to recycle ("The opportunities for you to recycle are none/plenty") less favorably (x-bar = 3.49) than the two knowledge items discussed above. But, the availability of recycling scheme items ("There are recycling schemes in your area" and "The necessary resources and facilities are available"; x-bar = 3.01 for both items) and the awareness of recycling scheme item fared the worst (x-bar = 2.94). Respondents thus feel that they do not have control over the act of recycling, especially in terms of where to recycle. Focused awareness-creation initiatives on the location of recycling drop-off infrastructure are thus needed.

5.7. Testing the Theory of Planned Behavior Model

The behavior, intention to recycle and attitude constructs show excellent reliability and internal consistency (Table 7). The reliability and internal consistency of the subjective norm and perceived behavioral control constructs are good. Thus, it was decided to keep all the items of all constructs.

Table 7. Reliability and internal consistency of the TPB constructs.

Construct	Instrument Quality	Coefficients *					
		α	λ_6	β	ω_h	ω_{lim}	ω_t
B	Excellent	0.965	0.965	0.944	0.883	0.906	0.974
IR	Excellent	0.931	0.938	0.897	0.811	0.856	0.948
A	Excellent	0.941	0.936	0.808	0.905	0.946	0.956
SN	Good	0.853	0.839	0.760	0.719	0.809	0.888
PBC	Good	0.873	0.884	0.711	0.720	0.772	0.932

* Where: α = Cronbach's alpha, λ_6 = Guttman's lambda 6, β = Revelle's beta, ω_h = McDonald's omega hierarchical, ω_{lim} = McDonald's omega asymptotic, ω_t = McDonald's omega total.

The results of the SEM analysis show a good fit of the survey data to the TPB theoretical model (Table 8).

Table 8. Goodness-of-fit statistics for the fitted SEM model.

Model Fit Criteria	Observed Value	Interpretation/Accepted Levels [47,71]	Comment
GFI (Goodness-of-fit index)	0.990	0 (no fit)—1 (perfect fit); values $\geq$ 0.95 suggest good fit	Good fit of model
AGFI (Adjusted goodness-of-fit index)	0.924	0 (no fit)—1 (perfect fit); values $\geq$ 0.9 suggest good fit	Good fit of model
RMSEA (Root-mean-square error of approximation)	0.112	0 = perfect fit	Good fit of model
Bentler–Bonnett NFI (Normed fit index)	0.990	0 (no fit)—1 (perfect fit); values $\geq$ 0.95 suggest good fit	Good fit of model
Tucker–Lewis NNFI (Non-normed fit index)	0.953	0 (no fit)—1 (perfect fit); values $\geq$ 0.95 suggest good fit	Good fit of model

The regression coefficients (βeta) are shown in Table 9. The proportion of variance explained for IR and B is shown in Table 10 and Figure 3.

Table 9. Parameter estimates for applying the TPB.

Path (Hypothesis)	Path Coefficients (β-Value)	Std Error	z-Value	Pr (>|z|)	Result of Hypothesis Hest
A $\rightarrow$ IR (H$_1$)	0.275	0.029	9.628	0.000	Statistically significant
SN $\rightarrow$ IR (H$_2$)	0.590	0.035	17.083	0.000	Statistically significant
PBC $\rightarrow$ IR (H$_3$)	-0.020	0.032	-0.610	0.542	Not significant
IR $\rightarrow$ B (H$_4$)	0.131	0.014	9.589	0.000	Statistically significant
PBC $\rightarrow$ B (H$_5$)	0.276	0.017	16.326	0.000	Statistically significant

Table 10. Model R^2-values.

Variable	Observed Variance	Estimated Variance	R^2	Std. Error Coefficient
B	0.881	0.648	0.264	0.858
IR	2.379	1.275	0.464	0.732

The TPB explain 46.4% of the variance in intention to recycle and 26.4% of the variance in recycling behavior (B) (Figure 3).

Fitting the TPB model to the survey data shows that the subjective norm (SN) (β = 0.590) has greater influence than either attitude (A) (β = 0.275) or perceived behavioral control (β = 0.020) on intention to recycle. The relatively strong and significant relationship (β = 0.590; p < 0.0001) between subjective norm and intention to recycle supports H$_2$. Overshadowed by the effect of subjective norm on intention to recycle, attitude results in having a smaller (β = 0.275), though significant (p < 0.0001), effect on intention to recycle (H_1). Perceived behavioral control shows an insignificant effect on intention to recycle ($\beta = -0.020$; $p = 0.54$) and thus rejects H_3. With $R^2 = 0.464$, the three variables attitude, subjective norm, and perceived behavioral control together account for 46.4% of the variance in intention to recycle. Both intention to recycle and perceived behavioral control influence recycling behavior, accounting for 26.4% of the variance in recycling behavior ($R^2 = 0.264$). Perceived behavioral control has a significant effect on recycling behavior ($\beta = 0.276$; $p < 0.0001$) (H_5).

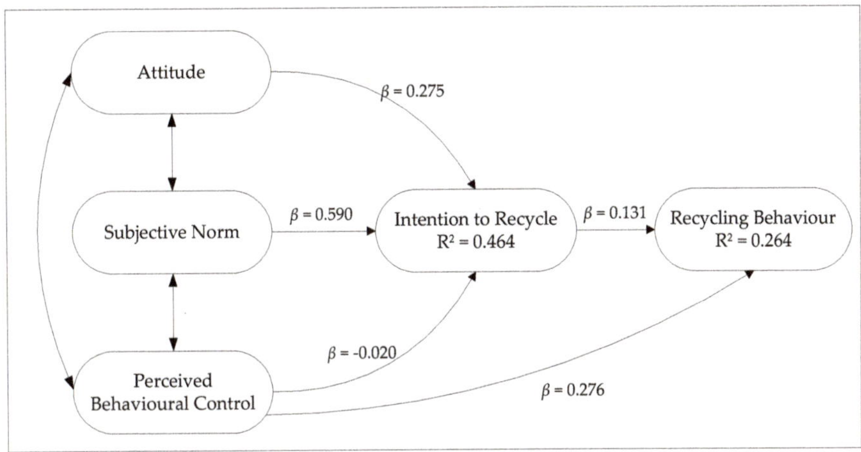

Figure 3. Path diagram of the TPB fitted to the unstandardized/raw latent variables.

The results from this study suggest that perceived behavioral control carries more weight (H_5) than intention to recycle (H_4) in explaining recycling behavior. Typically, in studies where the behavior is likely to not only be affected by personal motivation, but also by other factors such as the availability of resources and access to services, perceived behavioral control appears to have a greater influence on intention [33]. This is in line with the argument that the effect of perceived behavioral control varies with the availability of curbside recycling schemes [36].

The small influence of attitude on intention to recycle ($\beta = 0.275$) is in line with findings of Martin et al. (2006) that a positive attitude towards recycling does not guarantee recycling behavior [72]. The attitude-behavior link proves to be strong where there is no "resources and cooperation" needed [40]. The relatively weak link between attitude and intention to recycle, which was found in this study, could thus be ascribed to ancillary variables and situational factors not accommodated or explained through the TPB. Although A is not the variable with the strongest/largest effect, attitude does contribute to the intention to recycle. Awareness-raising initiatives to improve people's attitudes towards recycling have a better chance for success should it include a moral component.

With the TPB explaining 26.4% of the variance in recycling behavior, the results from this study compares well with the Armitage and Conner meta-analysis of 185 behavioral studies dated pre-1998 [45]. The meta-analysis found that, on average, the TPB explained 27% of the variance in behavior and 39% of the variance in intention to act [45].

Armitage and Conner (2001) point out that, over the years, researchers measured the IR construct in different ways [45]. The IR construct is a measure of "how hard people are willing to try or how much effort they would exert to perform the behavior" [43] (p. 181). Distinguishing between intentions and self-predictions of behavior, it is argued that self-predictions (the likelihood to perform a behavior) provide a better prediction of behavior than intentions [45]. The inclusion of likelihood to recycle statements in the IR construct is a possible explanation for this study's higher explanation value of intention to recycle (46.4%) than Armitage and Conner's meta-analysis average of 39% [45].

Respondents show a higher probability that they intend to recycle than what their self-reported behavior suggests. In the behavioral sciences this is one of the challenges of predicting behavior [40]. There are several external factors that influence the path between intention and action even though the best intentions might exist. Examples may include a family crisis, or just forgetting to put out the recyclables, or a breakdown of a motor vehicle which makes it impossible to take recyclables to the drop-off center. The best intention might also be deliberately suppressed or ignored, for example due to unfavorable weather conditions on recycling day. Thus, from a personal point of view a person may be a recycler, but the actual recycling behavior may be absent.

Similar to what other studies showed [30,55,73], this study with 26.4% of the variance explained by the TPB, suggests that there are other variables than those proposed in the TPB that appears to have an effect on recycling behavior.

The importance of perceived behavioral control as the construct with the largest effect on recycling behavior in the TPB model, confirms that people should feel in control of their ability to recycle. One manner in which to promote a sense of control is through buy-in, e.g., through allowing communities to co-design their recycling services, because a waste scheme that is acceptable and functional in one area might not be suitable for another area [74]. Thus, through co-designing of recycling schemes, the necessary buy-in and awareness of the recycling facilities can be created. Co-designing also creates opportunity for established co-responsibility—a moral imperative (injunctive norm) which has the potential to be more successful at creating social pressure than descriptive norms in a society where recycling behavior is very low. Co-designing of a recycling scheme would also serve as a direct communication of the importance of the communities' participation in recycling and indicate that the municipality/recycling company takes recycling seriously (descriptive norm). Another advantage of co-designing of recycling schemes would be that potential leaders of the recycling initiative in communities would be identified. These leaders could assist with future recycling related communications and also through playing the part of much-needed role-models. In addition, recycling needs to be reinforced as a normative behavior through, e.g., well-targeted recycling advertisements, awareness creation, and the deliberate visibility of recycling bins.

6. Conclusions and Recommendations

The study shows that at a point in time, November 2010, before the NEMWA was widely implemented, about a quarter (26.9%) of South Africa's city dwellers engaged in some form of recycling which include their paper and packaging waste as well as compostable garden and/or food waste. Only 3.3% of the respondents reported that they recycle about half or more of their recyclables on a frequent basis. While the TPB remains a useful model for examining the variables that affects recycling behavior, the TPB explains 26.4% of the variance in recycling behavior and 46.4% of the variance in intention to recycle. Compared to intention to recycle, which has a far smaller effect on recycling behavior than expected, perceived behavioral control appears to be the most important variable to explain recycling behavior. This confirms that there are other variables than those proposed in the TPB that appears to play an important role in recycling behavior.

It is encouraging that respondents are positive about their intention to recycle should they have a curbside collection for their recyclables, especially given the very low self-reported recycling score and the negativity that is overwhelmingly present in all the variables. The results also suggest that a less complicated and more convenient 2-bag waste collection system to accommodate the collection of recyclables at curbside has the greatest potential to be supported and thus encourage household recycling. The results strongly suggest that awareness raising that has the greatest chance to influence recycling behavior positively, should contain a balanced mix of moral values (injunctive norms) and information about available recycling schemes. However, raising awareness would be meaningless without a positive contribution to householders' perceived behavioral control over their ability to recycle. One way to alleviate perceived behavioral control is through the provision of tailor-made recycling schemes to fit communities' particular needs. While co-designing of recycling schemes would be beneficial for creating buy-in, such schemes should continue to operate as designed to ensure that the sense of having control over the act of recycling is obtained and maintained. For example, even the best designed recycling scheme will not show the envisaged results if such a scheme is failing or not well maintained.

The recommendation to policy-makers is to provide the most convenient and least complicated curbside collection for recyclables, such as a 2-bag collection system. However, it is important to acknowledge the need for diversity in designing recycling schemes, as well as the effect of continuity, reliability, and maintenance of recycling schemes on recycling behavior. Thus, the challenge for waste management is the ability to create an enabling environment in South Africa that would not only encourage recycling behavior, but also ensure continuation of recycling behavior. In particular recycling of the materials currently recycled in the lowest quantities, i.e., garden and food waste, metals, and glass, would benefit from improved recycling facilities.

The implication of this research is that over time, as South Africans become more aware of recycling, and recycling behavior thus would have shifted from the current baseline for recycling, that the strategies to further improve recycling behavior needs to be adapted to the changed needs at the particular point in time. It should also be acknowledged that South Africa is a country with much diversity which complicates generalization. It can thus be dangerous to be prescriptive on the best ways for improving recycling behavior without considering specific areas, its unique characteristics due to its local set-up, and operating within its local constraints. Furthermore, some of the findings need clarification which confirms the need for further research.

Funding: This research was funded by the Council for Scientific and Industrial Research (CSIR).

Acknowledgments: Ipsos-Markinor was contracted to conduct the survey on behalf of the CSIR and included the set of questions in their biannual survey. The analysis of the data was partly outsourced to Mark Difford at the Nelson Mandela Metropolitan University. Suzan Oelofse assisted with pilot study logistics on the day of pilot data collection. Several colleagues in employment of the CSIR at the time, assisted with the initial literature review. Heidi van Deventer compiled the map (Figure 2) for this publication.

Conflicts of Interest: The author declares no conflict of interest.

Appendix A

Table A1. Demographic variables of the sample group (*n* = 2004).

Parameter	Sample Distribution	
	n	**%**
Gender		
Male	1002	50
Female	1002	50
Total	2004	100
Age (years)		
16–17	69	3.4
18–19	85	4.2
20–24	271	13.5
25–29	215	10.7
30–34	222	11.1
35–39	256	12.8
40–44	205	10.2
45–49	176	8.8
50–54	133	6.6
55–59	95	4.7
60–64	103	5.1
65+	173	8.6
Refused	1	0

Table A1. *Cont.*

Parameter	Sample Distribution	
	n	**%**
Population		
Black	1338	67
White	290	14
Colored	255	13
Indian/Asian	121	6
Working status		
Full-time	665	33.2
Part-time	233	11.6
Not working: Housewife	126	6.3
Student	199	9.9
Retired	232	11.6
Unemployed: Looking for work	481	24.0
Not looking for work	68	3.4
Education (formal)		
No schooling	29	1.4
Some primary	113	5.6
Primary school	115	5.7
Some high school	725	36.2
Matric/Grade 12	678	33.8
Artisans certificate obtained	73	3.6
Technicon diploma/degree completed	148	7.4
University degree completed	87	4.3
Professional	15	0.7
Technical	15	0.7
Secretarial	2	0.1
Other	4	0.2
Type of dwelling		
House/cluster house	1213	60.5
Flat	85	4.2
Matchbox/RDP	348	17.4
Hostel/compound/boarding house	16	0.8
Room in backyard	57	2.8
Shack	276	13.8
Other	9	0.5
Financial status (household income)		
None	46	2.3
R1-<R2k	321	16.0
R2k-<R5k	434	21.7
R5k-<R10k	285	14.2
R10k-<R20k	175	8.7
R20k+	148	7.4
Refused	564	28.1
Do not know	31	1.6
Motor vehicle in household		
Yes	833	41.6
No	1171	58.4
Telephone in house (land line)		
Yes	432	21.6
No	1572	78.4

Table A1. *Cont.*

Parameter	Sample Distribution	
	n	%
Television set in household		
Yes	1919	95.8
No	85	4.2
Radio set in household (in working order): 1	1207	60.2
2	300	15.0
3	83	4.1
None	414	20.7

Table A2. Correlation values between items making up the recycling behavior construct.

Items	B	Responsibility	Frequency	Paper	Glass	Metal	Plastic	Compost
B	1	-	-	-	-	-	-	-
Responsibility	0.815	1	-	-	-	-	-	-
Frequency	0.882	0.829	1	-	-	-	-	-
Paper	0.833	0.651	0.723	1	-	-	-	-
Glass	0.805	0.584	0.648	0.607	1	-	-	-
Metal	0.709	0.468	0.496	0.493	0.528	1	-	-
Plastic	0.835	0.609	0.671	0.640	0.612	0.553	1	-
Compost	0.695	0.418	0.467	0.463	0.511	0.544	0.556	1

Table A3. Correlation values between latent variables.

Variables	B	IR	A	SN	PBC
B	1	-	-	-	-
IR	0.408	1	-	-	-
A	0.365	0.597	1	-	-
SN	0.494	0.662	0.739	1	-
PBC	0.480	0.523	0.658	0.766	1

References

1. Abdrabo, M.A. Assessment of economic viability of solid waste service provision in small settlements in developing countries: Case study Rosetta, Egypt. *Waste Manag.* **2008**, *28*, 2503–2511. [CrossRef] [PubMed]
2. Imam, A.; Mohammed, B.; Wilson, D.C.; Cheeseman, C.R. Solid waste management in Abuja, Nigeria. *Waste Manag.* **2008**, *28*, 468–472. [CrossRef] [PubMed]
3. Parrot, L.; Sotamenou, J.; Dia, B.K. Municipal solid waste management in Africa: Strategies and livelihoods in Yaoundé, Cameroon. *Waste Manag.* **2009**, *29*, 986–995. [CrossRef] [PubMed]
4. Kgosiesele, E.; Zhaohui, L. An evaluation of waste management in Botswana: Achievements and challenges. *J. Am. Sci.* **2010**, *6*, 144–150.
5. United Nations Environment Programme (UNEP). *Africa Waste Management Outlook*; United Nations Environment Programme: Nairobi, Kenya, 2018.
6. Department of Environmental Affairs (DEA). *White Paper on Integrated Pollution and Waste Management for South Africa. A Policy on Pollution Prevention, Waste Minimisation, Impact Management and Remediation*; Government Gazette Vol. 417, No. 20978, General Notice 227 of 17 March 2000; Government of South Africa: Pretoria, South Africa, 2000.
7. Nhamo, G.; Oelofse, S.; Godfrey, L.; Mvuma, G. *WasteCon'08 Workshop Report: Unpacking Governance Opportunities and Challenges for Integrated Municipal Waste Management in South Africa*; CSIR Report No. CSIRNRE/PW/IR/2009/0034/B; CSIR: Pretoria, South Africa, 2009.
8. United Nations Environment Programme (UNEP). *Global Waste Management Outlook (GWMO)*; UNEP DTIE International Environmental Technology Centre: Osaka, Japan, 2015.
9. Scarlat, N.; Motola, V.; Dallemand, J.F.; Monforti-Ferrario, F.; Mofor, L. Evaluation of energy potential of municipal solid waste from African urban areas. *Renew. Sustain. Energy Rev.* **2015**, *50*, 1269–1286. [CrossRef]

10. Hoornweg, D.; Bhada-Tata, P. *What a Waste: A Global Review of Solid Waste Management*; Urban Development Series Knowledge Papers; World Bank: Washington, DC, USA, 2012.

11. United Nations Human Settlements Programme (UN-HABITAT). *Solid Waste Management in the World's Cities: Water and Sanitation in the World's Cities 2010*; UN-HABITAT: Nairobi, Kenya, 2010.

12. Curtis, J.; Lyons, S.; O'Callaghan-Platt, A. Managing household waste in Ireland: Behavioural parameters and policy options. *J. Environ. Plan. Manag.* **2011**, *54*, 245–266. [CrossRef]

13. Organisation for Economic Co-operation and Development (OECD). *Greening Household Behaviour: Overview from the 2011 Survey (Revised Edition)*; OECD Studies on Environmental Policy and Household Behaviour; OECD Publishing: Paris, France, 2014.

14. Republic of South Africa (RSA). *National Environmental Management: Waste Act, 2008 (Act 59 of 2008)*; Government Notice 278 Government Gazette 32000 of 10 March 2009; Government of South Africa: Pretoria, South Africa, 2008.

15. Republic of South Africa (RSA). *National Environmental Management: Waste Amendment Act, 2014. (Act 26 of 2014)*; Government Notice 449 Government Gazette 37714 of 2 June 2014; Government of South Africa: Pretoria, South Africa, 2014.

16. European Environment Agency. *Earnings, Jobs and Innovation: The Role of Recycling in a Green Economy*; European Environment Agency (EEA): Copenhagen, Denmark, 2011.

17. United Nations Environment Programme (UNEP). *Guidelines for National Waste Management Strategies: Moving from Challenges to Opportunities*; United Nations Institute for Training and Research (Unitar); UNEP: Geneve, Switzerland, 2013. Available online: http://cep.unep.org/meetings/documents/41ff9f7eb276c7989419dca11245970c (accessed on 28 August 2018).

18. Department of Science and Technology (DST). *A National Waste R&D and Innovation Roadmap for South Africa: Phase 2 Waste RDI Roadmap. The Economic Benefits of Moving Up the Waste Management Hierarchy in South Africa: The Value of Resources Lost through Landfilling*; Department of Science and Technology: Pretoria, South Africa, 2014.

19. African Union Commission. *Agenda 2063: The Africa We Want*; Final Edition; African Union Commission: Addis Ababa, Ethiopia, 2015. Available online: https://au.int/sites/default/files/pages/3657-file-agenda2063_popular_version_en.pdf (accessed on 28 August 2018).

20. Godfrey, L.; Strydom, W.; Phukubye, R. *Integrating the Informal Sector into the South African Waste and Recycling Economy in the Context of Extended Producer Responsibility*; Briefing Note; CSIR: Pretoria, South Africa, 2016.

21. Oelofse, S.H.; Strydom, W.F. Trigger to Recycling in a Developing Country: In the Absence of Command-and-Control Instruments. In Proceedings of the Conference Waste 2010: Waste and Resource Management—Putting Strategy into Practice, Warwhickshire, UK, 28–29 September 2010.

22. Godfrey, L.; Oelofse, S. Historical review of waste management and recycling in South Africa. *Resources* **2017**, *6*, 57. [CrossRef]

23. Witmer, J.F.; Geller, E.S. Facilitating paper recycling: Effects of prompts, raffles and contests. *J. Appl. Behav. Anal.* **1976**, *9*, 315–322. [CrossRef] [PubMed]

24. Arbuthnot, J. The roles of attitudinal and personality variables in the prediction of environmental behavior and knowledge. *Environ. Behav.* **1977**, *9*, 217–232. [CrossRef]

25. Weigel, R.H. Ideological and demographic correlates of proecology behavior. *J. Soc. Psychol.* **1977**, *103*, 39–47. [CrossRef]

26. O'Riordan, T.; Turner, R.K. Recycling and householder attitudes: A survey of Norwich. *Resour. Policy* **1979**, *5*, 42–50. [CrossRef]

27. Ajzen, I.; Fishbein, M. *Understanding Attitudes and Predicting Social Behaviour*; Prentice-Hall: Upper Saddle River, NJ, USA, 1980.

28. Vining, J.; Ebreo, A. What Makes a Recycler? A comparison of recyclers and non-recyclers. *Environ. Behav.* **1990**, *22*, 55–73. [CrossRef]

29. Oskamp, S.; Harrington, M.J.; Edwards, T.C.; Sherwood, D.L.; Okuda, S.M.; Swanson, D.C. Factors Influencing Household Recycling Behavior. *Environ. Behav.* **1991**, *23*, 494–519. [CrossRef]

30. Gamba, R.J.; Oskamp, S. Factors influencing community residents' participation in commingled curbside recycling programs. *Environ. Behav.* **1994**, *26*, 587–612. [CrossRef]

31. Schwartz, S.H. Normative influences on altruism. *Adv. Exp. Soc. Psychol.* **1977**, *10*, 221–279.

32. Jackson, T. Motivating sustainable consumption. *Sustain. Dev. Res. Netw.* **2005**, *29*, 30.

33. Godin, G.; Kok, G. The theory of planned behavior: A review of its applications to health-related behaviors. *Am. J. Health Promot.* **1996**, *11*, 87–98. [CrossRef] [PubMed]

34. Trumbo, C.W.; O'Keefe, G.J. Craig Intention to conserve water: Environmental values, planned behavior, and information effects. A comparison of three communities sharing a watershed. *Soc. Nat. Resour.* **2001**, *14*, 889–899.

35. Knussen, C.; Yule, F.; MacKenzie, J.; Wells, M. An analysis of intentions to recycle household waste: The roles of past behaviour, perceived habit, and perceived lack of facilities. *J. Environ. Psychol.* **2004**, *24*, 237–246. [CrossRef]

36. Tonglet, M.; Phillips, P.S.; Read, A.D. Using the Theory of Planned Behaviour to investigate the determinants of recycling behaviour: A case study from Brixworth, UK. *Resour. Conserv. Recycl.* **2004**, *41*, 191–214. [CrossRef]

37. Davies, J.; Foxall, G.R.; Pallister, J. Beyond the intention–behaviour mythology an integrated model of recycling. *Mark. Theory* **2002**, *2*, 29–113. [CrossRef]

38. Ioannou, T.; Zampetakis, L.A.; Lasaridi, K. Using the theory of planned behavior for understanding the drivers of household recycling behavior. In Proceedings of the 3rd International CEMEPE & SECOTOX Conference, Skiathos, Greece, 19–24 June 2011; pp. 429–434.

39. Conner, M.; Armitage, C.J. Extending the theory of planned behavior: A review and avenues for further research. *J. Appl. Soc. Psychol.* **1998**, *28*, 1429–1464. [CrossRef]

40. Eagly, A.H.; Chaiken, S. *The Psychology of Attitudes*; Thomson Wadsworth: Belmont, CA, USA, 1993; p. 794.

41. Boldero, J. The Prediction of Household Recycling of Newspapers: The Role of Attitudes, Intentions, and Situational Factors 1. *J. Appl. Soc. Psychol.* **1995**, *25*, 440–462. [CrossRef]

42. Ajzen, I.; Madden, T.J. Prediction of goal-directed behaviour: Attitudes, intentions, and perceived behavioural control. *J. Exp. Soc. Psychol.* **1986**, *22*, 453–474. [CrossRef]

43. Ajzen, I. The theory of planned behavior. *Organ. Behav. Hum. Decis. Process.* **1991**, *50*, 179–211. [CrossRef]

44. Sniehotta, F.F.; Presseau, J.; Araújo-Soares, V. Time to retire the theory of planned behaviour. *Health Psychol. Rev.* **2014**, *8*, 1–7. [CrossRef] [PubMed]

45. Armitage, C.J.; Conner, M. Efficacy of the theory of planned behaviour: A meta-analytic review. *Br. J. Soc. Psychol.* **2001**, *40*, 471–499. [CrossRef] [PubMed]

46. Kollmuss, A.; Agyeman, J. Mind the gap: Why do people act environmentally and what are the barriers to pro-environmental behavior? *Environ. Educ. Res.* **2002**, *8*, 239–260. [CrossRef]

47. Oom do Valle, P.; Rebelo, E.; Reis, E.; Menezes, J. Combining behavioral theories to predict recycling involvement. *Environ. Behav.* **2005**, *37*, 364–396. [CrossRef]

48. Armitage, C.J. Time to retire the theory of planned behaviour? A commentary on Sniehotta, Presseau and Araújo-Soares. *Health Psychol. Rev.* **2015**, *9*, 151–155. [CrossRef] [PubMed]

49. Fishbein, M.; Ajzen, I. *Belief, Attitude, Intention and Behavior: An Introduction to Theory and Research*; Addison-Wesley Pub. Co.: Boston, MA, USA, 1975.

50. Ajzen, I. The theory of planned behavior. In *Handbook of Theories of Social Psychology*; SAGE Publications: London, UK, 2012; Volume 1, pp. 438–459.

51. Dulany, D.E. *Awareness, Rules, and Propositional Control: A Confrontation with SR Behavior Theory*; Prentice-Hall: Upper Saddle River, NJ, USA, 1968.

52. Salazar, H.A.; Oerlemans, L.; van Stroe-Biezen, S. Social influence on sustainable consumption: Evidence from a behavioural experiment. *Int. J. Consum. Stud.* **2013**, *37*, 172–180. [CrossRef]

53. Ajzen, I. From intentions to actions: A theory of planned behavior. In *Action Control from Cognition to Behavior*; Springer: Heidelberg, Germany, 1985; pp. 10–39.

54. Francis, J.; Eccles, M.P.; Johnston, M.; Walker, A.E.; Grimshaw, J.M.; Foy, R.; Kaner, E.F.; Smith, L.; Bonetti, D. *Constructing Questionnaires Based on the Theory of Planned Behaviour: A Manual for Health Services Researchers*; Centre for Health Services Research, University of Newcastle upon Tyne: Newcastle upon Tyne, UK, 2004.

55. Barr, S. Factors influencing environmental attitudes and behaviors: A UK case study of household waste management. *Environ. Behav.* **2007**, *39*, 435–473. [CrossRef]

56. Babbie, E.R.; Mouton, J. *The Practice of Social Research*; Oxford University Press: Cape Town, South Africa, 2001.

57. Kempton, W.; Boster, J.S.; Hartley, J.A. *Environmental Values in American Culture*; MIT Press: Cambridge, MA, USA, 1995.

58. Kish, L. A Procedure for Objective Respondent Selection within the Household. *J. Am. Stat. Assoc.* **1949**, *44*, 380–387. [CrossRef]
59. Page, C.; Meyer, D. *Applied Research Design for Business and Management*; McGraw-Hill Higher Education: Sydney, Australia, 2003; p. 331.
60. Brace, I. *Questionnaire Design: How to Plan, Structure and Write Survey Material for Effective Market Research*; Kogan Page Publishers: London, UK, 2018.
61. Likert, R. A technique for the measurement of attitudes. *Arch. Psychol.* **1932**, *140*, 55.
62. Guttman, L. A basis for analyzing test-retest reliability. *Psychometrika* **1945**, *10*, 255–282. [CrossRef] [PubMed]
63. Sijtsma, K. On the use, the misuse, and the very limited usefulness of Cronbach's alpha. *Psychometrika* **2009**, *74*, 107. [CrossRef] [PubMed]
64. Cronbach, L.J. Coefficient alpha and the internal structure of tests. *Psychometrika* **1951**, *16*, 297–334. [CrossRef]
65. Revelle, W. Hierarchical cluster analysis and the internal structure of tests. *Multivar. Behav. Res.* **1979**, *14*, 57–74. [CrossRef] [PubMed]
66. R Core Team. *R: A Language and Environment for Statistical Computing*; R Foundation for Statistical Computing: Vienna, Austria, 2012.
67. Difford, M. Nelson Mandela Metropolitan University, Port Elizabeth, South Africa. Addendum to Results of Statistical Analysis Contract Report. Personal communication. Unpublished report. 2010.
68. Cohen, J. A power primer. *Psychol. Bull.* **1992**, *112*, 155. [CrossRef] [PubMed]
69. Cialdini, R.B.; Kallgren, C.A.; Reno, R.R. A focus theory of normative conduct: A theoretical refinement and reevaluation of the role of norms in human behavior. *Adv. Exp. Soc. Psychol.* **1991**, *24*, 201–235.
70. Minato, W.L.; Curtis, A.L.; Allan, C. Understanding the role and influence of social norms: Lessons for NRM. *Local Environ.* **2012**, *17*, 863–877. [CrossRef]
71. Hooper, D.; Coughlan, J.; Mullen, M. Structural equation modelling: Guidelines for determining model fit. *Electron. J. Bus. Res. Method* **2008**, *6*, 53–60.
72. Martin, M.; Williams, I.D.; Clark, M. Social, cultural and structural influences on household waste recycling: A case study. *Resour. Conserv. Recycl.* **2006**, *48*, 357–395. [CrossRef]
73. Vicente, P.; Reis, E. Factors influencing households' participation in recycling. *Waste Manag. Res.* **2008**, *26*, 140–146. [CrossRef] [PubMed]
74. Korfmacher, K.S. Solid waste collection systems in developing urban areas of South Africa: An overview and case study. *Waste Manag. Res.* **1997**, *15*, 477–494. [CrossRef]

 *recycling*

Article

Barriers to Household Waste Recycling: Empirical Evidence from South Africa

Wilma F. Strydom

Natural Resources and the Environment, Council of Scientific & Industrial Research, P.O. Box 395, Pretoria 0001, South Africa; wstrydom@csir.co.za

Received: 30 July 2018; Accepted: 3 September 2018; Published: 6 September 2018

Abstract: A small percentage of South Africans regularly recycle most of their recyclables, which was only 4% and 7.2% in 2010 and 2015, respectively. This empirical quantitative study, the first study on this scale in South Africa, aimed to ascertain the reasons why people do not recycle. This paper reports the results from a survey conducted among a representative sample of 2004 respondents in eleven of South Africa's large urban areas. Each respondent selected three main reasons why people do not recycle from ten possible options as well as the one main reason. The results show that (i) insufficient space, (ii) no time, (iii) dirty and untidiness associated with recycling, (iv) lack of recycling knowledge, and (v) inconvenient recycling facilities are perceived as the main reasons why people do not recycle. Non-recycling households (74% of the respondents) give high priority to time and knowledge. Low recyclers—those that sporadically recycle few items—and young South Africans give high priority to services (inconvenient facilities and no curbside collection). Lack of knowledge is an important factor for people from dense settlements as well as the unemployed looking for work. Improved recycling services such as regular curbside collections have the potential to overcome time and space barriers. Recycling services as well as recycling knowledge will have to improve to encourage the youth, the unemployed, and those living in informal areas to recycle and realize the opportunities locked in the waste sector. The perceptions of respondents from non-recycling households differ from those from recycling households. The larger representation of non-recyclers in developing countries emphasize the importance of understanding local evidence when comparing and implementing results from developed countries. The learning from this study could also assist other developing countries to encourage household participation in recycling initiatives.

Keywords: quantitative survey; empirical study; developing country; household recycling behavior; reasons; barriers

---

1. Introduction

The waste management hierarchy has been adopted into South African waste legislation [1] as a means to support sustainable development. However, in 2011, an estimated 90% of all waste generated was still being disposed of in landfills [2]. Diverting recyclables from landfills is a worldwide concern in the drive towards a circular economy that strives to keep resources in use for as long as possible through reuse, recycling, and recovery [3–5]. South Africa has successfully grown a recycling economy over the past three decades [6]. In South Africa, about half of all post-consumer packaging waste was recycled in 2012 [7]. However, as is the case in most developing countries, the contribution of the active informal sector is acknowledged in achieving the encouraging recycling figures [8,9]. Growing the circular economy provides opportunities to improve the livelihoods and working conditions of waste pickers as well as to improve the methods used to collect materials for recycling [10]. The South African National Waste Management Strategy (NWMS) set targets for diversion of 25% of recyclables from landfills by 2016 [11], but, due to a lack of accurate waste data, there is no evidence suggesting that these targets were met. Agenda 2063, which is the 50-year strategic socio-economic transformation framework for Africa, set diversion targets for Africa [12]. The NWMS is in the process of being updated in line with the Agenda 2063 target of 50% diversion of waste from landfill by 2023 [12] and a target of 50% of households separating waste at source by 2023 [13]. For such ambitious targets to be reached, separation of recyclables at source becomes essential. In addition, separation at a household level is an opportunity to improve the quality of the recyclables as well as the working conditions of the informal sector.

Caution should be taken when comparing results from earlier studies when participation in recycling initiatives were mostly voluntary with results from later studies where curbside collection of "multiple materials" were more likely to be the norm [14] (p. 106). In South Africa, to date, household participation in recycling is still voluntary either by taking recyclables to buy-back or drop-off centers [15] or via street collection where such a recycling scheme is available. This voluntary household recycling resulted in 4.0% of South Africans living in the large urban areas recycling their paper and packaging on a regular basis in 2010 [16]. Although there was a decrease in the percentage households that do not recycle—from 74.0% in 2010 to 66.6% in 2015—only 7.2% of South African urban households reported in 2015 that they are dedicated recycling households, i.e., regularly recycling more than half of their recyclable waste [16].

This paper reports results from an empirical study conducted in 2010 on the reasons why households living in large urban areas think South Africans do not recycle. This study on a national level is a first for South Africa. The objective of this study is to ascertain the reasons why South Africans do not recycle, i.e., the barriers to recycling that they experience or perceive to experience. The results from this study will assist decision makers in the waste sector by pointing out possible interventions to encourage household participation in recycling initiatives to reach targets in line with national and African policy objectives.

2. Literature on Barriers to Household Recycling

Schultz and co-workers grouped variables having an effect on recycling behavior into personal and situational factors when they reviewed published empirical studies on recycling behavior [14]. Personal factors include attitude towards and beliefs about the environment, knowledge about recycling, taking responsibility, locus of control, and demographic variables (age, gender, income, education, etc.) [14]. Situational factors include the antecedents (e.g., collection method, goal-setting, normative factors, removing of any barriers, prompting) and consequence variables (e.g., rewards, feedback) [14].

Barr differentiated between environmental values, situational variables (which include the behavioral context such as recycling facilities and services, the socio-demographic variables such as age, gender, education, and income, and the knowledge and experience variables) and psychological factors (taking responsibility, altruism, intrinsic motivation, social norm, and self- efficacy or perceived behavioral control, which encompass time, space, and convenience among others) [17]. Results from Barr's study in the United Kingdom showed that a combination of two psychological factors (perceptions of convenience and acceptance of recycling as a social norm) and two situational factors (curbside collection for recyclables and knowledge of recycling services) are the most likely to increase both intention to recycle and recycling behavior [17].

Godfrey et al. grouped factors with the potential to influence recycling behavior into socio-demographic factors (e.g., age, gender, level of education), recycling facilities and services (also called the behavioral factors), and psychological factors (e.g., values, attitudes) [18].

The examples above show how the clustering of variables that affect recycling behavior differs between studies. For the purpose of this study, reasons for non-recycling obtained from the published literature are grouped into four main issues known as: situational factors at a household level, situational (knowledge), situational factors related to recycling facilities and services, and psychological factors (Table 1).

Table 1. Variables or factors affecting recycling behavior.

Issues or Factors	Published Research Findings
Situational factors at a household level	
It takes time to recycle [19–26]	There is an issue of time associated with recycling especially when having to take recyclable materials to a drop-off site. People with more time is more likely to take recyclables to drop-off facilities. Working people have less time to recycle than retired individuals.
Insufficient storage space for recyclables [19–21,23,24,26–28]	Storage space both inside and outside homes becomes a logistical issue. Larger gardens provide more space for storing recycling bins and bags while apartment buildings mostly have a limited space for storing recyclables.
Unpleasant odors, the feeling of untidiness recycling [19,21,23,24]	Possibilities of unpleasant odors and feelings of untidiness is associated with recycling. Perceptions exist that only clean materials can be recycled.
Situational (knowledge)	
Not enough awareness or knowledge about recycling and recycling initiatives [17,20,24,25,27,29–31]	Knowledge of the recycling program, awareness of the location of recycling facilities, and knowing how to and knowing what materials are recyclable is needed. Well targeted communications, door-to-door promotions, regular information and feedback including "how to" messages is needed to improve recycling. The visibility of recycling bins reminds people to put recyclables out.
Situational factors related to recycling facilities and services	
Perceptions of recycling (in)convenience [17,20,24,26,32–35]	Collection of recyclables is perceived as the most convenient ("least-effort") method for recycling and long distances to recycling centers discourage recycling. Simultaneously, binary sorting of waste is preferred to multi-sorting. Recycling schemes do not consider household preferences. Perceptions that it is inconvenient and requires effort to recycle [12,19,21] can be cancelled out by access to good recycling facilities.
Curbside collection [17,19,24–26,36]	Curbside collection motivates and encourages recycling. Preference for and higher participation rates obtained in collection schemes than in drop-off schemes, e.g., in Scotland, and in the Czech Republic. Curbside schemes counter the time it takes to recycle.
Lack of services/facilities [17,20,26,35]	Access to good recycling facilities encourages recycling behavior. A good recycling service is reliable, convenient, and easy to use. Improving recycling schemes to suit household preferences has greater potential to positively change householders' recycling behavior than either incentives or penalties.
Psychological factors	
Sense of responsibility [19,29,37]	Both non-recyclers and recyclers feel that recycling is a local authority responsibility. Household waste recycling is everybody's responsibility. Recyclers show a stronger sense of responsibility—attitudes of "duty"—towards recycling than non-recyclers. Feelings of responsibility is not influenced by the type of facility. It helps to foster recycling behavior if someone in the household constantly reminds everyone to recycle.
Disinterest/cannot be bothered [20,22,25]	Disinterest to recycle is a barrier to recycling and reported, amongst others, in London, Glasgow and Falkirk and Malaysia.
Doubts about whether it would make a difference [25,29,38]	Non-recyclers think it would not make a difference whether they recycle or not. Knowing a household's recycling contribution makes a difference in the bigger scheme of waste management and encourages recycling behavior. Individuals who feel indifferent to recycling are unlikely to recycle. The challenge is to replace indifference with concern.

3. Materials and Methods

3.1. Research Design

This paper ascertains the reasons why South Africans do not recycle. A descriptive quantitative research approach [39] with a fixed form survey [40] was followed. Recycling behavior of the household is obtained from self-reporting on questions relating to frequency, number of materials recycled, and a qualitative measurement of quantities recycled. The initial list of reasons why people do not recycle were obtained from literature (refer to Table 1).

3.2. Sampling Method and Data Collection

Ipsos was contracted to include a set of recycling behavior questions in their annual household survey. Using a standard random probability selection procedure, the survey targeted a representative sample of 2000 households in the large cities. Starting points were selected at random and a kish-grid was used to select individual respondents [41]. Due to the survey instrument and representative sample, the results can be generalized to the population [42].

3.3. Questionnaire Design

This study considered methods followed in previous studies. While some previous studies allowed respondents to formulate their own perceived barriers to recycling [28,43], others tested several situational and psychological variables, which are thought to have an effect on recycling behavior [17]. In another UK study, respondents indicated three reasons why they recycle or do not recycle and the reasons "ranked according to their popularity to allow comparisons to be made" [24] (p. 64). Kaciak and Kushner followed a qualitative approach to determine barriers to recycling by using open-ended questions such as: *"in your opinion, why do you think others don't recycle?"* [44].

Due to the envisaged sample size of 2000 respondents, the national quantitative survey followed the example of Barr et al. by providing 10 possible pre-selected reasons for non-recycling to choose from [17]. Following Kaciak and Kushner's example [44], the question *"what would you say are the reasons why people do not recycle?"* was phrased in an impersonal manner. First, to prevent a desirability bias which could be caused by feelings of guilt of non-recycling households and, second, to provide respondents that recycle the opportunity to also convey their perceptions about why they think non-recyclers do not recycle. It can be argued that those respondents that recycled at the time of the study would have noted difficulties with recycling they experience from day-to-day as well as the barriers they think others might experience.

The 10 reasons to include was selected as follows: First, a list of possible reasons for non-recycling was compiled from a review of international literature [17,21,28,45–47]. Second, a qualitative pre-study with open-ended questions was conducted [39] to identify gaps in the international literature, to test for relevance, and to adjust phrasing where appropriate for South African circumstances. Third, a set of 10 factors was tested in a pilot study ($n = 59$) where respondents could tick any number of statements with which they agree and circle the most important ones. The results from the pilot study were difficult to analyze because some respondents ticked all 10 options and others ticked one only. This learning was taken forward and the main study was adapted accordingly.

Each respondent was presented with a "show card" on which the 10 possible reasons why people do not recycle were listed. The respondents had the opportunity to study the options before selecting the three main reasons [24] and indicating the most important reason from the selected three reasons. The order of the factors was rotated on the "show card" to ensure that the order of the statements does not bias the response. The structured format of the question with options allows for uniformity in responses from a large sample size within a short period of time [48].

Recycling behavior of the household is calculated from self-reporting on questions relating to frequency, number of materials recycled, and a qualitative measurement of quantities recycled (Appendix A). Demographic information was obtained from the questions included in the standard Ipsos questionnaire.

3.4. Data Collection

Face-to-face interviews were conducted in November 2010 at the homes of the respondents. Respondents that refused to participate in the study were substituted using the same probability selection procedure. The participants in the study received a briefing pamphlet, which stated the purpose of the study, how the information will be used, on whose behalf the survey is conducted, and contact details where further information could be obtained. The participants also received assurance that they would stay anonymous. The data was captured in a manner that could not be linked back to the individual respondent.

3.5. Analyses

Each household's recycling behavior for paper and packaging waste is calculated by averaging the frequency of recycling (one variable) and how much of each of four recyclable materials, namely paper, plastic, glass, and metal (four variables), is recycled. The households are then grouped into clusters based on their recycling behavior [26,49]. One group comprises the households that reported no recycling behavior (B = 1). The recycling households (B > 1) are further grouped into three groups, according to their various levels of reported recycling behavior. Borrowing from both Thomas and co-workers' description of high, medium, and low recyclers [49] and Martin and co-workers' description of full (recycle five materials including plastic, paper, glass, cans, and cardboard), casual (recycle 1 to 4 materials), and non-recyclers (recycle no items) [26], for this study, the terms with definitions are adopted as follows:

- High recycling households with dedicated recycling activity: often or more regularly recycle more than half of their recyclables, i.e., behavior score greater than 4, up to and including 7, and selected to conform with results previously reported [16],
- Medium recycling households with casual recycling activity: the group between the sporadic and dedicated recycling households, i.e., behavior score of 2 to 4,
- Low recycling households with sporadic recycling activity: seldom or almost never recycle very little of one recyclable material only, i.e., behavior score greater than 1 but smaller than 2, and
- Non-recycling households with no recycling activity: never recycle i.e., behavior score of 1.

For the purpose of this paper, the segmentation into groups diverts from the Thomas et al. example [49] by also differentiating between the non-recyclers (B = 1) and the low (sporadic) recyclers. These two groups were analyzed separately because the large number of non-recycling households would overshadow the responses from the sporadic-recyclers. The author's classification also differs from the Martin et al. example [26] by splitting off a group called the sporadic recyclers (the households that almost never/seldom recycle very little/some of one material) from the casual recyclers (those households that recycle more materials and more frequently than the casual recyclers but less than the dedicated recyclers).

Using MS Excel, the total number of times each reason for non-recycling was selected was calculated and expressed as a percentage of the total number of reasons mentioned. First, the total number of times an item was selected as the one most important reason was expressed as a percentage of the total number of responses. Second, each of the three most important reasons selected was totaled and each statement expressed as a percentage of the total number of reasons mentioned. These calculations were repeated for the total sample group (n = 2004) as well as for each of the sub-groups as described above. According to the popularity of each selected reason, i.e., times selected, the factors were ranked [24] (p. 64).

3.6. Research Ethics

The impersonal phrasing of the question "what would you say are the reasons why people do not recycle?" has value in an ethical sense by not putting the respondents in a position where they feel guilty about their personal recycling behavior. It can also be argued that respondents would be more willing to share information and share it accurately with less desirability bias when a question is posed in a more impersonal manner. However, a more direct question: "why don't you recycle?" would have allowed a more confident comparison. Due to the impersonal phrasing of the question, interpretation of the results should thus be done with caution by keeping in mind that the results portrayed are the perceptions of the respective groups about the "people" and not their perceptions or knowledge about the specific group within which they are placed.

3.7. Assumptions

The assumption is made that the respondent as an individual (i) is an accurate proxy for his/her household's recycling behavior and (ii) answered the question from the perspective of someone with the specific recycling behavior.

4. Results and Discussion

The demographic representation of the sample is shown in Appendix B.

4.1. Recycling Behaviour

The results show that 26.0% of South African households living in large urban areas recycle varying quantities of materials at varying intervals while 74.0% of households never recycle (Table 2). Additionally, 4.0% of the respondents come from high recycling households with a dedicated recycling behavior. The respondents from households with medium or casual recycling activity comprise 14.2% of the total sample and those with low or sporadic recycling activity comprise 7.8% of households tested. The majority (74.0%) of the respondents represents households not recycling any of their household waste (non-recycling households). These sub-groups are used to portray the results in the following sections.

Table 2. Descriptive statistics for the total sample and sub-categories.

Respondents from	Sample Size (*n*)	Percentage of Sample (%)	Recycling Behavior Score (B) *	x-Bar
All households	2004	100%	1–7	1.46
High recycling households (dedicated recycling activity)	81	4.0%	5–7	5.11
Medium recycling households (casual recycling activity)	284	14.2%	2–4	2.89
Low recycling households (sporadic recycling activity)	157	7.8%	>1–<2	1.48
Non-recycling households (no recycling activity)	1482	74.0%	1	1

* for paper and packaging recyclables where 7 = best possible score for recycling activity and 1 = no recycling activity.

In the bottom group of recycling households—the low or sporadic recyclers—the behavior score band of between 1 and 2 is very narrow in comparison with the dedicated recyclers (5 to 7). This is because of the relatively small percentage of households that separate their recyclables in South Africa in comparison with many developed countries. For example, both the UK studies reported much higher percentages of recycling households [26,49] when compared to this South African study.

4.2. Perceived Barriers to Recycling—All Respondents

Results from the selected (i) one most important reason, and (ii) three main reasons why people do not recycle are discussed below.

(i) The One Most Important Reason

From 10 options, 28.1% of the respondents selected *no time* (ranked first) as the one most important reason why people do not recycle, followed by a *lack of knowledge* (14.0%, ranked second), *insufficient space* (12.5%, ranked third), *facilities inconvenient* (10.4%, ranked fourth) and *not responsible* (8.6%, ranked fifth) (Table 3). These five factors attracted almost three quarters (73.6%) of the responses. These results, drawn from a large percentage of the respondents, suggest that a combination of these five factors is required to enable households to recycle. The high percentage of respondents selecting *no time* as compared to the other factors shows that, although there are several barriers that could possibly prevent household recycling, the perception is that recycling is a time consuming activity.

This finding is not unique to this study, which suggests that factors deterring household recycling behavior are not country specific. In Bangladesh, where 25.6% of the respondents indicated that they recycle sporadically or more often despite not having a convenient recycling scheme, lack of time (38.49%) and no space in the home (37.2%) are the main reasons for not recycling [50]. A Malaysian study found that inadequate facilities (30.9% of the respondents), followed by inconvenience/no time (25.2%), lack of information (10.8%), too much effort needed (10.3%), and not interested (9.4%) are the main reason why people in Penang do not recycle [51].

Table 3. Perceptions why people do not recycle—order of priority (rank) of the one most important reason.

Reasons Why People Do Not Recycle: The One Most Important Reason	All Respondents (*n* = 2004)		Respondents from Recycling Households (*n* = 522)		Respondents from Non-Recycling Households (*n* = 1482)	
	%	Rank *	%	Rank *	%	Rank *
They do not have the time to recycle (no time)	28.1	1	24.3	1	29.4	1
They do not know what can and what cannot be recycled (lack of knowledge)	14.0	2	13.4	3	14.2	2
They lack space to keep the recyclables (insufficient space)	12.5	3	14.4	2	11.8	3
Recycling facilities are inconvenient (facilities inconvenient)	10.4	4	9.8	4	10.6	4
They are not responsible for recycling in their households (not responsible)	8.6	5	8.6	5	8.6	5
Keeping the materials until it is recycled is dirty and untidy (dirty and untidiness)	7.8	6	6.7	6	8.2	6
They do not have a curbside collection service for recyclables (no curbside service)	5.4	7	6.3	8	5.1	7
They think it will not make a difference whether they recycle or not (it makes no difference)	5.1	8	6.5	7	4.7	8
They cannot be bothered (not bothered)	4.8	9	5.9	9	4.4	9
Recycling services does not exist (no service)	3.3	10	4.0	10	3.0	10
Total percentage represented by the factors ranked 1 to 5	73.6	1–5	70.5	1–5	74.6	1–5

* Ranked from 1 to 10 including from the highest to the lowest percentage of respondents selecting each option.

The low percentage (3.3%, ranked 10th) of respondents selecting *no service* suggests that recycling services exist, but this does not ignore the fact that recycling facilities are perceived to be inconvenient (*facilities inconvenient*, ranked fifth) or is not the kind of service respondents would prefer (e.g., *no curbside collection*, ranked seventh).

Due to the high percentage (74.0%, Table 2) of non-recycling households in South Africa, the results are dominated by perceptions originating from non-recycling households. However, the results show that the same five reasons were selected the most by the respondents from both recycling and non-recycling households. However, respondents from recycling households give higher priority to *insufficient space* (with 14.4% ranked second following *no time* 24.3%, ranked first) than to *lack of knowledge* (13.4%, ranked third). This could imply that recycling households already have the knowledge that enable them to recycle, but the space the recyclables take up and the time it takes to recycle is something they are confronted with on an ongoing daily basis.

(ii) The Three Main Reasons

The results from the three main reasons (Table 4) show that the three situational factors at household level, which includes *no space, no time,* and *dirty and untidy,* are the main reasons why people do not recycle. The priority given to these three factors suggest that in-house barriers carry much weight when it comes to the recycling of household waste. *Lack of knowledge* of what can and cannot be recycled and *inconvenient recycling facilities* are also important reasons why people do not recycle. The results from the quantitative national survey suggest that factors that would assist with in-house management of separation at the source combined with the necessary knowledge and a convenient recycling facility would encourage household recycling. This finding is in line with the findings of international studies where lack of space [28], no time [23], dirty and unhygienic [43], lack of knowledge [27,29], and inconvenient recycling facilities [26,35] were identified as barriers to household recycling.

Table 4. Perceptions why people do not recycle—order of priority (rank) of the three main reasons.

Reasons Why People Do Not Recycle: Select the Three Main Reasons from the 10 Options	All Respondents Total Sample Group (*n* = 2004)		Respondents from Recycling Households (*n* = 522)		Respondents from Non-Recycling Households (*n* = 1482)	
	% **	Rank *	% **	Rank *	% **	Rank *
They lack space to keep the recyclables (insufficient space)	15.0	1	14.6	1	15.1	1
They do not have the time to recycle (no time)	14.9	2	14.2	2	15.1	1
Keeping the materials until it is recycled is dirty and untidy (dirty and untidy)	12.4	3	11.6	4	12.7	3
They do not know what can and what cannot be recycled (lack of knowledge)	12.3	4	11.8	3	12.5	4
Recycling facilities are inconvenient (facilities inconvenient)	10.8	5	10.8	5	10.8	5
They think it will not make a difference whether they recycle or not (makes no difference)	8.0	6	9.4	6	7.6	7
They do not have a curbside collection service for recyclables (no curbside collection)	7.9	7	7.8	7	7.9	6
They are not responsible for recycling in their households (not responsible)	7.0	8	6.5	9	7.2	8
They cannot be bothered (not bothered)	6.9	9	7.6	8	6.7	9
Recycling services do not exist (no service)	4.8	10	5.7	10	4.4	10
Total responses represented by the factors ranked 1 to 5	65.4	1–5	63.0	1–5	66.2	1–5

* Ranked from 1 to 10 from the highest to the lowest percentage of responses for each option. ** The three selected reasons totalled and expressed as a percentage of the total responses (three per respondent).

The combined effect of the three main reasons (Table 4) has a moderating or tempering effect on the results and shows which factors, apart from the most important reason (Table 3), contribute to why households do not recycle. Thus, the results from the selected three main reasons show which reasons, apart from the one which dominates, act as important barriers to household recycling. The small difference between the highest and lowest percentages (15% − 4.8% = 10.2%) could also be an indication that these 10 statements are all relevant and valid reasons why people do not recycle and that barriers to recycling could be context-specific.

Considering the responses from the total sample group, *insufficient space* attracted 15% of the responses (ranked first), followed by no time (14.9%, ranked second), dirty and untidy (12.4%, ranked third), lack of knowledge (12.3%, ranked fourth) and facilities inconvenient (10.8%, ranked fifth).

Similar to the results from the one most important reason, the results from both the recycling and non-recycling households show the same five factors as the main reasons why people do not recycle but with slight variation in ranking order. The main difference is that *lack of knowledge* (11.8%, ranked third) appears to be a more important factor among recycling households than *dirty and untidy* (11.6%, ranked fourth) but only marginally.

A comparison between the results from the one most important reason and the three main reasons shows that: *dirty and untidy,* ranked sixth in selecting the one most important reason, moved to third place when the three main reasons are considered. *Not responsible* (ranked fifth) as one of the top five reasons why people do not recycle moved to eighth place. In both, *not bothered* (6.9%) and *no service* (4.8%) are selected the least (ranked 9th and 10th, respectively) except for recycling households who acknowledge more so than non-recycling households that somewhat of a "bother" is needed (with 7.6% of the respondents selecting this factor, ranked eighth).

The five reasons receiving highest priority in this study are similar to Perrin and Barton's finding that inconvenience/no time, inconvenient or inadequate recycling facilities, storage/handling problems and lack of information were, in the absence of a curbside recycling scheme, the main reasons why people do not recycle [24]. Perrin and Barton conclude that convenience encompass several factors such as recycling scheme design, the level of knowledge a household needs to be able to participate in a recycling scheme, and the in-house handling of the recyclables [24]. All these convenience "barriers can be overcome with the introduction of a curbside recycling scheme which suits local circumstances" [24] (p. 65).

Personal responsibility normally motivates people to recycle more [52]. However, the data suggests that a sense of responsibility has a low priority among respondents from both recycling and non-recycling households in South Africa and, thus, is unlikely to be strong enough to overshadow the in-house difficulties as well as the difficulties experienced with services at this point in time in household recycling behavior in South Africa.

4.3. Perceived Barriers to Recycling—Recycling Households

Results from the separate analysis of the three sub-groups (high/dedicated, medium/casual, and low/sporadic) representing the recycling households bring nuances to the foreground, which would otherwise have been lost. Differentiating between the various levels of recycling households, results from the selected (i) one most important reason why people do not recycle and (ii) the three main reasons are shown and discussed below.

(i) The One Most Important Reason

The results show *no time* as having priority by all sub-groups as the one most important reason why people do not recycle (Table 5). The main differences between the sub-groups are that low recycling households give higher priority to *lack of knowledge* (ranked second) and *facilities inconvenient* (ranked third) than both the medium and high recycling households (ranked third and fourth, respectively). *Sufficient space* moves to the fifth ranked position among the low recycling households, likely because they recycle so sporadically that space to keep recyclables is outweighed by the other more pressing factors, which hamper recycling behavior such as *no time, lack of knowledge, facilities inconvenient,* and *not responsible. Insufficient space* also shares the fifth ranked place with *it makes no difference* and *no curbside service.*

Table 5. Comparing perceptions why people do not recycle across varying recycling levels—order of priority (rank) of the one most important reason.

Reasons Why People Do Not Recycle: The One Most Important Reason	All Recycling Households (*n* = 522)		High Recycling Households (*n* = 81)		Medium Recycling Households (*n* = 284)		Low Recycling Households (*n* = 157)	
	%	Rank *	%	Rank *	%	Rank *	%	Rank *
No time	24.3	1	25.9	1	23.2	1	25.5	1
Insufficient space	14.4	2	21.0	2	15.9	2	8.3	5
Lack of knowledge	13.4	3	13.6	3	14.1	3	12.1	2
Facilities inconvenient	9.8	4	7.4	4	9.9	4	10.8	3
Not responsible	8.6	5	7.4	4	8.8	5	8.9	4
Dirty and untidy	6.7	6	6.2	7	7.0	6	6.4	9
It makes no difference	6.5	7	7.4	4	5.3	9	8.3	5
No curbside service	6.3	8	3.7	8	6.0	7	8.3	5
Not bothered	5.9	9	3.7	8	5.6	8	7.6	8
No service	4.0	10	3.7	8	4.2	10	3.8	10

* Ranked from 1 to 10 from the highest to the lowest percentage of responses for each option.

These results suggest that, apart from the time and knowledge needed to be able to recycle, households' understanding of the difference their recycling efforts can make and that they need to take responsibility for recycling are important factors to trigger or encourage household recycling. In addition, a curbside collection for recyclables or any other equally convenient recycling facility has the potential to motivate households that never or sporadically recycle to start to recycle or to recycle more regularly. The results also suggest that respondents from high and medium recycling households ascribe lower priority (rank) to *no curbside service*, which could be an indication that these respondents perceive their recycling facilities to be more convenient than the low recycling households.

(ii) The Three Main Reasons

Facilities inconvenient received a higher priority among the low recycling households (12.3%, ranked fourth) compared to both the medium (10.7%, ranked fifth) and high recycling households (8.2%, ranked eighth) (Table 6). Similarly, *no curbside service* received a higher priority among the low recycling households (9.8%, ranked fifth) compared to both the medium (7.4%, ranked seventh) and the high recycling households (5.4%, ranked ninth). Although ranked ninth, *no service* attracted 6.6% of the responses from the low recycling households. In total, the service/facility related factors received 28.7% of the total responses from the low recycling households compared to 17.3% from the high recycling households. The large difference in priority between the low and high recycling households of the service/facility factors is in line with findings from UK studies, which suggest that low recyclers are more likely to report problems related to recycling services [49] and high recyclers are more satisfied with their recycling schemes [26]. These findings suggest that an improved recycling service is one of the major interventions needed to improve recycling behavior in South Africa. Several international studies highlight the importance of convenient recycling facilities especially for the low-recycling and non-recycling households [43,53]. However, it is not only access to a curbside recycling service but also awareness of other recycling facilities and especially the perceptions held of the convenience of these facilities that determine people's recycling behavior [17,54]. The reliability of a curbside scheme is important and especially the "lower-participating recyclers" prefer a weekly collection service, which could be a convenience preference or an indication of lack of space to store recyclables for longer periods [26] (p. 380).

While services received higher priority among the low recycling households, the results show that the combination of three factors known as *insufficient space, no time*, and *dirty and untidiness* attracted 44.1%, 42.1%, and 35.2% of the responses from the high, medium, and low recycling households, respectively. Thus, these in-house convenience factors appear to be more of an issue among the high recycling households. It is suggested that, as the low recycling households start to recycle more regularly and at higher volumes, these in-house convenience factors will also become more of an issue and might overshadow other relevant barriers, which is seen in the results from the high recycling households.

Table 6. Comparing perceptions why people do not recycle across varying recycling levels—order of priority (rank) of the three main reasons why people do not recycle.

| | Responses from Respondents Representing | | | | | | | |
| Reasons Why People Do Not Recycle: Select the Three Main Reasons from the 10 Options | All Recycling Households (n = 522) | | High Recycling Households (n = 81) | | Medium Recycling Households (n = 284) | | Low Recycling Households (n = 157) | |
	% **	Rank *	% **	Rank *	% **	Rank *	% **	Rank *
Insufficient space	14.6	1	14.8	2	15.4	1	13.2	1
No time	14.2	2	16.1	1	14.3	2	12.9	2
Lack of knowledge	11.8	3	12.3	4	11.0	4	12.9	2
Dirty and untidy	11.6	4	13.2	3	12.4	3	9.1	6
Facilities inconvenient	10.8	5	8.2	8	10.7	5	12.3	4
It makes no difference	9.4	6	8.6	6	9.7	6	9.1	6
No curbside service	7.8	7	5.4	9	7.4	7	9.8	5
Not bothered	7.6	8	9.1	5	7.1	8	7.9	8
Not responsible	6.5	9	8.6	6	6.1	9	6.2	10
No service	5.7	10	3.7	10	5.9	10	6.6	9

* Ranked from 1 to 10 from the highest to the lowest percentage of responses for each option. ** The three selected reasons totalled and expressed as a percentage of the total responses (three per respondent).

Lack of knowledge attracted responses as follows: low recycling households, 12.9% (ranked second), medium recycling households, 11.0% (ranked fourth), and high recycling households, 12.3% (ranked fourth). Although there is a difference in the ranking which indicates priority, the percentage responses are similar and show that a lack of knowledge is perceived as an important reason why people do not recycle by all recycling sub-groups. This is in line with findings by Clarke and Maantay [55] who recorded the lack of knowledge of what is recyclable in both areas with low and high participation in recycling initiatives and, thus, suggests that both recycling and non-recycling households perceive a lack of knowledge as a barrier to recycling.

With *not bothered* (9.1%) ranked fifth, and *not responsible* and *it makes no difference* sharing the sixth rank (8.6%), the high recycling households appear to be very aware that being bothered, taking responsibility for recycling in the household, and an understanding that every household can contribute is needed to activate and improve recycling behavior. *It makes no difference* is ranked sixth by sub-groups. Although not among the first five, this factor (*it makes no difference*) attracted more than 9% of all the recycling household responses. Perceptions that a household's recycling does not make a difference has the potential to be turned around with well-targeted communications to improve recycling behavior. The difference in the results between the recycling household sub-groups shows how people's perceptions could change once they become involved in and have more experience in household recycling.

What stands out in the comparison between the sub-groups is the small difference of seven percentage points (13.2% − 6.2% = 7.0%) between the responses of the lowest and the highest ranked factors of the low recycling households when compared to more than 12 percentage points (16.1% − 3.7% = 12.4%) of the high recycling households. The group of medium recyclers is in-between the high and low recycling households with 9.5 percentage points (15.4% − 5.9% = 9.5%). This suggests that low recycling households find all 10 options more relevant as barriers to their recycling behavior and more so than the high recycling households. It also suggests that dedicated recycling households have managed to overcome these barriers that still prevent low recycling households to recycle more and do not perceive these as barriers to household recycling anymore.

4.4. Demographic Variables and Perceived Barriers to Recycling

Considering the socio-demographics variables (Tables A1–A9 in Appendix C), analysis of the selected three main reasons why people do not recycle show that the same five factors known as *no time, insufficient space, dirty and untidiness, lack of knowledge,* and *facilities inconvenient* are selected the most. However, these five factors are not always in the same order of priority as the responses from the total sample group (as shown in Table 4) and there are a few exceptions where one of these five is replaced by another factor. While the data suggest that there is no major difference in perceptions about the reasons why people do not recycle, the largest deviations from the all respondents' order of priority are highlighted below.

- Age (Table A1): The respondents older than 60 appear to be, after *no time*, more concerned about the *dirty and untidiness* (ranked second) and *lack of knowledge* (ranked third), and less about *insufficient space* (ranked fourth). The age group 16–19 is more concerned about *facilities inconvenient* (ranked third) and less about the *dirty and untidiness* (ranked fifth).
- Gender (Table A2): The male respondents appear to be slightly more concerned about *insufficient space* (ranked first) and female respondents appear to be more concerned about the *time* and *knowledge* factors (ranked second and third, respectively).
- Type of dwelling (Table A3): The respondents living in flats are less concerned about *insufficient space* and *no time* (ranked third and fifth, respectively) and give higher priority to *lack of knowledge* (ranked first), *dirty and untidiness* (ranked second), and *facilities inconvenient* (ranked fourth). Those living in informal housing structures (squatter huts or shacks) also give higher priority to *lack of knowledge* (ranked second, after *insufficient space*, ranked first). The data suggests that those living in hostels, hotels, boarding houses, and compounds might not share the same perceptions as the majority, but the sample sizes of these groups are too small to be able to make meaningful deductions.
- Area type (urban formal and urban informal) (Table A4): Although these two groups share the same reasons among the five with the highest priority, the respondents from the urban informal areas appear to be more concerned about *insufficient space* (ranked first) and *lack of knowledge* (ranked second) and less about *no time* (ranked fourth) as reasons why people do not recycle.
- Employment status (Table A5): *Lack of knowledge* has higher priority among those working part-time (ranked second), the unemployed looking for work (ranked second), the housewives (ranked third), and the retired (ranked third). The unemployed not looking for work is more concerned about the *facilities inconvenient* factor (ranked third).
- Level of education (Table A6): The respondents without any schooling selected reasons why people do not recycle which, apart from the first three (insufficient space, no time, and lack of knowledge) differ from the order of priority of all the respondents. *Not bothered* and *it makes no difference* (both ranked fourth) have higher priority among those without any schooling. Among those with some primary schooling, *not responsible* has a higher priority (ranked fifth). Respondents with a completed diploma or degree ascribed less importance to *lack of knowledge* (ranked sixth) and gave higher priority to a *curbside service* (ranked fourth).
- Marital status (Table A7): The respondents that are single give higher priority to *lack of knowledge* (ranked third). *Lack of knowledge* is also a concern among those living together (ranked first). The widowed give higher priority to *no curbside service* (ranked fifth) while, among the divorced, both *facilities inconvenient* and *no curbside service* have higher priority (ranked second and fifth, respectively). Among the respondents that are separated, *no time* (ranked eighth) appears to be less of an issue, but they are more concerned about *it makes no difference* (ranked third). Whether the latter could be ascribed to a temporarily cynical outlook was not tested.
- Ethnic group (Table A8): Concern about the level of convenience appears to be the main difference between ethnic groups. The Indian/Asian and colored respondents give higher priority to the factor *facilities inconvenient,* (ranked second and fourth, respectively) while, among WHITE people,

no curbside service has a higher priority (ranked third). Being a representative sample, BLACK respondents makes up the largest percentage and thus dominate the order of priority of all the respondents.

- Occupation group (Table A9): The perception among those with a professional occupation is that *lack of knowledge* and *not bothered* are major factors that prevent people from recycling (both ranked third). Among the executives and managers, *lack of knowledge* appears to be less of an issue (ranked sixth) but *facilities inconvenient* and *no curbside service* have higher priority (ranked third and fourth, respectively). The semi-skilled are more concerned about *makes no difference* (ranked fifth). Among the unskilled respondents, on the ranking list, *lack of knowledge* shares the first place with *no time*. The factor *not responsible* also has higher priority (ranked fifth). The self-employed perceive *insufficient space* as less of a concern (ranked seventh) while *dirty and untidy* (ranked first) and *no curbside service* has higher priority (ranked fourth).

Given the rate of urbanization, the priority of *lack of knowledge* among those living in flats, informal housing structures, and in informal areas is an important finding. In addition, the data suggest that *not bothered* and *it makes no difference* as well as *not responsible* are factors that need attention among the unschooled and those with some primary school education. Oke and Kruijsen emphasize the need for the right kind of information [56], i.e., why it is important to recycle and why households should recycle in addition to the what, how, when, and where messages. The challenge would be to find alternative ways of communicating with the illiterate to get these messages across.

The *no time* factor is ranked the highest among the respondents working full-time and part-time. Kaciak and Kushner highlights the fact that inconvenience is a major barrier to recycling and that it has the ability to override the best personal intentions to recycle [44]. Therefore, in the absence of a street collection service and given the importance of time, the location and maintenance of recycling centers are crucial to encourage continued household recycling.

5. Conclusions and Recommendations

No time stands out as the one most important reason why people do not recycle, but analysis of the selected three most important reasons of the quantitative survey results show that (i) *insufficient space*, (ii) *no time*, (iii) the *dirty and untidiness* associated with recycling, (iv) *lack of knowledge*, and (v) *inconvenient recycling facilities* are the main reasons why people do not recycle. Being a representative sample of the South African urban population, the large percentage of non-recycling households dominates the results. Improved and more convenient services such as regular curbside collections might encourage low recyclers to recycle more while improved services could also be a solution for both the time and space problem that households currently either experience or envision to experience should they recycle.

The uniqueness of this study lies in the sample being representative of the South African population residing in large urban areas. Although desirability bias in answering the behavioral questions is possible, the chance of overrepresentation of recycling households is small in relation to compulsory postal or on-line surveys that focus on recycling behavior.

In the South African urban setting, household recycling is mostly voluntary and is even more so in rural areas. This empirical study highlights reasons why people do not recycle, which are useful for decision-making in South Africa. The fact that the findings of this study are in line with the findings from similar international studies underscores the value of this research for other countries especially developing countries that want to improve recycling participation.

The implication for developing countries is that recycling services as well as communications towards more knowledge will have to improve to get the buy-in from a larger percentage of the population to start recycling or to recycle more than they currently do. Knowledge about what is recyclable and the convenience of recycling facilities are important factors among the youth, the unemployed, and those in the informal urban areas. Thus, the waste sector has great potential to play a role towards job creation and poverty alleviation through recycling initiatives among the youth and the unemployed and especially in the informal urban areas. Communications for the semi-skilled and unskilled as well as the unschooled and semi-schooled should focus on why it is important to recycle —that recycling can make a difference—and why it is important to care and to take responsibility for recycling.

Equality in waste recycling services is still something to strive for in South Africa. Until such time that all households have access to recycling services of comparable convenience, it is and will remain difficult to compare recycling behavior among South Africans, between South Africa and other developing countries, and between South Africa and countries with mature recycling services. Since caution should be taken to compare results between studies with varying recycling scheme maturity [14], caution should also be taken to implement results and conclusions from developed countries and results in developing countries.

It is recommended that this quantitative study is repeated in 2020 to ascertain any change in recycling behavior and to also expand the survey to include rural towns to determine whether similar perceptions about why people from South Africa do not recycle exist in the smaller centra. Expanding the list of factors would be an improvement to the survey. However, there is a trade-off between functionality and length of the list of factors to choose from.

Funding: This research received no external funding.

Acknowledgments: This study forms part of a more comprehensive study to determine the baseline recycling behavior of households in South Africa and was funded by the Council for Scientific and Industrial Research (CSIR). The author would like to thank CSIR colleagues Linda Godfrey, Richard Meissner, and Suzan Oelofse who reviewed an earlier version and the two anonymous reviewers for their comments and suggestions, which helped to improve the quality of this paper.

Conflicts of Interest: The author has no vested interest in this research.

Appendix A

Questionnaire

Thinking of your household, would you say that your household recycles the recyclable materials from your household waste?	1. Never	1
	2. Almost never	2
	3. Seldom	3
	4. Sometimes	4
	5. Often	5
	6. Almost always	6
	7. Always	7
	8. Don't know (do not read out)	

For Each of the Recyclable Materials, Choose the Statement that Best Describes How Much Your Household Recycles

Statements of How Much Is Recycled	Recyclable Material			
	Paper	Glass	Metal	Plastic
My household recycles **nothing**	1	1	1	1
My household recycles **very little** of what can be recycled	2	2	2	2
My household recycles **some** things that can be recycled	3	3	3	3
My household recycles **about half** of everything that can be recycled	4	4	4	4
My household recycles **most** of everything that can be recycled	5	5	5	5
My household recycles **almost all** of what can be recycled	6	6	6	6
My household recycles **everything** that can be recycled	7	7	7	7

Which of the Following Would You Say Are the Reasons Why People Do Not Recycle and Which One Is the Most Important Reason Why People Do Not Recycle? (Choose the 3 Most Important Reasons and Mark in Order of Importance with One Being the Most Important Reason.)

One Mention Per Number Only	1st	2nd	3rd
1. They do not have the time			
2. They are not responsible for recycling in their households			
3. They do not know what can and what cannot be recycled			
4. Recycling facilities are inconvenient			
5. They lack space to keep the recyclables			
6. Keeping the materials until it is recycled is dirty and untidy			
7. They think that it will not make a difference whether they recycle or not			
8. They cannot be bothered			
9. Recycling services are poor or do not exist			
10. They do not have a pavement collection service for recyclables			

Note: order of options 1 to 10 is rotated on show cards.

Appendix B

Demographic composition of the sample:

Age: 16–19 y 7.6%, 20–29 y 24.2%, 30–39 y 23.9%, 40–49 y 19.0%, 50–59 y 11.3%, >60 y 13.7%.

Gender: male 50%, female 50%.

Type of dwelling: house/townhouse 60.5%, flat 4.2%, matchbox/RDP 17.4%, hostel 0.6%, hotel/boarding house <0.1%, compound 0.2%, room in backyard 2.8%, squatter hut/shack 13.8%, other 0.5%.

Area type: urban formal 84.0%, urban informal 15.9%, no answer/refused 0.1%.

Employment status: working full-time 33.2%, working part-time 11.6%, housewife 6.3%, student 9.9%, retired 11.6%, unemployed looking for work 24.0%, unemployed not looking for work 3.4%.

Education: no schooling 1.4%, some primary schooling 5.6%, primary school completed 5.7%, some high school 36.2%, Grade 12 33.8%, Artisans certificate completed 3.6%, diploma/degree completed 13.2%, other 0.3%.

Marital status: single 44.8%, married 36.5%, living together 8.8%, widowed 6.8%, divorced 2.2%, separated 0.9%.

Population: Black 66.8%, White 14.5%, Colored 12.7%, Indian/Asian 6.0%.

Occupation group: professional 2.5%, executive/managerial 3.8%, clerical/sales 12.0%, tradesman 7.1%, semi-skilled 8.4%, unskilled 6.9%, self-employed 2.2%, refused 1.8%, not answered (housewife/student/retired/unemployed) 55.2%.

Appendix C

Table A1. Reasons why people do not recycle as enumerated per age group (n = 2003 *).

Reasons Why People Do Not Recycle: Select the Three Main Reasons from the Ten Options	Times Selected Expressed as a Percentage of the Total of Each of the Respondent Groups											
	16–19 Years (n = 154)		20–29 Years (n = 486)		30–39 Years (n = 478)		40–49 Years (n = 381)		50–59 Years (n = 228)		60+ Years (n = 276)	
	%	Rank	%	Rank	%	Rank	%	Rank	%	Rank	%	Rank
Insufficient space	14.7	1	16.1	1	15.0	2	15.3	1	14.9	1	12.6	4
No time	14.5	2	15.9	2	15.1	1	14.3	2	13.5	2	14.6	1
Dirty and untidy	11.5	5	12.7	3	12.3	4	11.7	3	12.7	4	13.5	2
Lack of knowledge	12.1	4	12.5	4	13.2	3	11.2	4	11.3	3	12.9	3
Facilities inconvenient	13.2	3	10.1	5	10.2	5	10.9	5	11.5	5	11.0	5
It makes no difference	9.5	6	7.9	6	7.7	9	8.8	7	7.9	7	7.0	9
No curbside collection	7.1	7	6.7	9	8.3	6	9.0	6	8.2	6	7.6	7
No responsibility	6.7	8	7.1	7	6.8	8	6.7	9	7.7	8	7.4	8
Not bothered	6.3	9	7.0	8	6.8	8	6.9	8	6.3	9	8.1	6
No service	4.3	10	3.9	10	4.6	10	5.2	10	6.0	10	5.3	10

* One respondent refused age.

Table A2. Reasons why people do not recycle as enumerated per gender group (n = 2004).

Reasons Why People Do Not Recycle: Select the Three Main Reasons from the 10 Options	Times Selected Expressed as a Percentage of the Total of Each of the Respondent Groups			
	Male (n = 1002)		Female (n = 1002)	
	%	Rank	%	Rank
Insufficient space	15.2	1	14.7	2
No time	14.8	2	14.9	1
Dirty and untidy	11.9	3	12.9	3
Lack of knowledge	11.8	4	12.9	3
Facilities inconvenient	10.8	5	10.8	5
It makes no difference	8.8	6	7.3	7
No kerbside collection	7.8	7	7.9	6
No responsibility	7.2	9	6.8	8
Not bothered	7.4	8	6.5	9
No service	4.2	10	5.3	10

Table A3. Reasons why people do not recycle as enumerated per type of dwelling ($n = 2004$).

Reasons Why People Do Not Recycle: Select the Three Main Reasons from the 10 Options	Times Selected Expressed as a Percentage of the Total of Each of the Respondent Groups																	
	House/Cluster House/Townhouse ($n = 1213$)		Flat ($n = 85$)		Matchbox/Improved Matchbox/RDP ($n = 348$)		Hostel ($n = 12$) *		Hotel/Boarding House ($n = 1$) *		Compound ($n = 3$) *		Room in Backyard ($n = 57$)		Squatter Hut/Shack ($n = 276$)		Other (Mobile Home/Tent) ($n = 9$) *	
	%	Rank	%	Rank	%	Rank	%	Rank	%	Rank	%	Rank	%	Rank	%	Rank	%	Rank
Insufficient space	14.9	2	13.7	3	14.8	1	16.7	-	33.3	-	22.2	-	13.5	2	16.1	1	11.1	-
No time	16.1	1	11.8	5	11.9	2	16.7	-	0.0	-	22.2	-	20.5	1	13.0	3	7.4	-
Dirty and untidy	11.8	3	14.1	2	13.7	3	13.9	-	0.0	-	0.0	-	11.1	4	13.0	3	22.2	-
Lack of knowledge	11.1	4	14.9	1	13.5	4	8.3	-	0.0	-	11.1	-	12.9	3	15.5	2	11.1	-
Facilities inconvenient	10.7	5	12.5	4	10.8	5	2.8	-	0.0	-	22.2	-	9.9	5	11.1	5	11.1	-
It makes no difference	8.0	7	6.7	7	9.2	6	5.6	-	0.0	-	11.1	-	4.1	9	8.3	7	3.7	-
No curbside collection	8.1	6	9.8	6	7.3	7	22.2	-	33.3	-	11.1	-	5.8	8	6.3	8	11.1	-
No responsibility	6.6	9	6.3	8	7.2	8	5.6	-	0.0	-	0.0	-	9.4	6	8.6	6	3.7	-
Not bothered	7.4	8	6.3	8	6.6	9	5.6	-	0.0	-	0.0	-	9.4	7	5.2	9	7.4	-
No service	5.2	10	3.9	10	5.0	10	2.8	-	33.3	-	0.0	-	3.5	10	2.9	10	11.1	-

* Too few data points to derive meaningful conclusions.

Table A4. Reasons why people do not recycle as enumerated per area type: urban formal or urban informal ($n = 2002$ *).

Reasons Why People Do Not Recycle: Select the Three Main Reasons from the 10 Options	Times Selected Expressed as a Percentage of the Total of Each of the Respondent Groups			
	Urban Formal ($n = 1684$)		Urban Informal ($n = 318$)	
	%	Rank	%	Rank
Insufficient space	14.7	2	16.4	1
No time	15.4	1	11.8	4
Dirty and untidy	12.1	3	13.9	3
Lack of knowledge	11.9	4	14.6	2
Facilities inconvenient	10.9	5	10.2	5
It makes no difference	7.8	6	9.3	6
No curbside collection	7.8	6	7.9	7
No responsibility	7.2	8	6.1	8
Not bothered	7.1	9	5.9	9
No service	4.9	10	3.9	10

* Two respondents did not answer this question.

Table A5. Reasons why people do not recycle as enumerated per employment status group ($n = 2004$).

Reasons Why People Do Not Recycle: Select the Three Main Reasons from the 10 Options	Times Selected Expressed as a Percentage of the Total of Each of the Respondent Groups													
	Working Full-Time ($n = 665$)		Working Part-Time ($n = 233$)		Not Working: Housewife ($n = 126$)		Not Working: Student ($n = 199$)		Not Working: Retired ($n = 232$)		Unemployed: Looking for Work ($n = 481$)		Unemployed: Not Looking for Work ($n = 68$)	
	%	Rank	%	Rank	%	Rank	%	Rank	%	Rank	%	Rank	%	Rank
Insufficient space	15	2	13.6	3	15.7	1	16.1	1	12.2	3	16.6	1	12.7	4
No time	16.8	1	14.8	1	12.8	2	13.6	2	15.7	1	13	3	14.7	1
Dirty and untidy	11.5	3	12.6	4	12.2	4	12.7	3	12.9	2	13	3	14.2	2
Lack of knowledge	11.2	4	14.2	2	12.5	3	12.4	4	12.2	3	13.2	2	10.3	5
Facilities inconvenient	10.5	5	11.2	5	10.6	5	10.9	5	10.6	5	10.7	5	13.2	3
It makes no difference	7.9	7	8.7	6	7.4	8	7.9	6	7.2	9	8.8	6	6.4	9
No curbside collection	8.8	6	7	8	9.3	6	7.2	8	7.8	7	6.9	8	7.4	8
No responsibility	6.8	9	5.6	9	5.1	10	7.4	7	7.6	8	7.8	7	9.3	6
Not bothered	7.4	8	7.4	7	6.4	9	6.9	9	8.3	6	5.5	9	7.8	7
No service	4.2	10	4.9	10	8	7	5	10	5.5	10	4.4	10	3.9	10

Table A6. Reasons why people do not recycle as enumerated per education of the respondent ($n = 2004$).

Reasons Why People Do Not Recycle: Select the Three Main Reasons from the 10 Options	Times Selected Expressed as a Percentage of the Total of Each of the Respondent Groups															
	No Schooling ($n = 29$)		Some Primary School ($n = 113$)		Primary School Completed ($n = 115$)		Some High School ($n = 725$)		Grade 12 ($n = 678$)		Artisans Certificate Completed ($n = 73$)		Technicon Diploma/University Degree Completed ($n = 265$)		Other ($n = 6$) *	
	%	Rank	%	Rank	%	Rank	%	Rank	%	Rank	%	Rank	%	Rank	%	Rank
Insufficient space	17.2	1	13.0	1	15.7	1	15.2	1	15.1	2	16.9	1	13.8	2	11.1	-
No time	13.8	2	12.4	3	13.4	3	14.5	2	15.3	1	14.6	2	16.4	1	16.7	-
Dirty and untidy	9.2	6	13.0	1	14.8	2	12.4	4	12.1	3	11.9	5	12.6	3	11.1	-
Lack of knowledge	11.5	3	12.4	3	13.1	4	13.5	3	11.5	4	13.7	3	10.5	6	16.7	-
Facilities inconvenient	6.9	8	9.1	6	10.5	5	10.7	5	11.2	5	12.8	4	10.7	5	5.6	-
It makes no difference	10.3	4	8.8	7	7.6	7	8.4	6	7.8	7	8.2	6	7.3	7	5.6	-
No curbside collection	9.2	6	8.8	7	6.4	8	6.4	9	8.4	6	6.4	8	10.9	4	5.6	-
No responsibility	5.7	9	9.7	5	6.4	8	6.9	8	7.4	8	6.8	7	5.8	9	11.1	-
Not bothered	10.3	4	7.7	9	8.1	6	7.0	7	6.6	9	5.9	9	6.6	8	11.1	-
No service	5.7	9	5.0	10	4.1	10	5.0	10	4.5	10	2.7	10	5.4	10	5.6	-

* Too few data points to derive meaningful conclusions.

Table A7. Reasons why people do not recycle as enumerated per marital status ($n = 2004$).

Reasons Why People Do Not Recycle: Select the Three Main Reasons from the 10 Options	Times Selected Expressed as a Percentage of the Total of Each of the Respondent Groups											
	Single ($n = 898$)		Married ($n = 732$)		Living Together ($n = 176$)		Widowed ($n = 137$)		Divorced ($n = 44$)		Separated ($n = 17$)	
	%	Rank	%	Rank	%	Rank	%	Rank	%	Rank	%	Rank
Insufficient space	15.5	1	14.7	2	13.9	2	13.9	2	11.4	3	23.5	1
No time	15.3	2	15.0	1	11.2	4	15.6	1	18.2	1	3.9	8
Dirty and untidy	12.7	4	11.6	3	13.1	3	13.4	3	11.4	3	19.6	2
Lack of knowledge	12.9	3	11.2	4	15.4	1	12.2	4	6.8	8	11.8	4
Facilities inconvenient	10.9	5	11.1	5	10.6	5	7.5	8	15.2	2	7.8	5
It makes no difference	8.0	6	7.8	7	8.7	6	8.0	7	6.8	8	17.6	3
No curbside collection	7.1	7	9.0	6	6.5	9	9.0	5	9.8	5	0.0	10
No responsibility	6.9	8	6.6	9	8.0	7	8.8	6	8.3	6	3.9	8
Not bothered	6.6	9	7.6	8	6.1	10	7.1	9	7.6	7	5.9	6
No service	4.0	10	5.3	10	6.6	8	4.6	10	4.5	10	5.9	6

Table A8. Reasons why people do not recycle as enumerated per population/ethnic group ($n = 2004$).

Reasons Why People Do Not Recycle: Select the Three Main Reasons from the 10 Options	Times Selected Expressed as a Percentage of the Total of Each of the Respondent Groups							
	Whites ($n = 290$)		Blacks ($n = 1338$)		Indian/Asian ($n = 121$)		Colored ($n = 255$)	
	%	Rank	%	Rank	%	Rank	%	Rank
Insufficient space	13.2	2	15.0	1	14.6	1	16.6	2
No time	17.3	1	13.7	2	13.8	3	18.7	1
Dirty and untidy	10.5	4	13.3	3	12.4	4	10.2	5
Lack of knowledge	9.7	6	13.1	4	11.9	5	11.4	3
Facilities inconvenient	10.1	5	10.6	5	14.1	2	10.8	4
It makes no difference	6.9	8	8.0	6	9.9	6	8.6	6
No curbside collection	10.6	3	7.7	8	6.6	8	6.4	8
No responsibility	5.8	10	8.0	6	4.4	10	4.4	10
Not bothered	9.7	6	6.1	9	7.5	7	8.2	7
No service	6.2	9	4.5	10	4.7	9	4.6	9

Table A9. Reasons why people do not recycle as enumerated per occupation group (n = 898) **.

Reasons Why People Do Not Recycle: Select the Three Main Reasons from the 10 Options	Times Selected Expressed as a Percentage of the Total of Each of the Respondent Groups															
	Professional (n = 49)		Executive/Managerial (n = 77)		Clerical/Sales (n = 241)		Tradesman (n = 143)		Semi-Skilled (n = 168)		Unskilled (n = 139)		Self-Employed: (n = 45)		Refused (n = 36)	
	%	Rank	%	Rank	%	Rank	%	Rank	%	Rank	%	Rank	%	Rank	%	Rank
Insufficient space	17.7	1	15.2	2	14.8	2	16.8	1	15.9	1	12.0	4	8.1	7	11.4	2
No time	12.9	2	17.7	1	18.0	1	16.1	2	15.7	2	13.7	1	13.3	2	22.9	1
Dirty and untidy	11.6	5	11.3	4	11.6	4	11.0	4	11.7	3	12.5	3	14.8	1	11.4	2
Lack of knowledge	12.2	3	10.4	6	11.5	5	12.1	3	11.7	3	13.7	1	13.3	2	10.5	4
Facilities inconvenient	7.5	6	13.0	3	12.2	3	10.3	5	8.9	6	10.1	6	11.9	5	10.5	4
It makes no difference	6.8	9	7.4	7	7.9	6	7.0	8	11.1	5	6.7	9	8.9	6	7.6	7
No curbside collection	7.5	6	11.3	4	7.5	7	8.9	6	7.0	8	8.2	8	12.6	4	9.5	6
No responsibility	4.1	10	5.2	8	4.8	9	6.1	9	7.6	7	10.3	5	7.4	8	4.8	10
Not bothered	12.2	3	5.2	8	7.5	7	8.4	7	6.0	9	8.6	7	5.2	9	5.7	8
No service	7.5	6	3.5	10	4.3	10	3.5	10	4.4	10	4.3	10	4.4	10	5.7	8

** The respondents who selected not working (the housewives, students, retired) or unemployed (refer to Table A5) did not answer this question.

References

1. Republic of South Africa. *National Environmental Management: Waste Act, 2008 (Act 59 of 2008)*; Government Notice 278 Government Gazette 32000; Republic of South Africa: Cape Town, South Africa, 10 March 2009.
2. Department of Environmental Affairs. *National Waste Information Baseline Report*; Department of Environmental Affairs: Pretoria, South Africa, 2012.
3. Da Silva, C.L. Proposal of a dynamic model to evaluate public policies for the circular economy: Scenarios applied to the municipality of Curitiba. *Waste Manag.* **2018**, *78*, 456–466. [CrossRef]
4. Liu, L.; Liang, Y.; Song, Q.; Li, J. A review of waste prevention through 3R under the concept of circular economy in China. *J. Mater. Cycles Waste Manag.* **2017**, *19*, 1314–1323. [CrossRef]
5. Rada, E.C.; Cioca, L. Optimizing the Methodology of Characterization of Municipal Solid Waste in EU Under a Circular Economy Perspective. *Energy Procedia* **2017**, *119*, 72–85. [CrossRef]
6. Godfrey, L.; Oelofse, S. Historical review of waste management and recycling in South Africa. *Resources* **2017**, *6*, 57. [CrossRef]
7. *BMI Research Recycle Assessment Report*; BMI Research: London, UK, November 2013.
8. Department of Science and Technology (DST). *South African Waste Sector—2012. An Analysis of the Formal Private and Public Waste sector in South Africa*; Department of Science and Technology (DST): Pretoria, South Africa, 2012.
9. Wilson, D.C.; Velis, C.A.; Rodic, L. Integrated Sustainable Waste Management in Developing Countries. In Proceedings of the Institution of Civil Engineers: Waste and Resource Management; White Rose University Consortium, Universities of Leeds: Sheffield & York, UK, 2013; Volume 166, pp. 52–68.
10. Oelofse, S.; Nahman, A.; Godfrey, L. Chapter 6: Waste as a Resource: Unlocking Opportunities for Africa. In *Africa Waste Management Outlook*; United Nations Environment Programme (UNEP): Nairobi, Kenya, 2018; pp. 99–116.
11. Department of Environmental Affairs (DEA). *National Waste Management Strategy*; Department of Environmental Affairs: Pretoria, South Africa, November 2011.
12. African Union Commission. *Agenda 2063: The Africa We Want, Final ed.*; African Union Commission: Addis Ababa, Ethiopia, April 2015.
13. Department of Environmental Affairs (DEA). *Operation Phakisa: Chemicals and Waste Economy*; Briefing to Environmental Affairs Portfolio Committee, Parliament, Department of Environmental Affairs (DEA): Cape Town, South Africa, 17 October 2017.
14. Schultz, P.W.; Oskamp, S.; Mainieri, T. Who recycles and when? A review of personal and situational factors. *J. Environ. Psychol.* **1995**, *15*, 105–121. [CrossRef]
15. Department of Environmental Affairs and Tourism (DEAT). *National Waste Management Strategy Implementation South Africa—Recycling. Extended Producer Responsibility Status Quo Report, Report Number: 12/9/6, Annexure F*; Department of Environmental Affairs and Tourism (DEAT): Pretoria, South Africa, 4 April 2005.
16. Strydom, W.F.; Godfrey, L.K. Household waste recycling behaviour in South Africa-has there been progress in the last 5 years? In Proceedings of the 23rd WasteCon Conference and Exhibition, Johannesburg, South Africa, 17–21 October 2016.
17. Barr, S. Factors influencing environmental attitudes and behaviors: A UK case study of household waste management. *Environ. Behav.* **2007**, *39*, 435–473. [CrossRef]
18. Godfrey, L.; Scott, D.; Trois, C. Caught between the global economy and local bureaucracy: The barriers to good waste management practice in South Africa. *Waste Manag. Res.* **2013**, *31*, 295–305. [CrossRef] [PubMed]
19. MORI Social Research Institute. *Public Attitudes towards Recycling and Waste Management, Quantitative and Qualitative Review*; Research Study Conducted for The Strategy Unit, Cabinet Office: London, UK, 2002.
20. Robinson, G.M.; Read, A.D. Recycling behaviour in a London Borough: Results from large-scale household surveys. *Resour. Conserv. Recycl.* **2005**, *45*, 70–83. [CrossRef]
21. Ojala, M. Recycling and ambivalence: Quantitative and qualitative analyses of household recycling among young adults. *Environ. Behav.* **2008**, *40*, 777–797. [CrossRef]
22. Omran, A.; Mahmood, A.; Abdul Aziz, H.; Robinson, G.M. Investigating households attitude toward recycling of solid waste in Malaysia: A case study. *Int. J. Environ. Res.* **2009**, *3*, 275–288.

23. Miafodzyeva, S.; Brandt, N.; Olsson, M. Motivation recycling: Pre-recycling case study in Minsk, Belarus. *Waste Manage. Res.* **2010**, *28*, 340–346. [CrossRef] [PubMed]

24. Perrin, D.; Barton, J. Issues associated with transforming household attitudes and opinions into materials recovery: A review of two kerbside recycling schemes. *Resour. Conserv. Recycl.* **2001**, *33*, 61–74. [CrossRef]

25. McDonald, S.; Ball, R. Public participation in plastics recycling schemes. *Resour. Conserv. Recycl.* **1998**, *22*, 123–141. [CrossRef]

26. Martin, M.; Williams, I.D.; Clark, M. Social, cultural and structural influences on household waste recycling: A case study. *Resour. Conserv. Recycl.* **2006**, *48*, 357–395. [CrossRef]

27. De Young, R. Recycling as appropriate behavior: A review of survey data from selected recycling education programs in Michigan. *Resour. Conserv. Recycl.* **1990**, *3*, 253–266. [CrossRef]

28. McDonald, S.; Oates, C. Reasons for non-participation in a kerbside recycling scheme. *Resour. Conserv. Recycl.* **2003**, *39*, 369–385. [CrossRef]

29. Vicente, P.; Reis, E. Factors influencing households' participation in recycling. *Waste Manag. Res.* **2008**, *26*, 140–146. [CrossRef] [PubMed]

30. Gamba, R.J.; Oskamp, S. Factors influencing community residents' participation in commingled curbside recycling programs. *Environ. Behav.* **1994**, *26*, 587–612. [CrossRef]

31. Willman, K.W. Information sharing and curbside recycling: A pilot study to evaluate the value of door-to-door distribution of informational literature. *Resour. Conserv. Recycl.* **2015**, *104*, 162–171. [CrossRef]

32. Chung, S.S.; Poon, C.S. The attitudinal differences in source separation and waste reduction between the general public and the housewives in Hong Kong. *J. Environ. Manag.* **1996**, *48*, 215–227. [CrossRef]

33. Rousta, K.; Bolton, K.; Lundin, M.; Dahlén, L. Quantitative assessment of distance to collection point and improved sorting information on source separation of household waste. *Waste Manag.* **2015**, *40*, 22–30. [CrossRef] [PubMed]

34. Rousta, K.; Ordoñez, I.; Bolton, K.; Dahlén, L. Support for designing waste sorting systems: A mini review. *Waste Manag. Res.* **2017**, *35*, 1099–1111. [CrossRef] [PubMed]

35. Shaw, P.J.; Maynard, S.J. The potential of financial incentives to enhance householders' kerbside recycling behaviour. *Waste Manag.* **2008**, *28*, 1732–1741. [CrossRef] [PubMed]

36. Struk, M. Distance and incentives matter: The separation of recyclable municipal waste. *Resour. Conserv. Recycl.* **2017**, *122*, 155–162. [CrossRef]

37. Tucker, P. *Understanding Recycling Behaviour: A Technical Monograph. (Newspaper Industry Environmetal Technology Initiative)*; University of Paisley: Paisley, Scotland, UK, 2001; p. 179.

38. Shaw, P.J. *Kerbside Co-Mingled Recycling in the London Borough of Havering: Summary and Synthesis of Research, 2001 to 2004*; Report to the Cleanaway Havering Riverside Trust; Centre for Environmental Studies, School of Civil Engineering & the Environment, University of Southampton: Southampton, UK, November 2005.

39. Creswell, J.W. *Research Design: Qualitative, Quantitative, and Mixed Methods Approaches*; Sage Publications: Thousand Oaks, CA, USA, 2003.

40. Kempton, W.; Boster, J.S.; Hartley, J.A. *Environmental Values in American Culture*; MIT Press: Cambridge, MA, USA, 1995.

41. Kish, L. A Procedure for Objective Respondent Selection within the Household. *J. Am. Stat. Assoc.* **1949**, *44*, 380–387. [CrossRef]

42. Babbie, E.R.; Mouton, J. *The Practice of Social Research*; Oxford University Press: Oxford, UK, 2001.

43. Mosler, H.; Tamas, A.; Tobias, R.; Rodríguez, T.C.; Miranda, O.G. Deriving Interventions on the Basis of Factors Influencing Behavioral Intentions for Waste Recycling, Composting, and Reuse in Cuba. *Environ. Behav.* **2008**, *40*, 522–544. [CrossRef]

44. Kaciak, E.; Kushner, J. Determinants of residents' recycling behavior. *Int. Bus. Econ. Res. J.* **2009**, *8*, 1–12.

45. Oskamp, S.; Harrington, M.J.; Edwards, T.C.; Sherwood, D.L.; Okuda, S.M.; Swanson, D.C. Factors Influencing Household Recycling Behavior. *Environ. Behav.* **1991**, *23*, 494–519. [CrossRef]

46. Department of the Environment, Food and Rural Affairs (DEFRA). *Waste and Resources Research and Development Strategy 2004–2007*; Department of the Environment, Food and Rural Affairs (DEFRA): London, UK, 2004.

47. Tonglet, M.; Phillips, P.S.; Bates, M.P. Determining the drivers for householder pro-environmental behaviour: Waste minimisation compared to recycling. *Resour. Conserv. Recycl.* **2004**, *42*, 27–48. [CrossRef]

48. Bryman, A. *Social Research Methods*; Oxford University Press: Oxford, UK, 2012.

49. Thomas, C.; Yoxon, M.; Slater, R.; Leaman, J. Changing recycling behaviour: An evaluation of attitudes and behaviour to recycling in the Western Riverside area of London. In Proceedings of the Waste 2004 Integrated Waste Management and Pollution Control Conference, Stratford-upon-Avon, UK, 28–30 September 2004.

50. Afroz, R.; Hanaki, K.; Tuddin, R. The Role of Socio-Economic Factors on Household Waste Generation: A Study in a Waste Management Program in Dhaka City, Bangladesh. *Res. J. Appl. Sci.* **2010**, *5*, 183–190. [CrossRef]

51. Omran, A.; Schiopu, A. Reasons for non-participation in recycling of solid waste in northern Malaysia: A case study. *Environ. Eng. Manag. J.* **2015**, *14*, 233–243. [CrossRef]

52. Miliute-Plepiene, J.; Hage, O.; Plepys, A.; Reipas, A.R. What motivates households recycling behaviour in recycling schemes of different maturity? Lessons from Lithuania and Sweden. *Resour. Conserv. Recycl.* **2016**, *113*, 40. [CrossRef]

53. Vining, J.; Ebreo, A. What Makes a Recycler? A comparison of recyclers and non-recyclers. *Environ. Behav.* **1990**, *22*, 55–73. [CrossRef]

54. Lange, F.; Brückner, C.; Kröger, B.; Beller, J.; Eggert, F. Wasting ways: Perceived distance to the recycling facilities predicts pro-environmental behavior. *Resour. Conserv. Recycl.* **2014**, *92*, 246–254. [CrossRef]

55. Clarke, M.J.; Maantay, J.A. Optimizing recycling in all of New York City's neighborhoods: Using GIS to develop the REAP index for improved recycling education, awareness, and participation. *Resour. Conserv. Recycl.* **2006**, *46*, 128–148. [CrossRef]

56. Oke, A.; Kruijsen, J. The Importance of Specific Recycling Information in Designing a Waste Management Scheme. *Recycling* **2016**, *1*, 271–285. [CrossRef]

Article

Predictors of Recycling Intentions among the Youth: A Developing Country Perspective

Pradipta Halder [1],* and Harminder Singh [2]

[1]   Business School, University of Eastern Finland, Joensuu 80100, Finland
[2]   WAPCOS Ltd., Gurgaon, Haryana 122015, India; hmssingh2050@gmail.com
*   Correspondence: pradipta.halder@uef.fi; Tel.: +358-40-74-70711

Received: 20 June 2018; Accepted: 9 August 2018; Published: 16 August 2018

Abstract: India is currently facing a mounting challenge related to municipal waste management, due to an increasing urban population, and their high consumption lifestyles. India also has the world's highest number of young people in the 10–24 years age group. The study applied the theory of planned behaviour (TPB) model to predict school students' recycling intentions in Delhi, the capital of India and one of the highest producers of municipal solid wastes in the country. Data were collected from a school in New Delhi and the sample size consisted of 272 students from 9th and 10th grades. The TPB model explained 56% of the variance in the students' intentions to recycling. The predictor 'subjective norm' appeared to have the strongest impact on the students' recycling intentions, followed by 'attitude' and 'perceived behavioural control'. It indicated that social factors are driving the Indian youth's recycling intentions. It is important that the policymakers promote recycling as a social trend in India and provide adequate facilities to the public so that they can participate in recycling activities without facing difficulties. Schools also have a role in increasing students' awareness of recycling and motivating them to participate in household waste management practices.

Keywords: recycling; intentions; youth; India; theory of planned behaviour

---

1. Introduction

Recycling of ever-increasing urban waste has become a priority for sustainable environmental management and planning activities in both developed and developing countries. According to an estimate, the total generation of municipal solid waste (MSW) is expected to reach around 2.2 billion tonnes per annum globally by 2025 [1]. Waste generation has a strong linkage to production and consumption patterns in our societies. The current trend shows that the MSW-generation per capita has been growing steadily in the low and middle-income countries, whereas it has been gradually stabilising in the developed economies [2]. Unmanaged urban wastes affect biogeochemical cycles from local to global scales and hazardous wastes are particularly dangerous to the living organisms including humans [3–5]. Inadequate management of MSW has become a global concern that affects the quality of life in urban areas in many countries especially in the developing nations [6,7].

India currently produces much lower amount of MSW per day compared with the U.S. and China; however, it is among the top 10 MSW-generating countries due to its large size of urban population and their increasing adoption of high consumption lifestyles [8]. Current MSW-generation in India is about 170,000 tonnes per day (i.e., 62 million tonnes annually) and with an annual growth rate of 5% it is projected to reach about 436 million tonnes per year by 2050 [9]. The growth in urban waste generation is strongly linked with the growth in urban population as it has been projected that almost 50% of India's population will be residing in urban areas by 2050 from the present 31% urban dwellers in the country [9]. However, existing MSW-management systems in India are inefficient as 75–90% of the waste generated in urban areas is disposed of in open dumping sites, which affects public health, the quality of air, water, and soil, and the economy [10–14]. Therefore, it goes without saying that the development of an efficient and widely implemented urban waste management system in India is significant as it can provide economic benefits, conserve scarce resources, and decrease the quantities of waste ending at landfills [2,13].

A considerable number of studies have investigated waste management practices in various Indian cities [5,6,12,15]. The recycling of MSW is a tradition in many urban areas in India that involves both formal and informal sectors [13,16]. In the dominating informal waste recycling sector in India, waste pickers collect recyclables, segregate and transfer them to scrap dealers, which thereafter arrive at different enterprises for further processing such as recycling [16]. However, the quantity of recyclable materials in wastes is smaller in India than that of in developed countries. Wastes in the community bins in India mainly consist of organic materials, scrap papers and plastic materials [17].

In the recent years, Government of India (GoI) has introduced a wide range of policies to address the issues of urban waste management in the country. The latest policy prescription titled 'Solid Waste Management Rules' was introduced in 2016 that replaced the erstwhile 'Municipal Solid Wastes (Management and Handling) Rules' that came into being in 2000. Similarly, a new policy instrument called 'Plastic Waste Management Rules' emerged in 2016, replacing an earlier set of regulations on plastic waste management that was in place since 2011. Besides these measures, a completely new regulatory instrument called 'E-waste (Management) Rules' also appeared in 2015. Although the Ministry of Environment, Forests and Climate Change (under the GoI) is the nodal agency for implementing and monitoring these regulatory measures, their achievements have remained limited [13]. Additionally, the Prime Minister of India, Mr. Narendra Modi, initiated a 'Swachh Bharat Abhiyan (SBA)' or 'Clean India Mission' in 2014—the largest nationwide campaign to drive cleanliness in India by involving millions of government employees, the public, school and college students, and civil society organisations. The SBA has both urban and rural missions, and the urban mission known as the 'Swachh Bharat Urban' has prescribed guidelines for involving community members in urban waste management practices [18]. That is to say that the success of these programmes requires an active participation of citizens, local governments, and private entrepreneurs [14]. At present, almost no segregation of garbage takes place at source in India and the residents always dispose of garbage improperly [19]. The role of households is particularly important in the Indian context as "it will be nearly impossible for the civic body to provide better surroundings if residents do not make an effort to deposit waste into the bins and stop the practice of throwing garbage onto the road" Joseph [20].

1.1. Social Aspects in Recycling

Urban waste management and recycling practices are not simply technical matters and the technocrats do not entirely regulate them. According to Srivastava et al. [21], "a waste management programme that ignores the social aspects is doomed to failure". A number of studies appeared during the past two decades where scholars identified and explained the socio-psychological and situational factors influencing households' recycling behaviours [22]. However, research on social dimensions of MSW is still not expansive and therefore, it is crucial that researchers investigate the management of MSW from a social point of view [23]. In India, urban residents have to take on the majority of the responsibilities for segregating their household wastes and bringing them to community bins from where municipalities and private entrepreneurs will collect them for recycling and other uses [14]. Moving towards a recovery-centric approach of waste management (i.e., the 3R principles: Reuse, recycle, and reduce) from the current disposal-centric approaches requires partnership among the government, private sector, and citizens [14,16,24]. Citizens' participation in waste management could make it a decentralised and cost-efficient approach [14], and there is a need for civil society organizations and educational institutes to be actively involved in raising awareness of waste management among the masses in India [25].

Positive public opinion and their active participation are central to the long-term success of MSW-management program in the developing countries [26–28]. Therefore, waste management policies in many of these countries have tried to promote people-centric approaches in recycling [29]. In the industrialised countries, waste management and recycling are highly sophisticated. In these countries, recycling related studies mainly focus on technical applications such as design and innovation, policy and economic analysis, and explore socio-psychological influences on individuals' recycling behaviours. Recycling research in developing countries, on the contrary, has not been much oriented towards understanding the influences of indirect factors on individuals' recycling attitudes and behaviours [30]. However, a recent study by Ma and Hipel [23] has pointed out that social dimension studies in MSW-management have been increasing in the developing countries particularly in the Asian countries because of their high population density and rapid urbanization associated with fast economic growth, which have made MSW-management a daunting task for the local civic bodies in these countries.

1.2. Public Attitudes to Recycling in the Developing Countries

Among the studies related to public attitudes and behaviours concerning recycling in the developing countries, Bolaane [29] found from a study in Gaborone (Botswana) that the households' general awareness of recycling did not translate into actual recycling behaviours due to lack of financial incentives and absence of visible recycling facilities. In another study from Surabaya (Indonesia), Dhokhikah et al. [31] reported that the respondents did not sort and recycle household solid wastes due to lack of time, absence of tradition in separation of waste at source, shortage in collection facilities, inadequate knowledge, and apathy towards recycling as it lacks incentives. Similar attitudes to recycling were also observed among the urban residents of Hat Yai (Thailand), Mekelle (Ethiopia), Dar es Salaam (Tanzania), Putrajaya (Malaysia), and some other Indonesian cities [26,32–35]. In China, Xiao et al. [28] explored public willingness to participate in waste management practices in the city of Xiamen and they found that the residents' environmental knowledge and social motivation had the strongest positive effects on their willingness to participate in waste management activities.

A few studies have also explored public attitudes to recycling in different Indian cities. In this regard, Kumar and Nandini [36] found that most of the residents in Bengaluru, the third most populous city in India, were unaware of solid waste management, and they were disposing of their waste into open spaces. However, the authors reported that most of those respondents were willing to segregate their wastes into different bins if the local civic bodies provided such bins to them. A lack of awareness of solid waste management was also evident among the residents of both Kumbakonam, a small town in the state of Tamil Nadu [37] and Jalandhar, a medium-sized city in the state of Punjab [38]. The authors found that the residents of those two cities were disposing of their household wastes into open dumping sites, although many of them were aware of the potential health hazards arising from such a practice. Jayasubramanian et al. [39] reported that a considerable number of the residents in Coimbatore, the second largest city in the state of Tamil Nadu, were recycling their household wastes as they were aware of its benefits. However, the residents informed that the lack of time and inadequate waste disposal facilities in their vicinities were the main obstacles against recycling. There was only one study to our knowledge, which investigated school students' perceptions and attitudes regarding household waste management in India and reported that the school students in Thrissur city in the state of Kerala were aware of solid waste related issues and they demonstrated their willingness to participate in household waste management practices [40].

1.3. Importance of the Youth Participation in Recycling in India

Although the above studies shed light on the public perceptions and attitudes concerning solid waste management in India, they are by no means exhaustive. India has the world's highest number of people in the 10–24 years age group (ca. 242 million) [41]. According to the Census data of 2011, the country is expected to have about 34% share of youth (15–24 years) in the total population by

2020 from 19% in 2011 [42]. Therefore, the importance of youth-engagement in civic activities has started receiving greater attention in the Indian society and policies than ever before. There has been a major shift in the way societies now perceive the role of young people in India due to the realization that they are the most important section of the population and the country's future growth will be determined by the size of its youth and their ability to bring positive changes to the society [41]. However, no study has so far investigated the socio-psychological factors driving young citizens' such as school students' recycling intentions in the Indian context. In this regard, recycling intention can be defined as "an individual's self-commitment to engage in recycling behaviours" [43]. Consumption of electronic gadgets (e.g., mobile phones and computers) and fashionable items (e.g., clothing, bags, and accessories) has been growing rapidly among the youth in Indian cities and this trend is likely to continue in the future as urban households become wealthy. Students should be aware of the problems related to unmanaged urban wastes so that they can participate in waste management practices and become responsible citizens. They can also influence others, particularly the adults, who will follow sustainable waste management behaviours at home and thus can bring much desired behavioural changes to the Indian society related to waste management. The present study from this perspective aims to explain Indian students' recycling intentions by applying the framework of the Theory of Planned Behaviour (TPB) developed by Azjen [44], which is a widely applied socio-psychological theory explaining individuals' behavioural intentions.

1.4. The Theory of Planned Behaviour and Recycling Intention

The TPB is an extension of the Theory of Reasoned Action [45] and the TPB model comprises three independent variables—attitude (Attitude), subjective norm (SN), and perceived behavioural control (PBC), which together act as the predictor of a wide range of intentions (Intention, the dependent variable) [46]. According to the TPB model, an individual's intention to perform a given behaviour is determined by the positive evaluation of the behaviour (i.e., Attitude: e.g., recycling is useful, recycling is good, etc.), perceived social pressure (i.e., SN) from others who are important to them (e.g., family, friends, and colleagues) to behave (or not) in a certain manner (e.g., waste separation at source, bringing household wastes into community bins, etc.) and their motivations to comply with those views, and perceived ease of performing that behaviour (i.e., PBC; e.g., how difficult it is to perform recycling behaviour, how confident an individual is about performing waste separation behaviour, etc.) [47]. The TPB framework has been applied in several studies for explaining individuals' recycling intentions and behaviours. Among the most recent studies, Stoeva and Alriksson's [48] explored university students' waste separation intentions in Sweden and Bulgaria and reported that Attitude was the strongest predictor of the Swedish students' waste-separation intentions, whereas both Attitude and PBC were the most significant predictors of the Bulgarian students' intentions to separate household wastes. In another study, Pikturnienė and Bäumle [49] investigated public intentions for recycling in three Lithuanian cities and found only Attitude having statistically significant positive effect on the recycling intentions. In fact, Attitude appeared to be the most significant predictor of recycling intentions in several studies [50–54]. However, in some studies, PBC emerged as the most significant predictor of recycling intention. For example, PBC was the strongest predictor of recycling intention among the residents of Hong Kong [55], Australia [56], and Turkey [57]. There is also an instance where none of the TPB predictors had any effect on individuals' recycling intentions and additional factors such as 'moral obligation', 'past behaviour' and 'inconvenience' predicted individuals' recycling intentions in the Netherlands [58].

Traditionally, SN has appeared as the weakest predictor of intention in the TPB framework [46] and it was evident in many of the above-mentioned studies on recycling intentions. However, as an exception, SN emerged as the strongest predictor of university students' recycling intentions in Hong Kong [47]. Nevertheless, there are some examples in other contexts where SN emerged as the strongest predictor of school students' intentions to use bioenergy in India [59], Taiwanese citizens' intentions to visit green hotels [60], and Chinese entrepreneurs' intentions to adopt cleaner production technologies [61]. There is hardly any study, which has applied the TPB framework to investigate recycling intentions among the urban masses in India although there are some studies, which applied the TPB framework to explain Indian consumers' intentions to purchase green and environmentally sustainable products [62–65]. In those studies, the effects of both Attitude and PBC on individuals' purchase intentions of green products appeared to be statistically significant, whereas the effect of SN was inconsistent (i.e., either significant or insignificant). Therefore, it is evident from the above studies that though Attitude, PBC and SN were useful in explaining individuals' recycling and other pro-environmental intentions, their effects have been varying, which could be the result of different personal, contextual, and situational factors affecting individuals' pro-environmental intentions.

1.5. Objectives and Hypotheses

The present study was conducted in Delhi, the capital of India, the second most populous city in the country and the second largest megacity in the world [66,67]. MSW-generation in Delhi has been increasing at an alarming rate. It has increased from 8370 tonnes per day during 2014–2015 to 9260 tonnes per day during 2015–2016, i.e., almost 11% increase in one year [68]. It has been estimated to reach up to 18,000 tonnes per day by 2021 [69]. Door to door waste collection system exists in all the urban local bodies under the five municipal authorities in Delhi and the predominant waste management practice carried out in the city is landfills in four designated sites. A recent study by Singh et al. [70] has found that the landfills in Delhi are the major sources of emission of greenhouse gases, particularly the methane and therefore, there is an urgent need for better segregation of household organic wastes and to establish scientifically planned sanitary landfill sites. The first measure clearly requires the residents to be aware of MSW related issues and have positive attitudes to participating in waste management and recycling practices.

The present study is particularly relevant as it investigated the socio-psychological factors determining students' recycling intentions in Delhi by applying the standard TPB framework. In addition, the study explored students' awareness of recycling including the analysis of the effects of some of the demographical variables on the TPB constructs and students' perceptions of learning possibilities of recycling. Based on these findings, the study recommended some measures for developing students' awareness of and positive intentions to recycling. The main hypotheses of the study were as follows:

H1: *Students' attitude (Attitude) to recycling significantly influences their intentions (Intention) of recycling.*

H2: *Subjective Norm (SN) significantly influences students' intentions (Intention) of recycling.*

H3: *Perceived Behavioural Control (PBC) significantly influences students' intentions (Intention) of recycling.*

2. Materials and Methods

2.1. Data Collection

The present study was conducted as part of an international survey related to exploring recycling intentions among the youths in China, Greece and India. In Greece, the survey was conducted among university students. However, school students participated in the Chinese and Indian surveys. The findings from the Chinese and Greek surveys will be reported separately elsewhere. The main reason for not producing a comparative study with the data from these three countries was that the researchers aimed to present a detailed analysis of the social dimensions and policy frameworks concerning MSW-management from each country and relate them to the country-specific findings. The Indian data came from a high school based in New Delhi. A local collaborator contacted the Principal of the school for obtaining necessary permission for conducting the survey. The questionnaires were distributed to the students by the local collaborator and a class teacher was responsible for supervising the survey. Prior to distributing the questionnaires, the students were informed about the purpose of the survey and the use of the survey data. They were also assured of anonymity and confidentiality. The respondents spent an average twenty minutes for completing the survey and their participation was voluntary. After the survey, the local collaborator collected the questionnaires and performed an initial data entry in a statistical software package. The local collaborator sent the processed data to the researchers who conducted further analysis and interpreted the results.

2.2. Design of the Survey Instrument

The Indian and Chinese version of the survey instrument differed from the Greek version of the instrument in some respects. The Indian version of the survey instrument consisted of questions concerning students' demographic profiles and awareness of recycling, a seven-point Likert-type scale with fourteen items related to the TPB constructs, and questions regarding the students' perceptions of the possibilities to learn about recycling. There were also some questions, which asked the students to present their free comments on both recycling and the survey instrument. The survey instrument was developed in English, as the Indian school followed English as its medium of instructions. Therefore, there was no need for a back translation. The survey instrument can be obtained from the author upon request.

The seven-point Likert-type scale for the TPB constructs ranged from 'strongly disagree' to 'strongly agree' with a coding value from 1 to 7, respectively where the middle point was 'neither agree nor disagree', which was given a coding value of 4. Each of the TPB constructs had multiple statement-like items following the recommendation by Ajzen [44]. The constructs Intention, PBC and SN consisted of a few items from a study by Wan et al. [47]. Specifically, the construct Attitude consisted of five items that aimed to capture the students' evaluation of recycling, SN included three items that measured the students' perceptions of the social responses to their decisions about recycling, PBC contained three items that evaluated the students' perceived ease of practicing recycling, and the dependent variable Intention had three items that explored the students' planned behaviour towards recycling. In this type of coding, higher scores on the constructs indicated the students' stronger attitudes to recycling, greater perceptions of social pressure for recycling, higher perceived control over recycling, and stronger intentions to practice recycling. It should be noted that the items under the construct Attitude were formulated to measure students' attitudes to the relevance of recycling in a broader environmental context. They were somewhat different from the definition of the construct Attitude given by Ajzen [44]. A similar approach can also be found in the study by Wan et al. [47], which used an item 'Waste separation can create a better community environment' to formulate the construct Attitude. Two experts from India assessed the content validity of the questionnaire. A pilot test of the questionnaire was also carried out among a small group of students in a Delhi-based school, which helped to improve the clarity of some of the questions related to measuring the students' awareness of recycling.

2.3. Sample Characteristics, Data Screening and Measurement Model

About 306 students participated in the survey. However, only 272 students (ca. 88%) completed the questionnaires in all aspects with an equal number of male and female participants. The mean age of the students was 14.48 years (SD = 0.66). A reliability check of the TPB constructs was performed to evaluate the internal consistencies of the items corresponding to each latent construct. At first, Cronbach's alpha (α) values for all the TPB constructs appeared to be less than 0.7, which showed lack of internal consistency [71,72]. To improve the reliability, indicator variables AT5 under Attitude, SN3 under SN, PBC3 under PBC, and IN1 under Intention were removed from the analysis. This procedure raised the 'α' values for both Attitude and SN above 0.7 besides improving the 'α' values of the other two constructs although they remained below 0.7 (Table 1). The skewness and kurtosis value of the indicator variables was below ±3 and ±10, respectively as recommended by Kline [73].

Table 1. Theory of planned behaviour (TPB) constructs and their corresponding measurement items.

TPB Constructs	Cronbach's Alpha (α)	Measurement Items
Intention (IN) (M = 10.87; SD = 1.94)	0.44	IN2. I plan to take actions regarding recycling my recyclables regularly * IN3. I plan to take actions regarding recycling my recyclables every day for the next month
Attitude (AT) (M = 24.75; SD = 3.50)	0.72	AT1. Recycling can reduce the threat of global climate change AT2. Recycling can reduce the amount of waste that we produce AT3. Recycling can conserve natural resources AT4. Recycling can prevent pollution by reducing the need to collect new raw materials
Subjective Norm (SN) (M = 9.78; SD = 2.56)	0.78	SN1. My friends expect me to recycle my recyclables * SN2. My classmates expect me to recycle my recyclables *
Perceived Behavioural Control (PBC) (M = 11.49; SD = 2.14)	0.54	PBC1. I know what items can be recycled * PBC2. I know where to take my recyclables for recycling *

* source: Wan et al. [47].

The next step in the data analysis was to perform a structural equation modelling (SEM) using the maximum likelihood method to test the relationships between the predictors Attitude, SN, and PBC with the dependent variable Intention. SEM has two parts—the first one is constructing a measurement model or performing a confirmatory factor analysis (CFA), which assesses the validity and reliability of the measurement constructs while the second one is building a structural model, which determines the causal relationships among latent variables [74,75]. The goodness of fit of a CFA model is evaluated based on multiple indicators such as Chi-square normalised by degrees of freedom (χ^2/df), goodness off fit index (GFI), adjusted goodness of fit index (AGFI), comparative fit index (CFI), Tucker–Lewis index (TLI), and root mean square error of approximation (RMSEA). Normally, model fit is considered good (Table 2) when indices are $\geq$0.90, χ^2/df is between 2 and 5, and RMSEA is $\leq$0.08 [76–78]. The CFA model in this study showed an acceptable model fit (χ^2 = 82.42, χ^2/df = 2.75, GFI = 0.94, AGFI = 0.89, CFI = 0.90, TLI = 0.85, RMSEA = 0.08).

Table 2. The acceptable level of fit indices for the measurement model.

Fit Indices	Norm *
χ^2/df	2–5
RMSEA	<0.08
GFI	>0.9
AGFI	>0.8
CFI	>0.9
TLI	>0.9

* source: Bagozzi and Yi [78]. RMSEA: Root mean square error of approximation; GFI: Goodness of fit index; AGFI: Adjusted goodness of fit index; CFI: Comparative fit index; TLI: Tucker–Lewis index.

The reliability and convergent validity of the CFA model were checked by the values of average variance extracted (AVE) and composite reliability (CR) scores. According to Fornell and Larcker [79], the AVE and CR values should be more than 0.5 and 0.7, respectively although an AVE value, which is less than 0.05 for a construct is also acceptable if the CR value of that construct is more than 0.6. In this study, the AVE value of only SN was above 0.5 and the CR values of Attitude and SN were 0.7 (Table 3). Therefore, the results of the present study should be treated with some caution. Discriminant validity was ensured by comparing the value of square root of AVE of each construct with the correlation value of each construct. The square root of AVE of each construct was higher than its correlation's value

(Table 4), which ensured discriminant validity [80]. All the quantitative analyses were performed by the IBM SPSS Statistics 23 and IBM SPSS Amos 23 software packages.

Table 3. Composite reliability (CR) and average variance extracted (AVE) of the TPB predictors in the CFA model.

Constructs	CR	AVE
Intention	0.44	0.28
Attitude	0.73	0.41
SN	0.80	0.65
PBC	0.57	0.40

Table 4. Effects of the demographic variables on the TPB constructs and their correlations.

Constructs	One-Way ANOVA Test (*p*-Values)		Correlation Matrix			
	Gender	Age	Intention	Attitude	SN	PBC
Intention	0.03 *	0.00 **	*0.53*			
Attitude	0.47	0.58	0.18 **	*0.64*		
SN	0.10	0.000 ***	0.40 **	0.03	*0.82*	
PBC	0.43	0.00 **	0.25 **	0.22 **	21 **	*0.63*

Note: The bold diagonal values in italics represent the square root of AVE; * $p < 0.05$; ** $p < 0.01$; *** $p < 0.001$.

3. Results

3.1. Students' Awareness of Recycling

All the respondents informed that they had heard of recycling and their main information sources were textbooks, newspapers and teachers. Students' understanding of recycling was evaluated with a multiple-option question and almost all the students correctly selected the option that stated, "Recycling is a process of converting waste materials into reusable objects". The students were also asked to identify the recyclable materials from a list of items. About 75% of them selected paper, 47% selected biodegradable materials, 46% selected plastic and glass, 38% selected metal, 17% selected wood, 16% selected electronics, and 10% selected textile. In addition, the students were asked to identify the universal recycling symbol from a list of four images referring to the symbols of the United Nations, recycling, waste disposal, and compostable (Figure 1). It appeared that almost all the students were able to recognise the recycling symbol.

(a)   (b)   (c)   (d)

Figure 1. Symbols: (**a**) United Nations; (**b**) recycling; (**c**) waste disposal; and (**d**) compostable.

3.2. Test of the TPB Constructs and Their Correlations

One-Way ANOVA tests were conducted to determine whether the demographic variables had any significant effects on the TPB constructs (Table 4). It emerged that 'gender' had statistically significant effect on the students' recycling intentions. The male students appeared to be more positive than their female peers towards related to their recycling intentions although the effect size was small

(Cohen's d = 0.27). In terms of 'age', students in the age group of 13–14 years demonstrated greater positive intentions, stronger social pressure, and higher perceived control over behaviour concerning recycling than the students in the age group of 14–15 years. The effect size appeared to be small as Cohen's d values ranged between 0.33 and 0.46. A Pearson's correlation test was performed to explore the relationships among the TPB constructs and the results showed some statistically significant and positive correlations among all the constructs except between SN and Attitude though the strengths of such relationships were weak (Table 4).

3.3. Structural Model of the Students' Intentions to Recycle and Hypothesis Testing

The hypothesised TPB model was statistically significant ($p < 0.001$) and showed an acceptable fit to the data ($\chi^2 = 82.42$, $\chi^2/df = 2.75$, GFI = 0.94, AGFI = 0.89, CFI = 0.90, TLI = 0.85, RMSEA = 0.08). It explained about 56% of the variance ($R^2 = 0.56$) in the students' intentions to recycle (Figure 2). It appeared that the predictor SN had the strongest and statistically significant positive effect ($\beta = 0.64$, SE = 0.08, t = 5.49, $p < 0.001$) on the students' intentions of recycling. Attitude had the second highest effect on Intention ($\beta = 0.21$, SE = 0.11, t = 1.85, $p > 0.05$) followed by PBC ($\beta = 0.17$, SE = 0.11, t = 1.35, $p > 0.05$), although their effects were insignificant. All the factor loadings on the latent constructs were above 0.50. Among the three hypotheses, only H2 appeared to be acceptable as SN had a statistically significantly effect on Intention, whereas, the other two hypotheses (H1 and H3) were rejected due to their insignificant relationships with Intention.

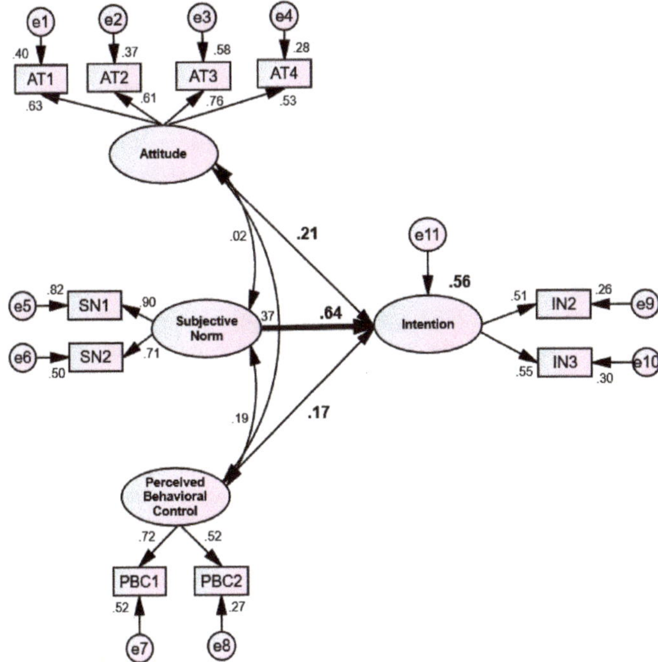

Figure 2. Structural model of the students' intentions to recycling ($N = 272$). The bold straight arrow from Subjective Norm to Intention shows statistically significant relationship.

3.4. Learning Possibility about Recycling and Students' Feedback

About 92% of the students informed that they could learn more about recycling from their school. When the students were asked about the ways they could learn more about recycling, nearly 75%

of them selected 'visiting a recycling facility', 62% selected 'teachers', 41% selected 'self-studying' and 29% selected both 'watching videos' and 'participating in debates'. About 74% of the students provided their free comments related to recycling and the survey procedure. Their comments reflected that they were interested to know more about recycling, as they appeared to be aware of its importance. There were comments from the students such as: "Recycling will make our country clean from waste material; recycling will make the world cleaner and greener; there are not many recycling bins in our county and therefore, some strict action should be taken regarding recycling of waste materials; a little amount of knowledge about recycling is not sufficient and therefore, schools need to take their students to recycling facilities for better understanding of the subject; and I believe that our natural resources are depleting at a faster rate than thought before and thus we must recycle everything as much as possible."

In addition, the students showed positive attitudes towards the survey as they thought that such a survey would enhance their awareness of recycling. Their comments were as follows: "This survey is quite inspiring and would promote the practice of recycling among young people; through the survey I am able to express my views on recycling; it has helped all the students to raise their awareness of recycling; some information on recycling should be provided by each school so that students can learn about it; the survey informed us about recycling, which we did not know beforehand; the survey was very useful as from our school we do not get much encouragement towards engaging in recycling; I will learn more about recycling by visiting a recycling facility; and being a nature lover, such environment related surveys are always appreciated by me but recycling facilities are very uncommon in India, which is a matter of concern."

4. Discussion and Conclusions

4.1. Synthesis of the Major Findings

The study investigated the socio-psychological determinants of school students' recycling intentions in India by applying the standard TPB framework. It also explored students' awareness of recycling and brought to the fore the educational aspects related to this issue. The findings of this study can be regarded as first-hand information on young citizens' recycling intentions. Results showed that almost all the students were aware of the concept of recycling. However, many did not perceive that plastic, metal, glass, wood, electronics and textile materials could be recycled. It also appeared that almost half of the students assumed biodegradable materials as recyclables.

The TPB model was able to capture a substantial variance in the students' recycling intentions, and thus the predictive utility of the model corresponded to a few earlier studies on this topic [47,49,55]. The emergence of SN as the most significant predictor of the students' recycling intentions contradicted the findings of most of the earlier studies where SN appeared to be the weakest among the TPB predictors although Wan et al. [47,55] found that SN had a significant role in determining individuals' recycling intentions in Hong Kong. The positive effect of PBC on recycling intentions corresponded to the findings of some earlier studies [55–57]. The insignificant effect of Attitude on the students' intentions of recycling was exceptional considering many studies reported Attitude as the strongest predictor of individuals' recycling intentions. However, this exception perhaps appeared due to the formulation of some of the items under the construct Attitude, which was somewhat different from the definition of Attitude given by Azjen [44].

These findings indicated that the social pressure would be driving Indian students' recycling intentions followed by their attitudes to recycling and their perceived ease of carrying out recycling activities. Jain et al. [81] have reported that since a *collectivist culture* prevails in India, SN tends to have a greater impact on intention than Attitude, and this could hold true in this study. The study found that the demographic variables had statistically significant effects on some of the TPB constructs although such effects were weak. The study by Oztekin et al. [57] found a gender difference in the Turkish university students' attitudes to recycling where female students were more positive than their male peers. However, the present study found an opposite result among the school students' attitudes to recycling in India. Overall, the school students demonstrated positive intentions and attitudes towards recycling and they were confident about their ability to participate in recycling activities. To some extent, these results corresponded to the study by Lucy et al. [40].

4.2. Implications and Recommendations

The study revealed that the students had some ambiguities about the materials that are recyclables. Such a lack of conceptual clarity about recyclable materials among the students can affect urban waste management program and make it costly for the municipalities. There is a high probability that due to this lack of knowledge among the youth, the segregation of wastes at source will be carried out improperly by them that will ultimately affect government's efforts towards urban waste management. Therefore, there is a clear need of raising awareness of recycling and recyclable materials among the youth in India. Wan et al. [55] have suggested that public authorities should position "recycling as a social trend and promote it in the society through encouraging messages from the celebrity personalities and showing the percentage or frequency of the local population performing recycling". It can be a relevant approach in the context of recycling, as Abrahamese and Steg [82] perceived that "socially desirable behaviours could be achieved in issues related to resource conservation by means of social influences, learning and comparison". Wan et al. [55] have also suggested that promoting recycling behaviour as a socially desirable trend instead of highlighting its benefits could be an effective strategy. It is evident that without increasing the knowledge of the benefits of recycling and improving individuals' waste separation abilities, socially desirable behaviours will be difficult to achieve in the context of urban waste management in India. Moreover, when it comes to developing students' recycling behaviours, schools can play an important role. Schools can initiate various environmental activities in their local communities involving students and residents, as the respondents in this study also suggested, which can enhance their awareness of waste separation and motivate them to take required practical actions both at the household and community levels [54]. There are already a few policies and initiatives existing in India related to urban waste management and therefore, it is the responsibility of the citizens, both young and old, to come forward and help implementing those policies successfully throughout the country.

4.3. Limitations and Future Research Needs

There were some limitations in the study, which are required to be addressed in the future research. The researchers collected data from only one school in Delhi and therefore, the sample cannot be considered a truly representative sample. Future studies should recruit a large sample of students from various Indian cities to improve the representativeness of the study as well as the generalizability of its findings. Moreover, besides school students, university students should be included in the future studies, as this will provide a comprehensive picture of the socio-psychological factors determining Indian youth's recycling intentions. Future studies could also explore the socio-psychological factors that affect recycling intentions of both the youth and adults to understand their similarities and differences across the two age groups, which could be useful in building relevant hypotheses. Apart from recycling intentions, past recycling behaviours should also be studied to observe possible gaps between intentions and behaviours. A number of studies also added constructs such as 'personal norm', 'awareness of consequences', 'environmental knowledge', 'past behaviour', 'self-identity', and 'situation factors' to the extended TPB frameworks to explain individuals recycling intentions, which appeared to be relevant [54,55,57,83–85]. Therefore, future studies could use such extended TPB models to predict young citizens' recycling intentions and behaviours in India. Future studies could also investigate how economic rewards (i.e., payment for recyclables) could influence young generation's intentions towards recycling as the lack of financial incentives was identified in many of the previous studies as one of the factors that affected individuals' recycling intentions.

4.4. Conclusions

In conclusion, it can be said that the TPB model was able to explain school students' intentions of recycling in the Indian context. All the predictors appeared to have positive effects on the students' intentions of recycling and the relevance of the TPB model in explaining students' intentions was apparent. SN appeared to be the strongest predictor of the Indian students' recycling intentions. Students' participation in recycling can be significantly enhanced by developing their positive attitudes to recycling, setting a social trend in recycling and making it easier for them to participate in recycling by providing better facilities. It appeared that the students were aware of the concept of recycling, although they lacked clarity about the recyclable materials. Students were also interested in receiving more information on recycling from their school. In addition, they recognised the need for having better waste separation facilities in their households and communities. The selection of the school students appeared to be relevant for this study because of the magnitude of the proportion of youth in India's population and their importance in addressing the mounting solid waste management problems in the country. The study was one of a kind in the Indian context, and thus could provide future directions in research on the social dimensions particularly on the behavioural aspects in recycling in India. Since India is a highly diverse country in terms of social and cultural settings, there is a need to take into consideration those driving factors while exploring young people's attitudes and behaviours related to recycling.

Author Contributions: Conceptualization, P.H.; Methodology, P.H. and H.S.; Data Curation, H.S.; Formal Analysis, P.H.; Writing-Original Draft Preparation, P.H.

Funding: This research received no external funding.

Acknowledgments: The authors greatly acknowledge the school and the students for participating in the survey. In addition, the valuable comments from the reviewers to improve an earlier version of the manuscript is also acknowledged.

Conflicts of Interest: The authors declare no conflict of interest.

References

1. Hoornweg, D.; Bhada-Tata, P. *What a Waste: A Global Review of Solid Waste Management*; Urban Development Series Knowledge Papers No. 15; The World Bank: Washington, DC, USA, 2012; 98p.
2. Wilson, D.C.; Rodic, L.; Modak, P.; Soos, R.; Rogero, A.C.; Velis, C.; Iyer, M.; Simonett, O. *Global Waste Management Outlook*; UNEP DTIE—International Environmental Technology Centre: Osaka, Japan, 2015; 332p.
3. Cointreau, S. *Occupational and Environmental Health Issues of Solid Waste Management Special Emphasis on Middle- and Lower-Income Countries*; The World Bank Group: Washington, DC, USA, 2006; 48p.
4. Grimm, N.B.; Faeth, S.H.; Golubiewski, N.E.; Redman, C.L.; Wu, J.; Bai, X.; Briggs, J.M. Global Change and the Ecology of Cities. *Science* **2008**, *319*, 756–760. [CrossRef] [PubMed]
5. Misra, V.; Pandey, S.D. Hazardous waste, impact on health and environment for development of better waste management strategies in future in India. *Environ. Int.* **2005**, *31*, 417–431. [CrossRef] [PubMed]
6. Das, S.; Bhattacharyya, B. Estimation of Municipal Solid Waste Generation and Future Trends in Greater Metropolitan Regions of Kolkata, India. *J. Ind. Eng. Manag. Innov.* **2014**, *1*, 31–38. [CrossRef]
7. Vergara, S.E.; Tchobanoglous, G. Municipal solid waste and the environment: A global perspective. *Ann. Rev. Environ. Resour.* **2012**, *37*, 277–309. [CrossRef]
8. Worldwatch Institute. Global Municipal Solid Waste Continues to Grow. 2012. Available online: http://www.worldwatch.org/global-municipal-solid-waste-continues-grow (accessed on 30 October 2017).
9. Planning Commission (Government of India). Report of the Task Force on Waste to Energy (Volume I) (In the Context of Integrated Municipal Solid Waste Management). 2014. Available online: http://planningcommission.nic.in/reports/genrep/rep_wte1205.pdf (accessed on 1 November 2017).
10. Biswas, A.K.; Kumar, S.; Babu, S.; Bhattacharyya, J.K.; Chakrabarty, T. Studies on environmental quality in and around municipal solid waste dumpsite. *Resour. Conserv. Recycl.* **2010**, *55*, 129–134. [CrossRef]

11. Hazra, T.; Goel, S. Solid waste management in Kolkata, India: Practices and challenges. *Waste Manag.* **2009**, *29*, 470–478. [CrossRef] [PubMed]

12. Jha, A.K.; Sharma, C.; Singh, N.; Ramesh, R.; Purvaja, R.; Gupta, P.K. Greenhouse gas emissions from municipal solid waste management in Indian mega-cities: A case study of Chennai landfill sites. *Chemosphere* **2008**, *71*, 750–758. [CrossRef] [PubMed]

13. Kumar, S.; Smith, S.R.; Fowler, G.; Velis, C.; Kumar, S.J.; Arya, S.; Rena; Kumar, R.; Cheeseman, C. Challenges and opportunities associated with waste management in India. *R. Soc. Open Sci.* **2017**, *4*, 160764. [CrossRef] [PubMed]

14. Narayana, T. Municipal solid waste management in India: From waste disposal to recovery of resources? *Waste Manag.* **2009**, *29*, 1163–1166. [CrossRef] [PubMed]

15. Sharholy, M.; Ahmad, K.; Vaishya, V.C.; Gupta, R.D. Municipal solid waste characteristics and management in Allahabad, India. *Waste Manag.* **2007**, *27*, 490–496. [CrossRef] [PubMed]

16. Talyan, V.; Dahiya, D.P.; Sreekrishnan, T.R. State of municipal solid waste management in Delhi, the capital of India. *Waste Manag.* **2008**, *28*, 1276–1287. [CrossRef] [PubMed]

17. Kumar, S.; Bhattacharyya, J.K.; Vaidya, A.N.; Chakrabarti, T.; Devotta, S.; Akolkar, A.B. Assessment of the status of municipal solid waste management in metro cities, state capitals, class I cities, and class II towns in India: An insight. *Waste Manag.* **2009**, *29*, 883–895. [CrossRef] [PubMed]

18. Ministry of Housing and Urban Affairs (Government of India). Guidelines for Community Engagement under Swachh Bharat Mission Urban. 2017. Available online: http://www.swachhbharaturban.in/sbm/home/lib/content/Community%20Engagement%20Guidelines.pdf (accessed on 2 November 2017).

19. Joshi, R.; Ahmed, S. Status and challenges of municipal solid waste management in India: A review. *Environ. Chem. Pollut. Waste Manag.* **2016**, *2*, 1–18. [CrossRef]

20. Joseph, K. Perspectives of solid waste management in India. In Proceedings of the International Symposium on the Technology and Management of the treatment and Reuse of the Municipal Solid Waste, Shanghai, China; 2002; p. 14.

21. Srivastava, P.K.; Kulshreshtha, K.; Mohanty, C.S.; Pushpangadan, P.; Singh, A. Stakeholder-based SWOT analysis for successful municipal solid waste management in Lucknow, India. *Waste Manag.* **2005**, *25*, 531–537. [CrossRef] [PubMed]

22. Varotto, A.; Spagnolli, A. Psychological strategies to promote household recycling. A systematic review with meta-analysis of validated field interventions. *J. Environ. Psychol.* **2017**, *51*, 168–188. [CrossRef]

23. Ma, J.; Hipel, K.W. Exploring social dimensions of municipal solid waste management around the globe—A systematic literature review. *Waste Manag.* **2016**, *56*, 3–12. [CrossRef] [PubMed]

24. Tripathi, D.K.; Shukla, J.P. Projection and Quantification of Municipal Solid Waste Management in Bhopal city M.P. India. *Int. J. Sci. Eng. Appl. Sci.* **2016**, *2*, 189–194.

25. Jha, A.K.; Singh, S.K.; Singh, G.P.; Gupta, P. Sustainable municipal solid waste management in low income group of cities: A review. *Trop. Ecol.* **2011**, *52*, 123–131.

26. Cheng, C.; Urpelainen, J. Who should take the garbage out? Public opinion on waste management in Dar es Salaam, Tanzania. *Habitat Int.* **2015**, *46*, 111–118. [CrossRef]

27. Dhokhikah, Y.; Trihadiningrum, Y. Solid waste management in Asian developing countries: Challenges and opportunities. *J. Appl. Environ. Biol. Sci.* **2012**, *2*, 329–335.

28. Xiao, L.; Zhag, G.; Zhu, Y.; Lin, T. Promoting public participation in household waste management: A survey based method and case study in Xiamen city, China. *J. Clean. Prod.* **2017**, *144*, 313–322. [CrossRef]

29. Bolaane, B. Constraints to promoting people centred approaches in recycling. *Habitat Int.* **2006**, *30*, 731–740. [CrossRef]

30. Troschinetz, A.M.; Mihelcic, J.R. Sustainable recycling of municipal solid waste in developing countries. *Waste Manag.* **2009**, *29*, 915–923. [CrossRef] [PubMed]

31. Dhokhikah, Y.; Trihadiningrum, Y.; Sunaryo, S. Community participation in household solid waste reduction in Surabaya, Indonesia. *Resour. Conserv. Recycl.* **2015**, *102*, 153–162. [CrossRef]

32. Charuvichaipong, C.; Sajor, E. Promoting waste separation for recycling and local governance in Thailand. *Habitat Int.* **2006**, *30*, 579–594. [CrossRef]

33. Malik, N.K.A.; Abdullah, S.H.; Manaf, L.A. Community participation on solid waste segregation through recycling programmes in Putrajaya. *Proc. Environ. Sci.* **2015**, *30*, 10–14. [CrossRef]

34. Sekito, T.; Prayogo, T.B.; Dote, Y.; Yoshitake, T.; Bagus, I. Influence of a community-based waste management system on people's behavior and waste reduction. *Resour. Conserv. Recycl.* **2013**, *72*, 84–90. [CrossRef]

35. Tadesse, T. Environmental concern and its implication to household waste separation and disposal: Evidence from Mekelle, Ethiopia. *Resour. Conserv. Recycl.* **2009**, *53*, 183–191. [CrossRef]

36. Kumar, M.; Nandini, N. Community attitude, perception and willingness towards solid waste management in Bangalore city, Karnataka, India. *Int. J. Environ. Sci.* **2013**, *4*, 87–95.

37. Indhira, K.; Senthil, J.; Vadivel, S. Awareness and attitudes of people perception towards to household solid waste disposal: Kumbakonam Town, Tamilnadu, India. *Arch. Appl. Sci. Res.* **2015**, *7*, 6–12.

38. Minhas, J. Solid Waste Management—Community Perception, Attitude and Participation. *Asian J. Res. Soc. Sci. Hum.* **2017**, *7*, 121–130. [CrossRef]

39. Jayasubramanian, P.; Saratha, M.M.; Divya, M. Perception of households towards waste management and its recycling in Coimbatore. *Int. J. Multidiscip. Res. Dev.* **2015**, *2*, 510–515.

40. Lucy, C.D.; Vivek, R.; Saritha, K.; Anies, T.K.; Josphina, C.T. Awareness, Attitude and Practice of School Students towards Household Waste Management. *J. Environ.* **2013**, *2*, 147–150.

41. United Nations. World Population Prospects: The 2015 Revision—Key Findings and Advance Tables. 2015. Available online: https://esa.un.org/unpd/wpp/publications/files/key_findings_wpp_2015.pdf (accessed on 12 December 2017).

42. Central Statistics Office (Ministry of Statistics & Programme Implementation, Government of India). Youth in India: 2017. Available online: http://mospi.nic.in/sites/default/files/publication_reports/Youth_in_India-2017.pdf (accessed on 12 December 2017).

43. Park, J.; Ha, S. Understanding consumer recycling behavior: Combining the theory of planned behavior and the norm activation model. *Fam. Consum. Sci. Res. J.* **2014**, *42*, 278–291. [CrossRef]

44. Ajzen, I. The theory of planned behavior. *Organ. Behav. Hum. Decis. Process.* **1991**, *50*, 179–211. [CrossRef]

45. Ajzen, I.; Fishbein, M. *Understanding Attitudes and Predicting Social Behavior*; Prentice-Hall: Englewood Cliffs, NJ, USA, 1980.

46. Armitage, C.J.; Conner, M. Efficacy of the Theory of Planned Behaviour: A meta-analytic review. *Br. J. Soc. Psychol.* **2001**, *40*, 471–499. [CrossRef] [PubMed]

47. Wan, C.; Cheung, R.; Shen, G.Q. Recycling attitude and behaviour in university campus: A case study in Hong Kong. *Facilities* **2012**, *30*, 630–646. [CrossRef]

48. Stoeva, K.; Alriksson, S. Influence of recycling programmes on waste separation behaviour. *Waste Manag.* **2017**, *68*, 732–741. [CrossRef] [PubMed]

49. Pikturnienė, I.; Bäumle, G. Predictors of recycling behaviour intentions among urban Lithuanian inhabitants. *J. Bus. Econ. Manag.* **2016**, *17*, 780–795.

50. Chan, K. Mass communication and pro-environmental behvior: Waste recycling in Hong Kong. *J. Environ. Manag.* **1998**, *52*, 317–325. [CrossRef]

51. Karim, G.W.A.; Rusli, I.F.; Biak, D.R.A.; Idris, A. An application of the theory of planned behaviour to study the influencing factors of participation in source separation of food waste. *Waste Manag.* **2013**, *33*, 1276–1281. [CrossRef] [PubMed]

52. Nigbur, D.; Lyons, E.; Uzzell, D. Attitudes, norms, identity and environmental behavior: Using an expanded theory of planned behaviour to predict participation in a kerbside recycling programme. *Br. J. Soc. Psychol.* **2010**, *49*, 259–284. [CrossRef] [PubMed]

53. Tonglet, M.; Phillips, P.S.; Read, A.D. Using the theory of planned behavior to investigate the determinants of recycling behaviour: A case study from Brixworth, UK. *Resour. Conserv. Recycl.* **2004**, *41*, 191–214. [CrossRef]

54. Zhang, D.; Huang, G.; Yin, X.; Gong, Q. Residents' Waste Separation Behaviors at the Source: Using SEM with the Theory of Planned Behavior in Guangzhou, China. *Int. J. Environ. Res. Publ. Health* **2015**, *12*, 9475–9491. [CrossRef] [PubMed]

55. Wan, C.; Shen, G.Q.; Choi, S. Experiential and instrumental attitudes: Interaction effect of attitude and subjective norm on recycling intention. *J. Environ. Psychol.* **2017**, *50*, 69–79. [CrossRef]

56. Chan, L.; Bishop, B. A moral basis for recycling: Extending the theory of planned behaviour. *J. Environ. Psychol.* **2013**, *36*, 96–102. [CrossRef]

57. Oztekin, C.; Teksöz, G.; Pamuk, S.; Sahin, E.; Kilic, D.S. Gender perspective on the factors predicting recycling behavior: Implications from the theory of planned behavior. *Waste Manag.* **2017**, *62*, 290–302. [CrossRef] [PubMed]

58. Philippsen, Y. Factors Influencing Students' Intention to Recycle. Master's Thesis, School of Management and Governance, University of Twente, Enschede, The Netherlands, 2015.

59. Halder, P.; Pietarinen, J.; Havu-Nuutinen, S.; Pöllänen, S.; Pelkonen, P. The Theory of Planned Behavior Model and Students' Intentions to Use Bioenergy: A Cross-Cultural Perspective. *Renew. Energy* **2016**, *89*, 627–635. [CrossRef]

60. Chen, M.F.; Tung, P.J. Developing an extended Theory of Planned Behavior model to predict consumers' intention to visit green hotels. *Int. J. Hosp. Manag.* **2014**, *36*, 221–230. [CrossRef]

61. Zhang, B.; Young, S.; Bi, J. Enterprises' willingness to adopt/develop cleaner production technologies: An empirical study in Changshu, China. *J. Clean. Prod.* **2013**, *40*, 62–70. [CrossRef]

62. Kumar, B. *Theory of Planned Behaviour Approach to Understand the Purchasing Behaviour for Environmentally Sustainable Products*; Working Paper No. 2012-12-08; Indian Institute of Management: Ahmedabad, India, 2012; 43p.

63. Paul, J.; Modi, A.; Patel, J. Predicting green product consumption using theory of planned behavior and reasoned action. *J. Retail. Consum. Serv.* **2016**, *29*, 123–134. [CrossRef]

64. Yadav, R.; Pathak, G.S. Young consumers' intention towards buying green products in a developing nation: Extending the theory of planned behavior. *J. Clean. Prod.* **2016**, *135*, 732–739. [CrossRef]

65. Yadav, R.; Pathak, G.S. Determinants of Consumers' Green Purchase Behavior in a Developing Nation: Applying and Extending the Theory of Planned Behavior. *Ecol. Econ.* **2017**, *134*, 114–122. [CrossRef]

66. Census of India. City Census 2011. Available online: http://www.census2011.co.in/city.php (accessed on 12 November 2017).

67. UN (United Nations). The World's Cities in 2016. Available online: http://www.un.org/en/development/desa/population/publications/pdf/urbanization/the_worlds_cities_in_2016_data_booklet.pdf (accessed on 11 November 2017).

68. CPCB (Central Pollution Control Board, Ministry of Environment, Forests & Climate Change, Government of India). Consolidated Annual Review Report Prepared in Compliance to the Provision 24(4) of the SWM Rules, 2016. Available online: http://cpcb.nic.in/MSW_AnnualReport_2015-16.pdf (accessed on 12 December 2017).

69. Chakraborty, M.; Sharma, C.; Pandey, J.; Singh, N.; Gupta, P.K. Methane Emission Estimation from Landfills in Delhi: A comparative Assessment of Different Methodologies. *Atmos. Environ.* **2011**, *45*, 7135–7142. [CrossRef]

70. Singh, S.K.; Anunay, G.; Rohit, G.; Shivangi, G.; Vipul, V. Greenhouse Gas Emissions from Landfills: A Case of NCT of Delhi, India. *J. Climatol. Weather Forecast.* **2016**, *4*, 1–6.

71. Kline, R.B. *Principles and Practice of Structural Equation Modeling*, 2nd ed.; Guilford Press: New York, NY, USA, 2005.

72. Nunnally, J.C.; Bernstein, I.H. *Psychometric Theory*; McGraw Hill: New York, NU, USA, 1994.

73. Kline, R.B. *Principles and Practice of Structural Equation Modeling*, 3rd ed.; The Guildford Press: New York, NY, USA, 2011.

74. Anderson, J.C.; Gerbing, D.W. Structural equation modeling in practice: A review and recommended two-step approach. *Psychol. Bull.* **1998**, *103*, 411–423. [CrossRef]

75. Arbuckle, J.L. *Amos7.0 User's Guide*; SPSS: Chicago, IL, USA, 2006.

76. Browne, M.W.; Cudek, R. Alternative ways of assessing model fit. In *Testing Structural Equation Models*; Bollen, K.A., Long, J.S., Eds.; Sage: Thousand Oaks, CA, USA, 1993; pp. 136–162.

77. Hair, J.F.; Anderson, R.E.; Tatham, R.L.; Black, W.C. *Multivariate Data Analysis*, 5th ed.; Prentice-Hall: Upper Saddle River, NJ, USA, 1998.

78. Bagozzi, R.P.; Yi, Y. On the evaluation of structural equation models. *J. Acad. Market. Sci.* **1988**, *16*, 74–94. [CrossRef]

79. Fornell, C.; Larcker, D. Evaluating structural equation models with unobservable variables and measurement error. *J. Market. Res.* **1981**, *18*, 39–50. [CrossRef]

80. Chin, W.W.; Gopal, A.; Salisbury, W.D. Advancing the theory of adaptive structuration: The development of a scale to measure faithfulness of appropriation. *Inf. Syst. Res.* **1997**, *8*, 342–367. [CrossRef]

81. Jain, S.; Mohammed Khan, M.N.; Mishra, S. Understanding consumer behavior regarding luxury fashion goods in India based on the theory of planned behavior. *J. Asia Bus. Stud.* **2017**, *11*, 4–21. [CrossRef]

82. Abrahamse, W.; Steg, L. Social influence approaches to encourage resource conservation: A meta-analysis. *Glob. Environ. Chang.* **2013**, *23*, 1773–1785. [CrossRef]
83. Pakpour, A.H.; Zeidi, I.M.; Emamjomeh, M.M.; Asefzadeh, S.; Pearson, H. Household waste behaviours among a community sample in Iran: An application of the theory of planned behaviour. *Waste Manag.* **2014**, *34*, 980–986. [CrossRef] [PubMed]
84. Graham-Rowe, E.; Jessop, D.C.; Sparks, P. Predicting household food waste reduction using an extended theory of planned behaviour. *Resour. Conserv. Recycl.* **2015**, *101*, 194–202. [CrossRef]
85. Khalil, M.S.; Abdullah, S.H.; Manaf, L.A.; Sharaai, A.H.; Nabegu, A.B. Examining the Moderating Role of Perceived Lack of Facilitating Conditions on Household Recycling Intention in Kano, Nigeria. *Recycling* **2017**, *2*, 18. [CrossRef]

Article

Release of Trace Elements from Bottom Ash from Hazardous Waste Incinerators

Tran Thi Thu Dung [1,2,*], Elvira Vassilieva [1], Rudy Swennen [1] and Valérie Cappuyns [1,3]

[1]   Department of Earth and Environmental Sciences, KU Leuven, 3001 Leuven, Belgium;
     elvira.vassilieva@ees.kuleuven.be (E.V.); rudy.swennen@kuleuven.be (R.S.);
     valerie.cappuyns@kuleuven.be (V.C.)
[2]   Faculty of Environment, University of Science Vietnam National University Ho Chi Minh City,
     Ho Chi Minh City 700000, Vietnam
[3]   Center for Economics and Corporate sustainability, KU Leuven, Brussels B-1000, Belgium
*   Correspondence: tttdung@hcmus.edu.vn; Tel.: +84-8-3835-3295

Received: 11 July 2018; Accepted: 9 August 2018; Published: 14 August 2018

Abstract: Bottom ash is the major by-product of waste incineration and can contain trace elements (As, Cd, Co, Cu, Cr, Mo, Ni, Pb, and Zn) with concentrations up to thousands of $mg \cdot k^{-1}$. In this study, a combination of different extractions and leaching tests (i.e., CH_3COOH and ammonium-EDTA (Ethylenediaminetetraacetic acid) extractions and pH_{stat} leaching tests) was used to investigate the potential release of trace elements from bottom ash samples derived from hazardous waste incineration plants. Although large variations have been found in the release of trace elements by different extractions, in general, the highest concentrations of most trace elements (except As and Mo) were released with the CH_3COOH extraction, whereas the release of As and Mo was highest with the ammonium-EDTA extraction. Kinetics of element release upon acidification based on a pH_{stat} leaching test at pH 4 could be related to the solid-phase speciation of some selected trace elements. The relatively high-potential mobility and elevated total concentrations of some trace elements imply a threat to the environment if these bottom ashes are not treated properly. Results of the present study may be useful to develop potential treatment strategies to remove contaminants and eventually recover metals from bottom ash.

Keywords: bottom ash; hazardous waste; kinetic release; potential mobility; trace elements; waste management

---

1. Introduction

When society moves towards more sustainable material cycles, recovery of metals from industrial waste can be an opportunity to turn waste into a valuable resource. It is estimated that numerous metals will be consumed in less than 50 years (Zn) or 100 years (Co, Cu, Ni, Mo, and Pb) if current rates of extraction are maintained [1]. However, sustainable reserves of these metals will remain for the future if they are recovered and the remaining reserves are used more efficiently [2]. Waste containing significant amounts of trace elements offers a potential for the recovery of these elements, since trace elements in anthropogenic matrices are often more labile compared to natural matrices (e.g., ores, rocks) [3]. Generally, the term "major elements" is used for elements which have concentrations greater than 1% and term "trace elements" has been used in geochemistry for chemical elements that occur in the Earth's crust in amounts less than 0.1% (1000 $mg \cdot k^{-1}$) [4]. In this study, the term "trace elements" is used to indicate particularly the elements As, Cd, Cr, Co, Cu, Mo, Ni, Pb, and Zn while the term "major element" is used to indicate the following elements: Al, Ca, Fe, K, Mg, P, S, and Si.

Waste incineration has been a preferred alternative in solid waste management since landfilling became more difficult to site due to high costs, diminishing land availability, and stricter regulations [5].

The bottom ashes (BAs) from waste incineration are often treated to recover the metals or they are reused in the cement and concrete industries to produce road construction materials in Europe and developed countries [6,7]. However, without any pretreatment, the residues which contain high amounts of leachable potentially toxic elements are also classified as hazardous wastes [8].

One of the key factors when evaluating the risks related to the management (e.g., disposal/reuse) of solid waste regarding environmental health and safety is the release of pollutants to the receiving environment. Co-disposal of different kinds of waste without knowing their leaching characteristics may pose a threat to environment, for example, the disposal of ash in an acidic environment increased the leaching of heavy metals and contaminated the ground water at the disposal site [9]. For this reason, leaching/extraction tests are commonly applied because they provide information about the potential release of constituents from solid materials to the liquid phase [10]. There are several extraction/leaching tests for solid materials, each with different purposes and implications. Among these tests, an ammonium-EDTA (ethylenediaminetetraacetic acid) extraction is usually used to determine the potential mobility of trace elements in soils and sediments as a consequence of complexation. Different types of EDTA (EDTA free acid, sodium-EDTA, and ammonium-EDTA) are used as extraction solutions [11]. Acetic acid (CH_3COOH) in different concentrations (0.43 mol$\cdot$L^{-1} or 0.11 mol$\cdot$L^{-1}) is often used to determine the fraction of elements in a soil or sediment that is susceptible to changes in acidity of the environment (e.g., through acid rain) [12]. Beside acetic acid, nitric acid is currently employed in pH-dependence or pH_{stat} tests to analyze the leachability of an element under acidic conditions since acetic acid is a weak acid that may form complexes.

For the sake of harmonization, ammonium-EDTA 0.05 mol$\cdot$L^{-1} and CH_3COOH 0.43 mol$\cdot$L^{-1} were selected as extraction solutions by the Standards Measurement and Testing (SM&T) Program to indicate respectively the "mobilizable/potentially available fractions" or the "acid extractable fractions" of elements in sediments and soils [13]. Since the release of trace elements from soils, sediments and waste materials is strongly influenced by pH, different types of tests are available to assess the release of elements due to changes in pH. While single extractions (e.g., the extraction with CH_3COOH) allow estimating element release at a single pH value, which is determined by the reagent and by the acid neutralizing capacity of the sample, pH_{stat} leaching tests allow investigating element release at a pre-defined pH value. In the standard pH_{stat} test [14], the release of trace elements is only assessed at one moment in time (after 48 h), which does not allow addressing the kinetics of the release of elements [14]. Hence, a detailed batch leaching test where the pH is kept at a preset value by using an automatic titration over time in conjunction with the analysis of the leachates at various times allows determining the release kinetics of elements due to changes in pH. This kind of test provides a more detailed assessment of time-dependent leaching behavior of trace elements.

Element leaching from BA has been the subject of many studies in different countries. An overview of some selected papers dealing with trace element leaching from BA which are relevant for the present study is provided in Table 1. It was found that leaching of trace elements from BA is pH-dependent and affected by accelerated ageing [15–17]. In developing countries, BA is sometimes treated differently from what is stipulated in legislation. For instance, some hazardous waste incinerators leave the BA covered at a dump site next to the plants (e.g., Ghana [18] and Vietnam). As for most developing countries, rapid industrial development has led to an increase in the generation of various types of industrial waste in Vietnam in recent years. However, data on the composition of BA from hazardous waste incinerators in Vietnam, as well as the potential release of trace elements from these materials are rather scarce, even though such studies provide essential information to select the most sustainable management options for this kind of waste, and eventually also contribute to the protection of the environment.

Table 1. Total concentrations (dry weight basis) of major and trace elements, mineral composition, organic carbon content (OC), and pH, and of studied BA compared to other studies. Average ± standard deviation of 2 replicates.

Type of Bottom Ash	Unit	This Study [19]		[20]	[21]	[22]	[18]
		Hazardous Waste Incinerators		Industrial Wastes	Municipal Solid Waste Incinerators (MSWI)	Coal, Peat and Forest Residues	Hospital Medical Waste Incinerators
Element		AS1 *	AS2 **				
Al	%	6.06 ± 0.33	2.90 ± 0.12	-	6.4	1.22	
As	$mg{\cdot}kg^{-1}$	7 ± 1	77 ± 75	21.4	21	3.9	
Ca	%	1.91 ± 0.02	3.85 ± 0.18	6.26	9.7	-	
Cd	$mg{\cdot}kg^{-1}$	1 ± 0.03	2 ± 0.1	<0.3	14	<3.0	7.54
Co	$mg{\cdot}kg^{-1}$	928 ± 23	63 ± 11	7.8	67	2.9	
Cr	$mg{\cdot}kg^{-1}$	573 ± 45	804 ± 185	50.6	1158	10.9	99.30
Cu	$mg{\cdot}kg^{-1}$	1126 ± 196	818 ± 188	33.5	7743	16.9	
Fe	%	3.92 ± 0.15	23.82 ± 3.53	-	8.9	0.84	
K	$mg{\cdot}kg^{-1}$	3800 ± 200	2400 ± 2	2910	9000	-	
Mg	$mg{\cdot}kg^{-1}$	1900 ± 6	7900 ± 617	5850	15,000	-	
Mn	$mg{\cdot}kg^{-1}$	377 ± 3	3597 ± 259	1450	1000	425.0	
Mo	$mg{\cdot}kg^{-1}$	278 ± 7	42 ± 1	1.1	99	<1.0	
Ni	$mg{\cdot}kg^{-1}$	1373 ± 77	233 ± 22	24.3	356	6.3	
P	$mg{\cdot}kg^{-1}$	2000 ± 47	1000 ± 41	3430	4000	-	
Pb	$mg{\cdot}kg^{-1}$	63 ± 4	817 ± 85	5.1	1022	<3.0	143.80
S	$mg{\cdot}kg^{-1}$	1130 ± 56	3820 ± 348	1580	4950	59.4	
Zn	$mg{\cdot}kg^{-1}$	930 ± 2	1461 ± 291	340	7732	256.0	16,417.69
OC	%	3.24 ± 0.01	0.95 ± 0.04	-		<0.5	
pH		8.64 ± 0.05	9.40 ± 0.70	-			
Calcite	%	1.0	3.6	-	0.5		
Corundum	%	6.9	-	-	4.5		
Hematite	%	-	7.2	-	1.9		
Magnetite	%	3.1	13.1	-	4.2		
Mullite	%	1.9	-		-		
Quartz	%	10.3	7.8		31.4		
Rutile	%	-	2.6		0.8		

* = BA sample 1; ** = BA sample 2.

In the present study, BA samples that were previously partly characterized using chemical, mineralogical, and physical approaches [19] were further studied through a complementary extraction/leaching approach with focus on the potential release of trace elements under influence of acidification and complexation. Single extraction with ammonium-EDTA ($0.05\ mol{\cdot}L^{-1}$) and CH_3COOH ($0.43\ mol{\cdot}L^{-1}$), which are commonly applied to soils and sediments, were performed. The results of the single extractions were compared with the release of elements during pH_{stat} titration leaching at pH 4. This pH was chosen since it represents the worst-case scenario when the samples would become exposed to acidification at the disposal site (e.g., acid rain conditions, mixing with other acid wastes, etc.). The comparability of different extractions and leaching tests is determined by the main characteristics of the tests (the chemical reagent used, the duration of the tests, the liquid/solid (L/S) ratio). The tests used in this study are all conventional batch tests in which the leaching solution is not renewed, and the mixing is performed over a relatively short time period (hours to days) with the aim of reaching equilibrium conditions. The comparison of potential mobility of trace elements by different and extractions is not completely straightforward since operational conditions and reagents are different. The estimation of the potential trace elements' mobility is thus operationally defined by the extracting agents used [23].

The aim of the study is to investigate the usefulness of a different extraction/leaching approach focusing on the potential release of trace elements from BAs from hazardous waste incinerators under influence of acidification and complexation. The kinetic of release of the various elements under acidic conditions is also discussed and related to the solid-phase composition of the BA. It is not the purpose of the present paper to investigate the metal recycling from BA, since this would also require a detailed economic and technological assessment. However, strategies for disposal or recycling of solid waste are also often based on the understanding of the leaching behavior of these materials [10]. In this study, the information deduced from the different methods was compared and evaluated. Results of the present study are helpful for the first steps in the evaluation the potential treatment strategies for these BAs. Results from extraction methods applied in the present study might be indicative for evaluation of options for metal recovery. Additionally, the investigation of release kinetics of trace elements under acidic conditions by a detailed batch leaching test is helpful to gain a better understanding of acid neutralization capacity as well as the solid-phase composition of the BA. Since this is just the first estimation for the usefulness of the CH_3COOH and ammonium-EDTA extraction, we use standardized conditions, using more diluted extraction agents and high liquid/solid (L/S) ratios instead of severe conditions with more concentrated agents and low L/S ratios. Moreover, mild extraction conditions are used since the chemical consumption should be minimized during BA washing for multiple reasons. Several studies have dealt with bottom ash from municipal solid waste incineration; however, bottom ash from hazardous waste incineration has not been the subject of many studies before. This paper will improve our understanding of the release of trace elements under different external factors that BAs may encounter during their processing or management. Extraction and leaching methods applied in the present study are standardized tests that have been developed for soils and sediments. Developing novel extractions for waste materials is not the purpose of the present study. However, the approach used in this study is useful to evaluate the fraction of metals that can be leached out from BA, in view of metal recovery and treated BA disposal/use. This approach is necessary for the development of appropriate waste management options, especially in countries facing inefficient waste and waste water treatment technologies, resulting in residual waste materials with considerable concentrations of valuable elements.

2. Materials and Methods

2.1. Material Characterization

The BA samples in the present study were collected from two hazardous waste incinerators in Ho Chi Minh City, South Vietnam. At the time of sampling, source material from the plant from which sample AS1 was collected included sludge from the waste water treatment of textile and printing ink production factories, and chemical containers. The general input for the incineration facility where AS2 was collected was sludge from waste water treatment of a textile dying factory, cloths containing chemical and lubricants, and out of date chemicals. After sampling, the BA materials were stored in plastic bags and brought to KU Leuven, Belgium for further treatment and laboratory analysis. More details about sampling and sample pretreatment and the determination of chemical and mineralogical composition of the samples can be found elsewhere [19]. In general, the composition of BA varies greatly and depends on the input material and the applied incineration technology. The BA samples in the present study were collected from two-stage incinerators having a quite low combustion temperature (550–650 °C) which can influence the complete burning of organic matter and the forming of metal oxides. Moreover, inputs of hazardous waste incinerators vary widely depending on the type of plants.

Some general characteristics (pH, total elemental concentrations, organic carbon content and mineralogical components) of these samples compared to other studies are summarized in Table 1. The fraction >2 mm determined by sieving was 12% for both samples. Mineralogical characterization was carried out by X-ray diffraction (XRD). According to XRD results, the studied BAs were mainly composed of oxides and Ca-, and Si-minerals. Magnetite (Fe_3O_4), quartz (SiO_2), and calcite ($CaCO_3$) were found in both samples. Corundum (Al_2O_3) and mullite ($Al_6Si_2O_{13}$) were detected in sample AS1 and hematite (Fe_2O_3) and rutile (TiO_2) were found in sample AS2 (Table 1). Total element concentrations in these samples were compared to the composition of other BA from different sources including industrial wastes, MSWI, coal, peat and forest residues, and hospital medical waste incinerators. Both samples were characterized by a low concentration of Ca and P compared to the composition of other BA. Iron concentration in AS2 was much higher compared to its concentration in AS1 and other BA (Table 1). The total element concentrations of some trace elements in the two BA samples were compared to the Flemish limit values for recycling of granular material in construction applications to evaluate whether the studied BAs would be suitable for being used in construction applications. Results indicated that Cu (sample AS1 and AS2), Ni (sample AS1) and Zn (sample AS2) exceeded the Flemish limit values for recycling of granular material in construction applications [24]. Moreover, according to the Vietnamese National Technical Regulation on Hazardous Waste Thresholds, As and Pb (sample AS2) exceeded the limit values [25]. This might raise concerns regarding the potential hazard of using this BA as a construction material if no pretreatment is applied.

2.2. Extractions/Leaching Test

Extractions and leaching test were performed at room temperature (20 °C). The ammonium-EDTA extraction followed the protocol of the Standards, Measurement and Testing (SM&T) Program [26]. Ammonium-EDTA 0.05 mol·L^{-1} solution was added to the dry sample with a liquid/solid (L/S) ratio of 10 (L·kg^{-1}). The suspension was shaken for 1 h in a reciprocal shaker, centrifuged (3500 rpm, 10 min), decanted and filtered (0.45 µm, Chromafil® PET-45/25, Macherey, Düren, Germany). The acid extractable fraction (generally considered to consist of exchangeable elements and elements bound to carbonates) was also based on the procedure described by the SM&T programme using an extraction with CH$_3$COOH 0.43 mol·L^{-1} [27]. The CH$_3$COOH 0.43 mol·L^{-1} solution was added to the dry sample to obtain a liquid/solid (L/S) ratio of 40 (L·kg^{-1}). The suspension was shaken for 16 h in a reciprocal shaker, centrifuged (3500 rpm, 10 min), decanted off and filtered (0.45 µm, Chromafil® PET-45/25, Macherey, Düren, Germany). After extraction, the pH of the extracts was measured with a pH Hamilton single-pore electrode.

The pH$_{stat}$ leaching test was carried out employing an automatic multititration system (Titro-Wico Multititrator, Wittenfield and Cornelius, Bochum, Germany). 80 g of dried sample was put in an Erlenmeyer flask together with 800 mL of Milli–Q water (L/S ratio = 10 L·kg^{-1}). Element release was measured at regular time intervals (0, 1, 3, 6, 12, 24, 48, 72 and 96 h). Bottom ash samples were first shaken for 30 min at the natural BA-pH (without acid addition) before the effective pH$_{stat}$- experiment was started. A sample of 5 mL of the suspension was taken over a filter (0.45 µm, Chromafil® PET-45/25, Macherey-Nagel GmbH & Co. KG, Düren Germany) by means of a syringe attached to a flexible tube at regular time intervals. This pH$_{stat}$ test corresponds to the CEN/TS 14997 test [14], except that the leachate is not only sampled after 48 h, but at regular time intervals. Moreover, in the present study, the pH$_{stat}$ tests were only performed at pH 4. The objective of this test was to examine the kinetic release of trace elements under acidic conditions, while the pH of the suspension (bottom ash suspended in water) is kept at a constant value (pH = 4) by continuous titration with HNO$_3$ (1 mol·L^{-1}). Due to the variety of processes occurring at landfills or open dumps such as sulfide oxidation, microbial activity, acidic deposition and reaction with atmospheric CO$_2$, the pH of ash in landfills may drop to a value as low as 3 to 5 [28]. Hence pH 4 was chosen to address a worst-case scenario of acidification.

2.3. Analysis and Quality Control

Single extractions with ammonium-EDTA 0.05 mol·L^{-1} and CH$_3$COOH 0.43 mol·L^{-1} were done in duplicate and the results are presented as a mean value. Blank solutions were also inserted in each batch of extractions. A certified reference material (BCR 483) was also analyzed for quality control (Table 2).

Table 2. Comparison of the results of the ammonium-EDTA 0.05 mol·L^{-1} extraction and CH$_3$COOH 0.43 mol·L^{-1} extraction of BCR 483 (average ± standard deviation of 2 replicates) and certified values.

Element	Ammonium-EDTA 0.05 mol·L^{-1} (This Work)	Ammonium-EDTA 0.05 mol·L^{-1} (Certified Values)	Acetic Acid 0.43 mol·L^{-1} (This work)	Acetic Acid 0.43 mol·L^{-1} (Certified Values)
Cd	20.7 ± 0.7	20.4 ± 1.3	19.6 ± 0.5	18.3 ± 0.6
Cr	37.6 ± 9.8	28.6 ± 2.6	26.3 ± 1.6	18.7 ± 1.0
Cu	192 ± 7	215 ± 11	38.2 ± 1.2	33.5 ± 1.6
Ni	24.2 ± 1.4	28.7 ± 1.7	25.5 ± 1.5	25.8 ± 1.0
Pb	189 ± 17	229 ± 8	1.9 ± 0.2	3.1 ± 0.25
Zn	529 ± 16	612 ± 19	659 ± 34	620 ± 24

The extracts/leachates were acidified immediately after the experiments, with concentrated HNO$_3$ to bring the pH < 2. The EDTA extracts were kept at 4 °C and not acidified prior to analysis to prevent precipitation of EDTA salts at low pH. Elemental concentrations (Al, Ca, Fe, K, Mg, P, S, As, Cd, Co, Cr, Cu, Mn, Mo, Ni, Pb and Zn) were measured by ICP-OES (Varian 720-ES). A Varian 720-ES instrument supplied with double-pass glass cyclonic spray chamber, concentric glass nebulizer SeaSpray and "extended high solids" torch was used. Solutions were introduced into the spectrometer using the Varian SPS3 sample preparation system. Calibration solutions were prepared from certified multi-element ICP standard stock solutions and from Plasma HIQU (High Quality) single element solutions from CHEM-LAB (Belgium). Blanks were also included in the calibration. All solutions were prepared from 18 MΩ·cm^{-3} ultra-pure water supplied from Millipore system and stabilized with ultra-pure nitric acid (CHEM-LAB). Sensitivity, linear dynamic range, and freedom from spectral interferences were taken into consideration during wavelength selection for each element. Each measurement was carried out with three replicates.

3. Results

In this study, although results and discussion mainly focus on trace elements (As, Cd, Co, Cr, Cu, Ni, Mo, Pb, and Zn), major elements (Al, Ca, Fe, K, Mg, P, S, and Si) are sometimes mentioned because of their relevance for interpretation of release mechanisms of elements of interest.

3.1. Single Extractions

The results of ammonium-EDTA and CH$_3$COOH 0.43 mol·L^{-1} extractions for the studied BA samples are displayed in Table 3. Extractability is expressed in percent of an element extracted with ammonium-EDTA and CH$_3$COOH extraction relative to its total concentration in the (solid) sample. Ammonium-EDTA extraction was used to determine the potential mobility of trace elements as a consequence of complexation and used as an estimation of the "pool" of a specific element that can deliver elements from the solid phase to the solution [13]. Moreover, it can also give a rough indication on the bioavailability of some trace elements and it is sometimes used to assess the availability of trace elements to plants [29,30].

Among the examined trace elements, Cd (in AS1) and Pb (in AS2) showed the highest EDTA extractability (27 and 31% of the total content in the samples, respectively) while EDTA extractability of some other trace elements such as Co (3%), Ni (2–3%), Cr (0.2–0.6%) and Mo (10%) is rather similar despite of their difference in total concentrations in both samples (Table 3).

Table 3. Extractability (as% of total concentration) of elements from ammonium-EDTA and acetic acid extractions.

Element	AS1 (%)		AS2 (%)	
	Major elements			
	EDTA	CH$_3$COOH	EDTA	CH$_3$COOH
Al	0.2	4	1	20
Ca	87	100	45	87
Fe	1	7	0.1	2
K	5	10	8	19
Mg	25	47	7	50
Mn	9	23	1	11
P	13	13	5	12
S	98	100	36	37
	Trace elements			
As	14	< 6	2	< 0.5
Cd	27	65	< 3	< 3
Co	3	17	3	25
Cr	0.6	3	0.2	4
Cu	22	33	11	29
Mo	10	6	10	< 0.9
Ni	2	15	3	27
Pb	9	< 1	31	42
Zn	20	43	6	29

Although Cd reached the highest extractability in AS1, its concentration in the EDTA extract was below limit of quantification (LOQ) in AS2. Arsenic also had a low EDTA-extractable fraction (2%) in sample AS2, but in sample AS1, 14% of the total As concentration was extracted. Similar to As and Cd, Zn also displayed a higher extractability in sample AS1 (20%) compared to sample AS2 (6%). Results of As and Cd were in accordance with Ca and S, since both Ca and S reached a high extractability in AS1 (87–98%), and a slightly lower extractability in AS2 (36–45%). A study of ash from coal combustion, Nugteren (2008) [31] reported that As, Cd, and Mo are belonging to the group of elements which are associated with calcium oxides and sulfates. However, FEG-EPMA analyses of the bottom ash samples indicated that Mo (in AS1) was related to Fe-bearing phases, while As and Cd could not be observed during solid-phase characterization with FEG-EPMA [19]. The extractability of Cu (AS1 and AS2) and Pb (AS2) was quite high compared to other trace elements. The high EDTA extractability of Cu and Pb (respectively 22% (in AS1) and 31% (in AS2) of their total content in the samples) might be explained by the high complexation constants for these two elements with EDTA (log K = 17.8 and 18.3, respectively) [12].

The concentrations of elements extracted with CH$_3$COOH varied between the two BA. Besides Cd (in AS1) and Pb (in AS2), Zn and Cu showed a high extractability (29–65% of their total concentration). In contrast, Cr, As and Mo display a low extractability ($\leq$6%). It should be mentioned that Pb and Cu are characterized by a higher stability for mononuclear monoligand and biligand complex systems with CH$_3$COOH compared to other metals [32,33]. However, the high extractability of Cd, Pb, Zn, and Cu can be a combination of both high stability of the acetate complexes and the drop of pH during extractions. In both samples, Co and Ni were released in similar amounts (15–17% for AS1 and 25–27% for AS2, respectively) possibly because they originate from the same host phases, namely Fe-alloys or Fe-oxides [19].

For both single extractions, Ca and S (in AS1) showed the highest extractability among the examined elements. Ca and S in AS1 were totally extracted during the CH$_3$COOH 0.43 mol·L^{-1} extraction. In general, major cations present in the solid samples may be one of the factors affecting trace element extraction efficiency due to their competition to form complex compounds with EDTA [34].

The dissolution of calcite can consume EDTA in calcareous soils, lowering the extraction efficiency for trace elements [35]. In the present study, important amount of Ca (87%) was extracted with ammonium-EDTA (Table 3), possibly affecting the extraction efficiency of the reagent.

3.2. Acid Neutralization Capacity (ANC) and Trace Element Release at pH 4

The $ANC_{pH4, 96 h}$ (i.e., the amount of acid added to maintain a pH of 4 until 96 h after the start of the pH_{stat} titration) of sample AS2 (936 mmol·kg^{-1}), was nearly double to the $ANC_{pH4, 96h}$ of sample AS1 (510 mmol·kg^{-1}). The higher amount of calcite as determined by XRD in sample AS2 (3.2%) most likely explains the difference in ANC between both samples. Although the pH change during the extraction with CH_3COOH 0.43 mol·L^{-1} might provide an indication of the ANC of the two BA samples, the difference in the final pH of the CH_3COOH extracts was not that high (final pH of 3.26 and 3.44 for AS1 and AS2 respectively). The initial pH of the CH_3COOH solution was 3.02. The reason for this small difference in pH, despite the important difference in ANC, might be the short duration of the CH_3COOH extraction test (16 h) in which the slow buffering reactions are not fully considered [36]. Hence, pH_{stat} leaching tests, performed for a longer period (96 h in the present study) likely allow a better estimation of ANC from BA samples. It should also be mentioned that weathering of BA (natural or artificial) is responsible for increasing the buffering capacity of the BA [15]. Several studies have shown that leaching for several metals appears to be less important than from fresh BA after weathering [37,38]. However, a study about the carbonation (artificial weathering) of BA from municipal solid waste incinerator, Van Gerven et al. (2005) [39] reported an increase in the leaching of Cr and a constant leaching of Mo and Sb after carbonation of BA. Therefore, ANC of a BA is very important and should be better estimation to investigate the release of trace metal upon external addition of H$^+$.

The evolution of ANC and the release of Ca with time in both samples during pH_{stat} leaching are displayed in Figure 1.

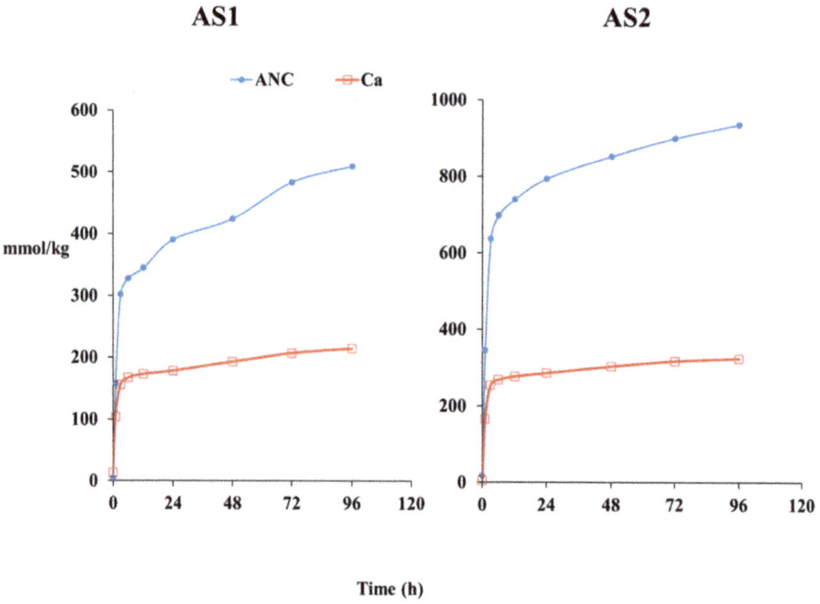

Figure 1. Evolution of ANC and Ca during pH_{stat} leaching test (pH 4).

We performed XRD phase analysis on residual BAs after pH_{stat} leaching (at pH = 4) to assess leaching related to changes in major solid phases. XRD phase analysis on the residual BAs after the pH_{stat} leaching (at pH = 4) showed that some peaks of calcite ($CaCO_3$) decreased in intensity (Figure 2). This suggests that main mineral phases in the BAs were stable at pH 4, except small change was observed for calcite. Possibly, the dissolution of other mineral phases was too small to be detected by the XRD technique.

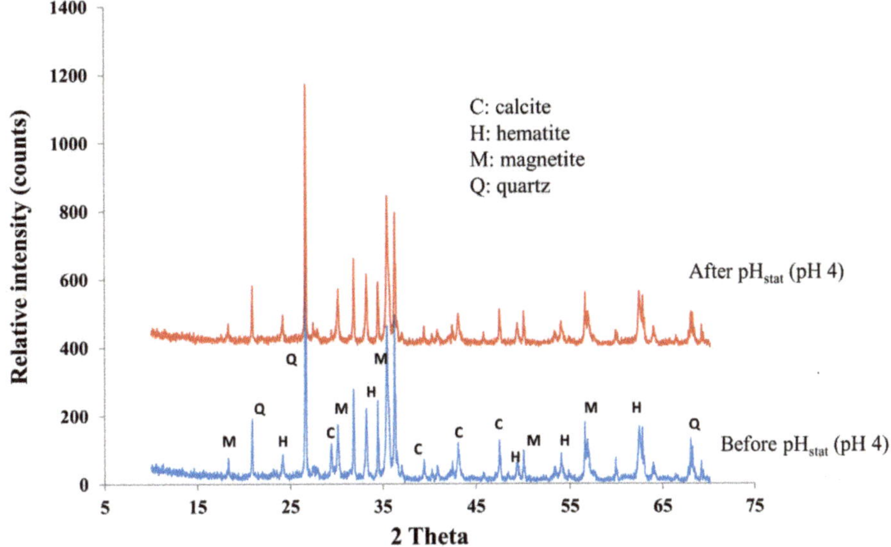

Figure 2. XRD patterns of original sample and sample after the pH_{stat} test (pH = 4, sample AS2).

In the following section, leachability refers to the concentration in the final leachate (after 96 h) expressed in percent of an element leached relative to its total concentration in the (solid) sample (Table 4) except for Mo in sample AS2 since its concentration in the leachates decreased to values below the LOQ from 3 h onward. For the latter, the concentration in the leachate after 1 h was used.

In both BA samples, despite the high total concentration, Al Fe, and P exhibit very low leachability (<0.5%) compared to other major elements, such as Ca, K, Mg, Mn, and S (>2%). This suggests that no significant dissolution of Al-Fe-P containing minerals occurred during the pH_{stat} leaching test. Release of some selected major and trace elements during pH_{stat} leaching are displayed in Figures 3 and 4. Cadmium concentrations in the leachates from both samples were below LOQ likely due to the low total Cd-concentrations ($\leq$1.5 mg·kg^{-1}).

Table 4. Leachability (as% of total concentration) of elements in the pH_{stat} test (after 96 h or, except for Mo in sample AS2, after 1 h).

Heading	AS1 (%)	AS2 (%)
Major Elements		
Al	0.1	0.4
Ca	45	34
Fe	0.3	0.02
K	5	7
Mg	14	14
Mn	5	2

Table 4. *Cont.*

Heading	AS1 (%)	AS2 (%)
P	0.03	0.04
S	54	22
Trace Elements		
As	<1	<1
Cd	<6	<3
Co	3	6
Cr	0.04	0.02
Cu	3	2
Mo	1	0.3
Ni	3	6
Pb	<0.2	1
Zn	9	3

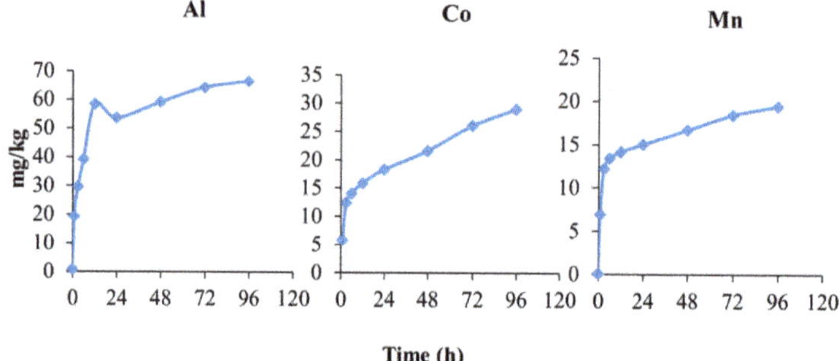

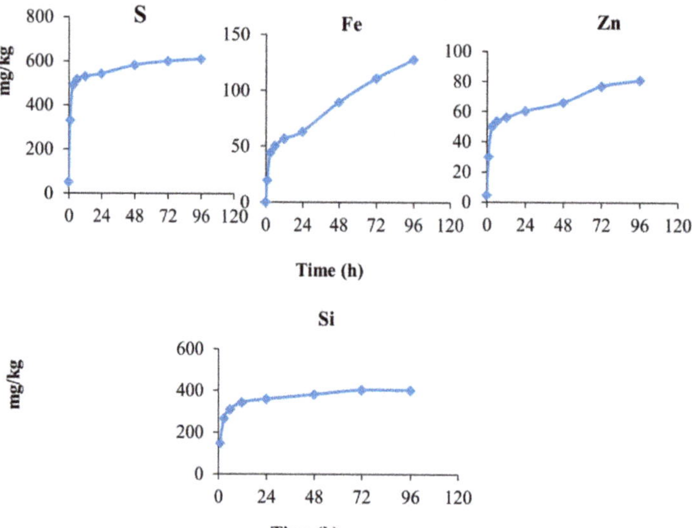

Figure 3. Release of major elements and some selected trace elements from AS1 during pH_{stat} leaching (at pH 4).

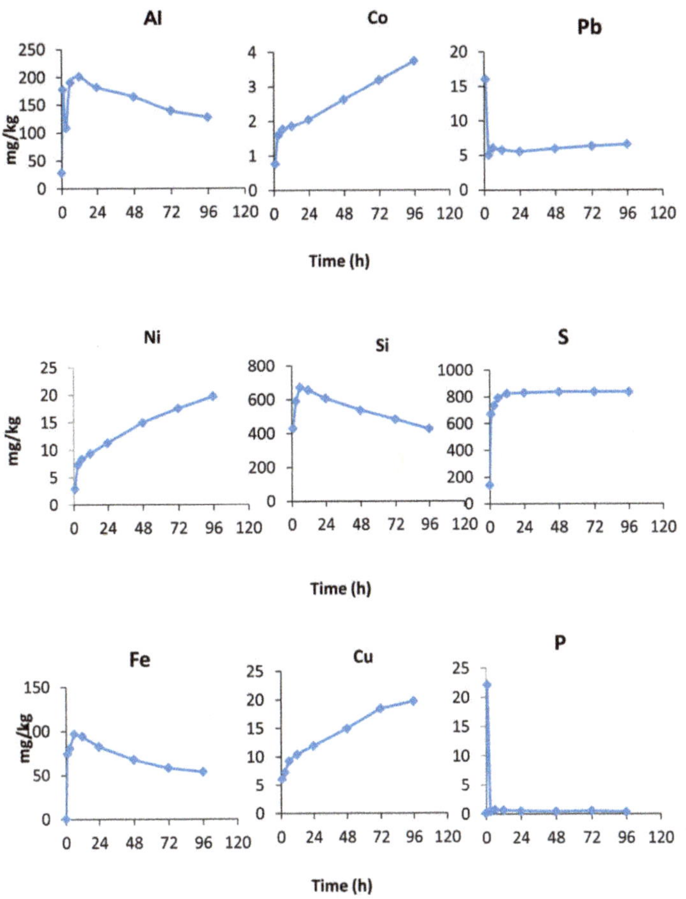

Figure 4. Release of major elements and some selected trace elements from AS2 as a function of time during pH$_{stat}$ leaching (at pH 4).

In the leachate of sample AS1, As and Pb were below LOQs while released Cr concentration was very low (0.04% or 0.3 mg·kg^{-1} was released at pH 4). The leachability of Co, Cu, Mo, and Ni varied between 1–3%. The highest leachability was observed for Zn (9%).

In sample AS2, most of trace elements were only detected in the leachates after 1 h except Cr, Mo and Zn which were already released immediately from the start of the experiment. Arsenic concentrations in the leachate varied just around LOQ while Mo was released at the beginning of the pH$_{stat}$ experiment, but the concentration decreased below LOQ after 3 h of leaching. Nickel exhibits the highest leachability (8%), while other trace elements such as Co, Cu, Pb and Zn show a moderate leachability (1–6%). Chromium exhibited a low leachability (0.02%) despite its high total concentration (804 mg·kg^{-1}). This indicates that most of the trace elements in the BAs do not occur in readily soluble forms, even if the external pH is lowered to a value of 4.

4. Discussion

4.1. Potential Release of Trace Elements Based on Different Extractions/Leaching Test

The results after 48 h of pH_{stat} leaching in this study corresponds to the standard pH_{stat} test CEN/TS 14997 test [14] at pH 4. Therefore, comparing the release of trace elements by the CH_3COOH extraction, ammonium-EDTA extraction and after 48 h of pH_{stat} leaching provides information on the potential release of trace elements in BA samples based on standardized extraction/leaching tests. The release of some selected trace elements from CH_3COOH 0.43 mol·L^{-1} and ammonium-EDTA 0.05 mol·L^{-1} extraction compared to the amount of these trace elements extracted with pH_{stat} leaching test (determined with 48 h) is displayed in Figure 5.

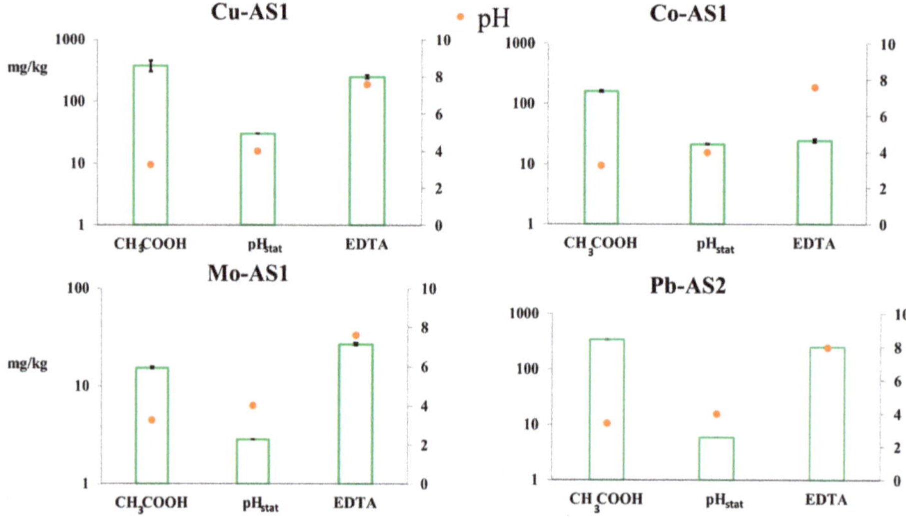

Figure 5. Release of Cu-Co-Mo (in AS1) and Pb (in AS2) after acetic acid 0.43 mol·L^{-1} and ammonium-EDTA 0.05 mol·L^{-1} extraction compared to the amount of these trace elements released by the pH_{stat} leaching test (after 48 h).

Most of trace elements showed the highest extractability with CH_3COOH, while As and Mo were most effectively extracted with ammonium-EDTA. The high amount of trace elements (except As and Mo) that were extracted by CH_3COOH can be explained by the low pH (pH = 3.26–3.44) of the CH_3COOH extract. Acidification has a pronounced effect on the release of trace elements from the studied BAs. Removal or recovery of trace elements from ash by acid leaching has been studied to find out the most effective leaching agent [24,40,41]. However, this recycling option suffers from some drawbacks such as the use of large amounts of acid due to the high ANC of ashes and the generation of wastewater. The more important extraction of As and Mo by ammonium-EDTA compared to CH_3COOH extraction and pH_{stat} leaching at pH 4 is related to the fact that they may occur as oxyanions which are known to be leached more under alkaline conditions than in acidic conditions. The pH of the CH_3COOH extract is 3.26 (AS1) and 3.44 (AS2), whereas the pH of the solution of the pH_{stat} leaching test was continuously kept at 4. Contrarily, the pH of the ammonium-EDTA extract is neutral to slightly alkaline with values of 7.59 (AS1) and 7.98 (AS2). Some trace elements such as Cd, Cu, Ni, Pb, and Zn can form complexes with dissolved organic carbon (DOC) which may enhance the leaching of these elements. Increased Cu leaching from MSWI bottom ash by DOC complexation was observed by [42,43]. Cu in AS1 exhibited a slightly higher mobility at alkaline pH values (9–11) compared to neutral pH, which can be related to the fact that Cu forms organometal complexes with dissolved organic matter in the leachates of sample AS1 at high pH [19]. Unfortunately, data of DOC in the leachates are not available; hence, the possible effects of complexation with DOC could not be evaluated in the present study.

The relatively high-potential mobility of Cu, Pb and Mo deduced from the ammonium-EDTA extraction, in combination with elevated total concentrations of Cu and Mo (in AS1) and Pb (in AS2) indicate a higher possibility of recovery of these metals from the BA. Up to 250 mg Cu/kg and 27 mg Mo/kg (in AS1) and 251 mg Pb/kg (in AS2) were released by the ammonium-EDTA extraction.

Removal of contaminants from ashes by washing with chelating agents (e.g., citrate buffer, EDTA or oxalate buffers) has been suggested since chelating agents can remove significant amounts of contaminants present in mobile forms in the outer layer of the ash particles [22]. Pre-washing with water, increasing the concentration of EDTA and increasing the extraction time even increase the extractability [41]. A high extractability of Cu (100%), Pb (94%) and Zn (40%) was observed from the fly ash with pre-water washing followed by a 24 h extraction with EDTA 0.1 mol·L^{-1} (end pH 8.2 and L/S 5.0 L·kg^{-1}) [41]. However, it is worth noting here that results for fly ash are not directly transferable to the BA samples in this work due to the differences in particle size, chemical composition, and mineralogy. Extraction efficiency obtained in the present study can probably be improved, by increasing the concentration of the EDTA solution, adapting the extraction time, L/S ratio, temperature, etc. It was not the purpose of the present study to investigate the optimal condition for maximal element recovery from the BA samples. However, the result show that the extractions used in the present study can be used for a relatively cheap and fast screening of the potential of element recovery from waste materials.

4.2. Kinetics of Trace Element Release during Leaching Test at pH 4

Understanding the kinetics of trace elements release is important for predicting the environmental risks associated with these elements over time. Kinetic leaching experiments performed on solid materials have shown that element leaching can be described by two steps, representing a fast release process followed by a slow process [44]. In this study, time-dependent leaching behavior of trace elements could be divided into three types (Figure 6), considering fast and slow release processes.

Type 1 includes elements which are released slowly, and steady state is not reached (e.g., Co and Ni in both samples, as well as Cu in AS2 and Zn in AS1). This type of element release is related to the desorption of elements that are strongly bound to solid phases or to the slow dissolution of solid phases. Similar leaching patterns of Co and Ni from both samples support the hypothesis that they might come from the same host phase or that they are retained by similar binding mechanisms. Compositional analysis by EPMA-EDS indicated that Fe, Co, and Ni co-existed in the analyzed spots in sample AS1 [19]. The slow release of Ni and Co is most likely due to the fact that Ni and Co are occluded in stable phases such as Fe-alloys and Fe-oxides and thus showed a slow mobilization under acidic environmental conditions. Similar time-dependent leaching patterns of Fe, Co, and Ni (Figure 3) from sample AS1 support this hypothesis. Although the Fe content of sample AS2 was much higher (24%) compared to that of sample AS1 (4%), it should be mentioned here that the total concentration of Co and Ni in AS2 is not that high (63 mg Co/kg and 233 mg Ni/kg), and Ni and Co did not show a release pattern similar to Fe in sample AS2. In sample AS1, Zn occurs in Si-rich phases which might be related to quartz, mullite or amorphous phases [19]. The release pattern of Zn during the pH$_{stat}$ leaching test in this study (Figure 3) is quite similar to the release of Zn from glass phases as observed by [45]. Moreover, quartz and mullite are known as stable phases under neutral and mildly acid conditions while Zn was observed to leach up to 9% in pH$_{stat}$ leaching test (Table 4). Therefore, Zn might be associated with amorphous or poorly crystalline phases, which are more easily dissolved than stable phases.

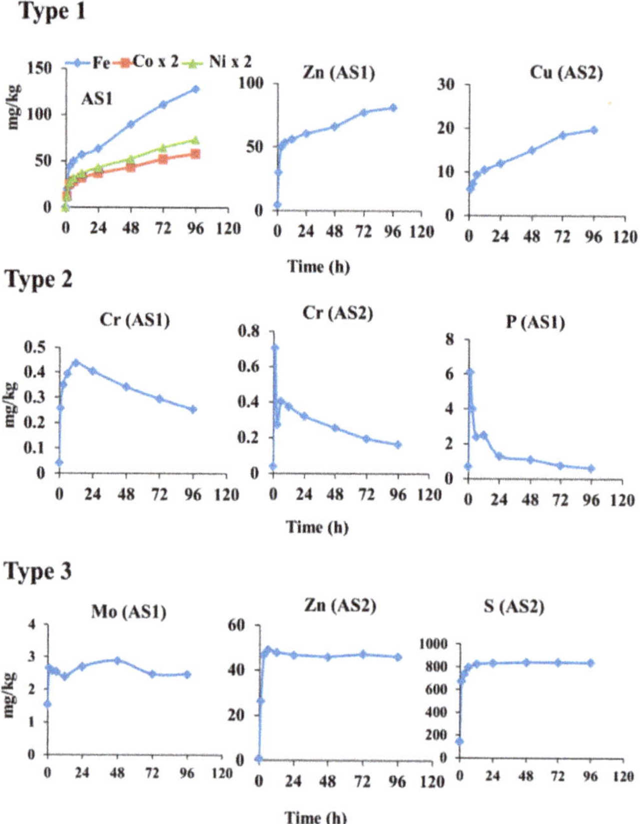

Figure 6. Release of Fe (AS1), P (AS1), S (AS2) and some selected trace elements illustrating three different types of time-dependent leaching behavior during pH$_{stat}$ leaching (at pH 4). Co and Ni concentrations were multiplied by a factor of 2 for a better visualization.

Release of elements according to "type 2" is characterized by an initial release at the beginning of the pH$_{stat}$ leaching test, followed by a decrease in dissolved concentrations over time. This type of release is related to precipitation reactions or re-adsorption onto solid phases (e.g., Cr and P in both samples). This is the case for example when elements forming oxyanions (e.g., chromate) are re-adsorbed on positively charged reactive surfaces at acidic pH. This phenomenon could not be assessed in the single extraction test (e.g., CH$_3$COOH extraction) due to the short duration of the extraction and the fact that only the final leachate was analyzed.

The last type of leaching pattern ("type 3") is related to elements which are released rapidly, and steady state seems to be reached after more or less 6 h (e.g., Cu–Mo in AS1 and Zn in AS2). Mo was also observed in Fe-rich phases in sample AS1 by FEG-EPMA; however, its leachability was higher compared to other elements associated with Fe-rich phases. The release kinetics of Mo (AS1) during pH$_{stat}$ leaching is also faster than Co and Ni (Figure 3) suggesting that this element might not be incorporated into crystalline phases but is distributed on the surface of Fe-rich phases. Although Zn-bearing phases were not identified by XRD and FEG-EPMA in sample AS2 [19], the release behavior of Zn during pH$_{stat}$ leaching was similar to S (Figure 6), suggesting that Zn may also exist in S-rich phases or that Zn and S are bound to a solid phase in a similar way.

It can be deduced from the pH$_{stat}$ test that release mechanism of elements from BA are probably related to the surface processes on the solid material such as desorption and re-adsorption, or to solubility of mineral phases. When pH is lower, desorption happens since the presence of hydrogen ions displaces metals that bound on the solid surfaces [46]. Precipitation or re-adsorption onto solid phases is observed for some elements which form oxyanions such as Cr and P. It was concluded by [47] that solid solution formation is frequently believed to be a controlling mechanism for oxyanion leaching, for example for Cr (VI) in MSWI bottom ash.

5. Conclusions

In the present study, two BA samples characterized by a different mineralogical and chemical composition and with a different ANC were investigated applying some standardized extractions/leaching tests (CH$_3$COOH extraction, ammonium-EDTA extraction, and pH$_{stat}$ leaching test). A high extractability of trace elements (except As and Mo) was observed in the CH$_3$COOH extraction which also resulted in the lowest pH among the experiments. pH$_{stat}$ leaching allows investigating the kinetics of element release under acidic conditions and can be linked to the solid-phase characteristics of some selected trace elements. Nickel and Co are occluded in stable phases such as Fe-alloys and Fe-oxides while Zn might be associated with amorphous or poorly crystalline phases. Moreover, Zn may also exist in S-rich phases. Based on the results obtained in this study, some preliminary treatment to remove trace elements (e.g., by washing) or immobilize trace elements from the ash should be applied to the studied BAs before landfilling. This is necessary to avoid contamination of the environment, both as consequence of the high total concentrations and because of the relatively high mobility (as deduced from single extractions) of Cu-Mo (in AS1) and Pb (in AS2). Additionally, the potential recovery of some metals should be explored. The extractions and leaching tests applied in this study, which are generally used for preliminary screening in assessment of elemental mobility, could also be used for a first estimate of the potential recovery of valuable elements from waste materials. The approach used in this study is helpful for the development of appropriate waste management options, especially in countries facing inefficient waste and waste water treatment technologies, resulting in residual waste materials with considerable concentrations of valuable elements.

Author Contributions: T.T.T.D. designed and performed the experiments. T.T.T.D. interpreted the results with significant contribution of Valérie Cappuyns. T.T.T.D. wrote the paper and significant contribution to writing was made by V.C. and R.S.E.V. performed ICP-OES analysis.

Funding: This research was funded by [BTC] grant number [10VIE/0015].

Acknowledgments: We acknowledge the members of Geology Division, Department of Earth and Environmental Sciences, KU Leuven for their support. We also thank the personnel at hazardous waste treatment plants for providing samples.

Conflicts of Interest: The authors declare no conflict of interest.

References

1. Mineral Commodity Summaries 2013. Available online: https://minerals.usgs.gov/minerals/pubs/mcs/2013/mcs2013.pdf (accessed on 10 July 2018).
2. Hunt, A.J.; Farmer, T.J.; Clark, J.H. Elemental sustainability & the importance of scarce element recovery. In *Element Recovery and Sustainability*; Hunt, A.J., Ed.; RSC Publishing: Cambridge, UK, 2013; Volume 22, pp. 1–28.
3. Chlopecka, A.; Bacon, J.R.; Wilson, M.J.; Kay, J. Forms of cadmium, lead and zinc in contaminated soils from southwest Poland. *J. Environ. Qual.* **1996**, *25*, 69–79. [CrossRef]
4. Kabata-Pendias, A.; Mukherjee, A.B. *Trace Elements from Soil to Human*; Springer: Berlin/Heidelberg, Germany; New York, NY, USA, 2007; pp. 1–549.
5. Veli, S.; Kirli, L.; Alyuz, B.; Durmusoglu, E. Characterization of bottom ash, fly ash, and filter cake produced from hazardous waste incineration. *Polish J. Environ. Stud.* **2008**, *17*, 139–145.

6. Cho, J.H.; Eom, Y.; Park, J.M.; Lee, S.B.; Hong, J.H.; Lee, T.G. Mercury leaching characteristics of waste treatment residues generated from various sources in Korea. *Waste Manag.* **2013**, *33*, 1675–1681. [CrossRef] [PubMed]

7. CEWEP. Bottom Ash Fact Sheet 2016. Available online: http://www.cewep.eu/wp-content/uploads/2017/09/FINAL-Bottom-Ash-factsheet.pdf (accessed on 4 June 2017).

8. Nowak, B.; Rocha, S.F.; Aschenbrenner, P.; Rechberger, H.; Winter, F. Heavy metal removal from MSW fly ash by means of chlorination and thermal treatment: Influence of the chloride type. *Chem. Eng. J.* **2012**, *179*, 178–185. [CrossRef]

9. Pani, G.K.; Rath, P.; Maharana, L.; Barik, R.; Senapati, P.K. Assessment of heavy metals and rheological characteristics of coal ash samples in presence of some selective additives. *Int. J. Environ. Sci. Technol.* **2016**, *13*, 725–731. [CrossRef]

10. Kosson, D.S.; Van der Sloot, H.A.; Sanchez, F.; Garrabrants, A.C. An integrated framework for evaluating leaching in waste management and utilization of secondary materials. *Environ. Eng. Sci.* **2002**, *19*, 159–204. [CrossRef]

11. Zou, Z.; Qiu, R.; Zhang, W.; Dong, H.; Zhao, Z.; Zhang, T.; Wei, X.; Cai, X. The study of operating variables in soil washing with EDTA. *Environ. Pollut.* **2009**, *157*, 229–236. [CrossRef] [PubMed]

12. Sahuquillo, A.; Rigol, A.; Rauret, G. Overview of the use of leaching/extraction tests for risk assessment of trace metals in contaminated soils and sediments. *Trends Anal. Chem.* **2003**, *22*, 152–159. [CrossRef]

13. Ure, A.M. 1996 Single extraction schemes for soil analysis and related applications. *Sci. Total Environ.* **1996**, *178*, 3–10. [CrossRef]

14. CEN/TS 14997. *Characterization of Waste—Leaching Behavior Tests—Influence of pH on Leaching with Continuous pH-Control*; CEN (Comité Européen de Normalisation): Brussels, Belgium, 2006.

15. Polettini, A.; Pomi, R. The leaching behavior of incinerator bottom ash as affected by accelerated ageing. *J. Hazard. Mater.* **2004**, *113*, 209–215. [CrossRef] [PubMed]

16. Cornelis, G.; Van Gerven, T.; Vandecasteele, C. Antimony leaching from MSWI bottom ash: Modelling of the effect of pH and carbonation. *Waste Manag.* **2012**, *32*, 278–286. [CrossRef] [PubMed]

17. Van Zomeren, A.; van der Sloot, H. *Systematic Leaching Behaviour of Worldwide MSWI Bottom Ashes in Spite of Their Variability in Content Sustainable Landfilling*; ECN Report; CISA Publisher: Padova, Italy, 2013.

18. Adama, M.; Esena, R. Heavy metal contamination of soils around a hospital waste incinerator bottom ash dumps site. *J. Environ. Pub. Health* **2016**, *2016*, 1–6. [CrossRef] [PubMed]

19. Dung, T.T.T.; Vassilieva, E.; Golreihan, A.; Phung, N.K.; Swennen, R.; Cappuyns, V. Potentially toxic elements in bottom ash from hazardous waste incinerators: an integrated approach to assess the potential release in relation to solid-phase characteristics. *J. Mater. Cycles Waste Manag.* **2017**, *19*, 1194–1203. [CrossRef]

20. Nurmesniemi, H.; Poykio, R.; Kuokkanen, T.; Ramo, J. Chemical sequential extraction of heavy metals and sulphur in bottom ash and in fly ash from a pulp and paper mill complex. *Waste Manag. Res.* **2008**, *26*, 389–399. [CrossRef] [PubMed]

21. Bayuseno, A.P.; Schmahl, W.W. Understanding the chemical and mineralogical properties of the inorganic portion of MSWI bottom ash. *Waste Manag.* **2010**, *30*, 1509–1520. [CrossRef] [PubMed]

22. Manskinen, K.; Pöykiö, R. Comparison of the total and fractionated heavy metal and sulphur concentrations in bottom ash and fly ash from a large-sized (120 MW) power plant of a fluting board mill. *Chemija* **2011**, *22*, 46–55.

23. Cappuyns, V.; Swennen, R. The application of pH(stat) leaching tests to assess the pH-dependent release of trace metals from soils, sediments and waste materials. *J. Hazard. Mater.* **2008**, *158*, 185–195. [CrossRef] [PubMed]

24. Van Gerven, T.; Cooreman, H.; Imbrecht, K.; Hindrix, K.; Vandecasteele, C. Extraction of heavy metals from municipal solid waste incinerator (MSWI) bottom ash with organic solutions. *J. Hazard. Mater.* **2007**, *140*, 376–381. [CrossRef] [PubMed]

25. Solid Waste—National Technical Regulation on Hazardous Waste Thresholds. Available online: http://vea.gov.vn/vn/vanbanphapquy/layykiengopy/Documents/QCVNnguongCTNH.doc (accessed on 10 July 2018).

26. Quevauviller, P. Operationally defined extraction procedures for soil and sediment analysis I. Standardization. *Trends Anal. Chem.* **1998**, *17*, 289–298. [CrossRef]

27. Quevauviller, P.; Rauret, G.; Rubio, R.; Lopez-Sanchez, J.F.; Ure, A.; Bacon, J.; Muntau, H. Certified reference materials for the quality control of EDTA-and acetic acid-extractable contents of trace elements in sewage sludge amended soils (CRMs 483 and 484). *Fresenius J. Anal. Chem.* **1997**, *357*, 611–618. [CrossRef]

28. Li, L.Y.; Ohtsubo, M.; Higashi, T.; Yamaoka, S.; Morishita, T. Leachability of municipal solid waste ashes in simulated landfill conditions. *Waste Manag.* **2007**, *27*, 932–945. [CrossRef] [PubMed]

29. Jamali, M.K.; Kazi, T.G.; Arain, M.B.; Afridi, H.I.; Jalbani, N.; Kandhro, G.A.; Shah, A.Q.; Baig, J.A. Heavy metal accumulation in different varieties of wheat (*Triticum aestivum* L.) grown in soil amended with domestic sewage sludge. *J. Hazard. Mater.* **2009**, *164*, 1386–1391. [CrossRef] [PubMed]

30. Soriano-Disla, J.M.; Gómez, I.; Navarro-Pedreño, J.; Lag-Brotons, A. Evaluation of single chemical extractants for the prediction of heavy metal uptake by barley in soils amended with polluted sewage sludge. *Plant Soil* **2010**, *327*, 303–314. [CrossRef]

31. Nugteren, H. Limitations of combustion ashes: 'From threat to profit'. In *Combustion Residues: Current, Novel and Renewable Applications*; Cox, M., Nugteren, H., Janssen-Jurkovičvá, M., Eds.; John Wiley & Sons: West Sussex, UK, 2008; pp. 137–198.

32. Martell, A.E.; Smith, R.M. *Critical Stability Constants. Other Organic Ligands*, 3rd ed.; Plenum Press: New York, NY, USA, 1977.

33. Martell, A.E.; Smith, R.M. *Critical Stability Constants. First Supplement, 5*; Plenum Press: New York, NY, USA, 1982.

34. Kim, C.; Lee, Y.; Ong, S.K. Factors affecting EDTA extraction of lead from lead-contaminated soils. *Chemosphere* **2003**, *51*, 845–853. [CrossRef]

35. Papassiopi, N.; Tambouris, S.; Kontopoulos, A. Removal of heavy metals from calcareous contaminated soils by EDTA leaching. *Water Air Soil Pollut.* **1999**, *109*, 1–15. [CrossRef]

36. Ganne, P.; Cappuyns, V.; Vervoort, A.; Buvé, L.; Swennen, R. Leachability of heavy metals and arsenic from slags of metal extraction industry at Angleur (eastern Belgium). *Sci. Tot. Environ.* **2006**, *356*, 69–85. [CrossRef] [PubMed]

37. Meima, J.A.; van der Weijden, R.D.; Eighmy, T.T.; Comans, R.N.J. Carbonation processes in municipal solid waste incinerator bottom ash and their effect on the leaching of copper and molybdenum. *Appl. Geochem.* **2002**, *17*, 1503–1513. [CrossRef]

38. Kaibouchi, S.; Germain, P. Comparative study of physico-chemical and environmental characteristics of (MSWI) bottom ash resulting from classical and selective collection for a valorization in road construction. In *Progress on the Road to Sustainability, Fifth International Conference on the Environmental and Technical Implications of Construction with Alternative Materials, San Sebastian, Spain, 4–6 June 2003*; Ortiz de Urbina, G., Goumans, H., Eds.; 2003; pp. 645–653.

39. Van Gerven, T.; Van Keer, E.; Arickx, S.; Jaspers, M.; Wauters, G.; Vandecasteele, C. Carbonation of MSWI-bottom ash to decrease heavy metal leaching, in view of recycling. *Waste Manag.* **2005**, *25*, 291–300. [CrossRef] [PubMed]

40. Liu, F.; Liu, J.; Yu, Q.; Jin, Y.; Nie, Y. Leaching characteristics of heavy metals in municipal solid waste incinerator fly ash. *J. Environ. Sci. Health A* **2005**, *40*, 1975–1985. [CrossRef] [PubMed]

41. Fedje, K.K.; Ekberg, C.; Skarnemark, G.; Steenari, B.M. Removal of hazardous metals from MSW fly ash—An evaluation of ash leaching methods. *J. Hazard. Mater.* **2010**, *173*, 310–317. [CrossRef] [PubMed]

42. Olsson, S.; van Schaik, J.W.J.; Gustafsson, J.P.; Kleja, D.B.; van Hees, P.A.W. Copper (II) binding to dissolved organic matter fractions in municipal solid waste incinerator bottom ash leachate. *Environ. Sci. Technol.* **2007**, *41*, 4286–4291. [CrossRef] [PubMed]

43. Arickx, S.; Van Gerven, T.; Boydens, E.; L'Hoëst, P.; Blanpain, B.; Vandecasteele, C. Speciation of Cu in MSWI bottom ash and its relation to Cu leaching. *Appl. Geochem.* **2008**, *23*, 3642–3650. [CrossRef]

44. Meima, J.A.; Comans, R.N.J. Reducing Sb-leaching from municipal solid waste incinerator bottom ash by addition of sorbent minerals. *J. Geochem. Explor.* **1998**, *62*, 299–304. [CrossRef]

45. Kirby, C.S.; Rimstidt, J.D. Interaction of municipal solid waste ash with water. *Environ. Sci. Technol.* **1994**, *28*, 443–451. [CrossRef] [PubMed]

46. Pickering, W.F. Metal ion speciation- soils and sediments (a review). *Ore Geol. Rev.* **1986**, *1*, 83–146. [CrossRef]
47. Cornelis, G.; Johnson, C.A.; Van Gerven, T.; Vandecasteele, C. Leaching mechanisms of oxyanionic metalloid and metal species in alkaline solid wastes: A review. *Appl. Geochem.* **2008**, *23*, 955–976. [CrossRef]

Article

Study of Temperature Fields and Heavy Metal Content in the Ash and Flue Gas Produced by the Combustion of Briquettes Coming from Paper and Cardboard Waste

Harouna Gado Ibrahim [1,*], Salifou K. Ouiminga [2], Arsène Yonli [2], Oumar Sanogo [3], Tizane Daho [2] and Jean Koulidiati [2]

1   Département de Physique, Faculté des Sciences et Techniques, Université Dan Dicko Dankoulodo de Maradi, BP, 465 MARADI, Niger
2   LPCE, Département de Physique, Université Joseph Ki-Zerbo de Ougadougou,
    03 BP 7021 Ouagadougou 03, Burkina Faso; salif0477@yahoo.com (S.K.O.); yarsene@hotmail.com (A.Y.);
    tizane_daho@yahoo.fr (T.D.); j.koulidiati@yahoo.fr (J.K.)
3   IRSAT, Institut de Recherche en Sciences Appliquées et Technologies,
    03 BP 7047 Ouagadougou, Burkina Faso; sanogo_oumar@hotmail.com
*   Correspondence: gado_foga@yahoo.fr

Received: 16 May 2018; Accepted: 12 July 2018; Published: 13 July 2018

Abstract: The present study focused on the combustion of four types of briquettes made from paper and cardboard waste produced in Ouagadougou (Burkina Faso). Rotary and tubular kilns were used to study the combustion. The combustion mean temperatures, nitrogen, phosphorus and potassium (NPK) content in the ash and heavy metals content in the ash and the flue gas were analyzed. The combustion steady phase mean temperatures ranged from 950 °C to 750 °C were obtained according to briquettes type. The temperature favored the transfer of the heavy metal in the flue gas comparatively to the ash mainly for Hg, Cd and Pb. The Pb, Hg and Mn content in flue gas and the ash are higher than their content in the parent wood used for paper production due to the additive during the manufacturing process. The results showed a high content of heavy metal in flue gas produced by combustion of briquette made with office paper and in the ash for the briquette made of corrugated cardboard. Furthermore, the low heavy metal contain in the ash allow their use for soil amendment. However, ash contained a low proportion of NPK (less than 2%) which does not allow their usage as fertilizer alone.

Keywords: briquettes; ash; nitrogen-phosphorus-potassium; heavy metals

1. Introduction

More than two billion people around the world do not have access to modern energy. In Burkina Faso, most of the energy needs are provided by wood (more than 80%). The consequence is an increasing scarcity of wood, hence the need to find other sources of energy. This fully justifies the importance of energy recovery from waste, especially paper and cardboard waste. Indeed, the typology of waste from the city of Ouagadougou, the biggest city of Burkina Faso, reveals that paper/cardboard accounts for 10% of waste, which is more than 20,000 tonnes per year according to the work of Tezanou et al. [1]. This quantity can be valorized energetically or recycled. According to the life cycle analysis conducted by Karna et al. [2], the energy recovery of paper waste contributes less to global warming than recycling. However, the study conducted by Merrild et al. [3] showed that a high-performance recycling process is more beneficial than energy recovery. In the case of Burkina Faso, energy recovery is possible because of the absence of a suitable recycling system. Indeed, a little fraction of waste composed of metal, hard plastic and bottles (made of glass or plastic) are recycled by the informal sector. The waste paper and cardboard ends in the landfills. Thus, waste paper and cardboard can be used as fuels in the form of briquettes as proposed in our previous study [4]. The environmental advantages of replacing wood or fuel oil with paper/cardboard briquettes are important (reducing deforestation, reducing the global warming associated with the use of fuel oil, etc.). In addition to recovering the energy released by the combustion of the briquettes, the ashes obtained can also be used to amend the soil. Note that, these ash quantities can achieve 28% of the mass of the raw paper and cardboard waste [5]. The ash may be used to return nutrients to the soil, thus closing the nutrient cycle and enhancing soil fertility [6,7]. However, the burning of waste paper and cardboard briquettes presents major challenges. Indeed, depending on their nature and their uses, waste paper and cardboard are soiled. In general, the waste paper and cardboard present physical impurities such as staples, sand, plastics, inks... But also chemical and toxic matters as heavy metals [8,9]. The heavy metals are also present in the biomass as a micronutrient [10]. These heavy metals may be present in the paper at the end of the manufacturing process. The chemical additives and ink used during the manufacturing process and the usage of the paper and cardboard are probable sources of heavy metals. Indeed, colorants and other additive used to improved paper color and quality content zinc, cadmium, lead and chromium [11]. The ink is composed up to 30% of pigment which contents many types of heavy metals according to its formulation [12–14]. Thus, during the combustion process the heavy metals can be found in the bottom ash, the fly ash and the flue gas [15]. Heavy metals could cause air, water and soil pollution according to Kovacs et al. [15]. As a result, the combustion and valorization of residual ash are subject to multiple regulations, mainly on the content of heavy metals. The main heavy metals targeted are cadmium (Cd), arsenic (As), lead (Pb) and Zinc (Zn), [16,17]. This imposes the need for the evaluation of energy performances and environmental impacts with a view to energy recovery of briquettes and ash. For this purpose, the ash compositions of many type of biomass were determined by Jenkins et al. [8]. These authors found a great composition variation of biomass ash; furthermore, the manufacturing process could impact the proportion and the composition of ash coming from biomass like paper and cardboard waste.

Waste paper and cardboard are considered for briquetting process by many studies but there is a need to study a possible environmental risk through the heavy metal emission, in the flue gas and ash, produced during the combustion of the briquettes. During the present work, a study of the energy and environmental performance of the combustion of briquettes was carried out. Four types of briquettes obtained in our previous study were used in this study [4]. The study consisted in analyzing the temperature fields, heavy metals in the flue gas and in the ash during the combustion of the four types of briquettes. The possibility of using ash as an amendment to soil was also studied.

2. Materials and Methods

2.1. Characteristics of Briquettes

The paper and cardboard were collected directly in the waste produced in the city of Ouagadougou. The briquettes were produced and characterized in the laboratory during our previous work [4]. The characteristics of the briquettes based on waste paper/cardboard used are summarized in the Table 1. The concerned characteristics are the density, the higher heating value (HHV) and the moisture content of the briquettes.

Table 1. Characteristics of briquettes used in this study [4].

Briquette	Density (kg/m^3)	HHV (MJ/kg)	Moisture Content (%)
Type 1 (corrugated cardboard)	486	15.83	7.5
Type 2 (office paper)	550	14.10	5.7
Type 3 (mixture of paper waste)	490	15.22	5.1
Type 4 (mixture of paper and green waste)	430	15.54	6.4

2.2. Study of Combustion

The study of briquette combustion was conducted using two devices in order to analyze the temperature fields and the heavy metals content in the flues gas and ash.

2.2.1. Analysis of Temperature Fields

The device shown in Figure 1 was used for the study of the combustion of briquettes. The rotary kiln is a cylinder slightly inclined with a diameter of 88 cm and 108 cm of length. The refractory and insulation material of 4 cm and 5.6 cm of thickness respectively were placed in the rotary kiln. The flue gases were evacuated through a chimney of 8 m of length. The rotary kiln allowed the study of temperature fields during the combustion of briquettes. Two kilograms (2 kg) of briquettes were weighed and then arranged on a horizontal grid as shown in Figure 1. Five thermocouples were placed along a metal rod for temperature measurement. The thermocouples of type K were used for this purpose. The first thermocouple was placed close to the kiln gate and the four other thermocouples are regularly placed at 10 cm in order to cover the briquette bed formed on the grate. The measured temperatures were collected by a data logger connected to a computer. Inflammation of the fuel charge was done at the front of the oven (outward). Thus, the flame propagates from the front to the inside of the oven.

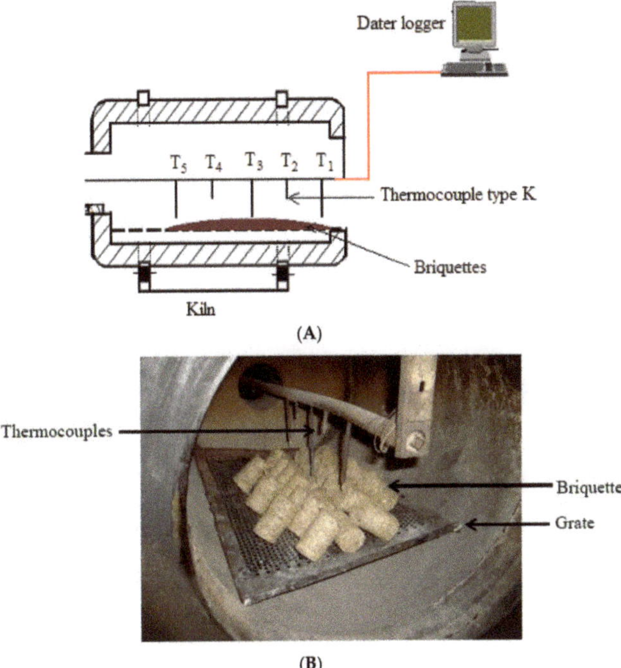

Figure 1. Experimental device of the briquettes combustion. (**A**) Schematic Diagram of the device; (**B**) Arrangement of briquettes in the rotary kiln.

The established phase (or homogeneous) of the combustion is the main concern of the present study. The combustion is called in the established phase when all the reaction intermediates are released and are in the oxidation phase: this is the part of the combustion where we have the maximum of chemical reactions that occur. This step is characterized by the stabilization of the temperatures in the oven. The duration of the established phase of the combustion was estimated by considering the time spent between the beginning and the end of the combustion indicated by the rise of the temperature and the drop of the temperature respectively. In addition, the temperature of the established phase of the combustion was determined based on the temperature fields. Three tests were carried out for each type of briquette. The ash from each test was collected and weighed. The ash samples were digested according the method described by ISO 16968 [18] and the heavy metals content were determined by ICP-MS. The determination of nitrogen was done by mineralized the samples with H_2SO_4. The nitrogen content in the ash was determined by the Kjeldahl method. For the phosphorus content determination, $NaHCO_3$ and the formed phosphate ions were treated with an ascorbic acid. The phosphorus (P_2O_5) content was determined by dosing the emitted blue color with a spectrophotometer at 660 nm of wavelengths as described by Olsen et al. [19]. For the potassium (K_2O) content determination, the ash sample was digested according to the method describe by the Part 2 of standard ISO 16967 [20]. Then the potassium content was determined by flame emission spectroscopy at a wavelength of 766.5 nm. The heavy metal (i) content in the ash $(HM_{ash})_i$ is given by the Equation (1).

$$(HM_{ash})_i = 100\frac{M_{ash}C_{i,ash}}{M_{sample}} \tag{1}$$

M_{ash}, $C_{i,ash}$ and M_{sample} are respectively the masse of collected ash in kg, the concentration of the heavy metal determined in the ash sample in mg per kg and the masse of the briquette sample in kg.

2.2.2. Flue Gas Analysis

The tubular kiln was used for the study of the fumes resulting from the combustion of the briquettes (Figure 2). A test bench has been created for this purpose. The tubular kiln of 88 cm represents the central part of the bench. A quart reactor of 120 cm long and 7 cm of internal diameter was placed inside the tubular kiln. The reactor temperature can be set up to a maximum of 1250 °C. The air flow was fixed by a digital control box (Brooks Microprocessor Control & Read Out Unit Models 0154) connected to a Brooks flow meter of type 5850E. Mass samples of 500 mg are weighed and placed inside the preheated oven at the set temperature. The air flow was set at 1.07 liter/min. The combustion takes place in a few minutes. The flue gases were bubbled to a bottle containing 250 mL of solution of 5% of HNO_3. The resulted ash from the combustion was weighed after each trial.

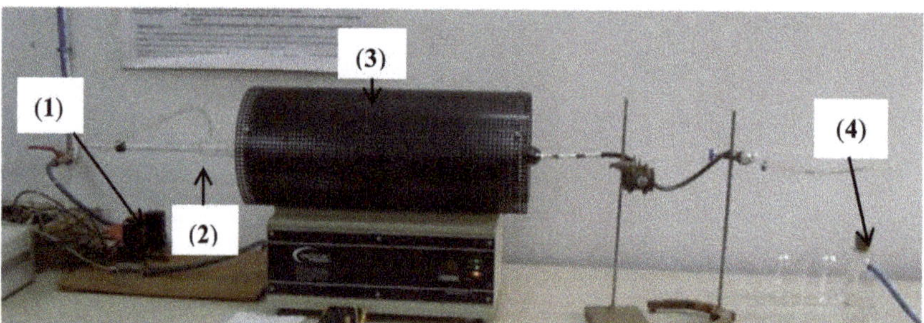

Figure 2. Experimental device for the briquettes combustion in a tubular furnace: (**1**) Flowmeter; (**2**) Quart reactor; (**3**) Tubular kiln; (**4**) Solution of 5% HNO_3.

For each type of briquette three combustion tests were performed to ensure better reliability of the results. The ash fractions (*Ash*) obtained in the tubular and rotary kiln devices were calculated on wet basis by the Equation (2).

$$Ash = 100\frac{M_{ash}}{M_{sample}} \tag{2}$$

The bubbling in the acid solution of the flue gas at the outlet of the reactor was carried out in order to trap the heavy metals. The analysis of the contents of these metals was carried out by

ICP-MS. The following heavy metals were analyzed: Arsenic (As), Cadmium (Cd), Mercury (Hg), Lead (Pb) and Manganese (Mn). This study was conducted to simulate the combustion under the temperature obtained during the combustion in the rotary kiln. The heavy metal (*i*) content in the flue gas $\left(HM_{flue\ gas}\right)_i$ is given by the Equation (3).

$$\left(HM_{flue\ gas}\right)_i = 100\frac{V_{solution}C_{i,solution}}{M_{sample}} \tag{3}$$

$V_{solution}$ and $C_{i,solution}$ are respectively the volume of the acid solution in liters and the concentration of the heavy metal determined in the solution in mg per liter.

3. Results and Discussion

3.1. Temperature Fields and Ash Fractions

The temperature fields during the combustion of the four types of briquettes were shown in the Figure 3.

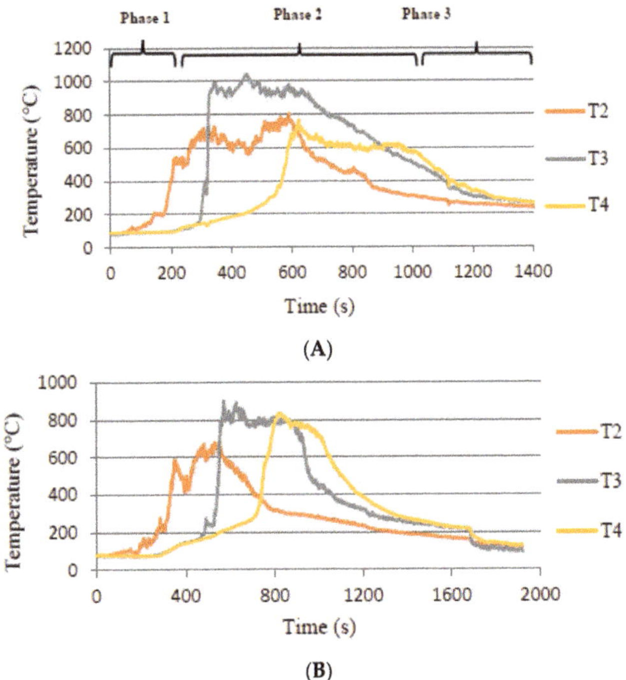

Figure 3. *Cont.*

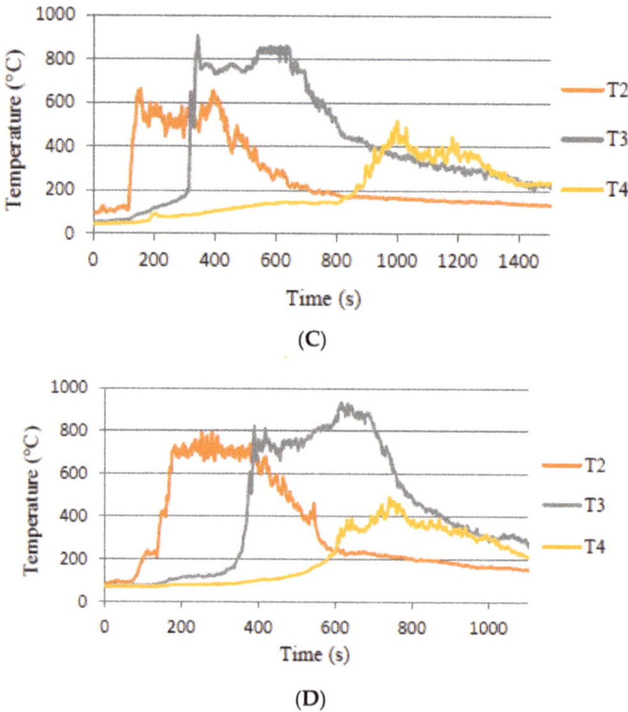

Figure 3. Temperature variation versus time for: (**A**) Type 1 briquettes (corrugated cardboard); (**B**) Type 2 briquettes (office paper); (**C**) Type 3 briquettes (mixture of paper waste); (**D**) Type 4 briquettes (mixture of paper and green waste).

Only the thermocouple temperatures T_2, T_3 and T_4 were represented. Indeed, these thermocouples were placed above the solid bed. This was not the case of T_1 which was placed in combustion initiation zone and T_5 which was placed at the combustion end zone. The temperature rises were shifted chronologically from T_2 to T_4 (Figure 3); this is explained by the fact that the propagation of the flame was done along the solid bed from outside to inside the furnace. The rise of temperatures characterized the phase 1 of the combustion which represents the phase of the initiation (Figure 3A). The established phase of the combustion is also observed on the temperature fields during the combustion of each type of briquette (Phase 2). The temperature drop characterizes the end of combustion (Phase 3). The average temperature of the established phase of the combustion and the duration of the combustion obtained during the combustion are given by the Table 2. The results showed a better combustion for the briquette of Type 1 made of corrugated cardboard for which the highest mean temperature was recorded. This result is predictable because briquettes made of corrugated cardboard have the highest calorific value, 15.83 MJ/kg. After briquettes made of corrugated cardboard, come the briquettes of Type 3 (mixture of papers) and Type 4 (mixture with green waste) which had the same mean temperature since their HHV are similar (15.22 MJ/kg and 15.54 MJ/kg respectively).

Table 2. Summary of combustion times and average temperatures of the established phase of combustion.

Briquette Type	Type 1	Type 2	Type 3	Type 4
Combustion time (min)	10.5	11.0	18.5	12.5
Temperature (°C)	950	750	800	800

The briquettes which produced the lowest mean temperature are the briquettes made from office paper. This fact is also predictable because office papers have the lowest HHV (14.10 MJ/kg). Furthermore, the briquettes made of corrugated cardboard burned quickly than the other briquettes. The high cellulose content of cardboard is the base of the rapid degradation since cellulose degrades rapidly and at low temperature according to Sorum et al. [5]. The duration of the combustion was quite the same for briquettes of Types 1, 2 and 4. The duration of the combustion was higher for briquettes of Type 3 (18 min) comparatively to other briquettes. This can be explained by the fact that the papers in the category "commercial printing paper" were composed of additives giving them a more flame retardant character than the others.

The ash fractions produced by the combustion in the rotary and tubular furnace were analyzed in the present work (Figure 4). In general, the ash proportions were almost the same regardless of the device used (rotary kiln for large-scale tests and tubular kiln for small-scale tests). The ash content varied from 10% to 16%, which will lead to a large production of ash during the combustion of briquettes in industrial plants. The ash content obtained for the paper/cardboard are higher than those obtained for the raw wood.

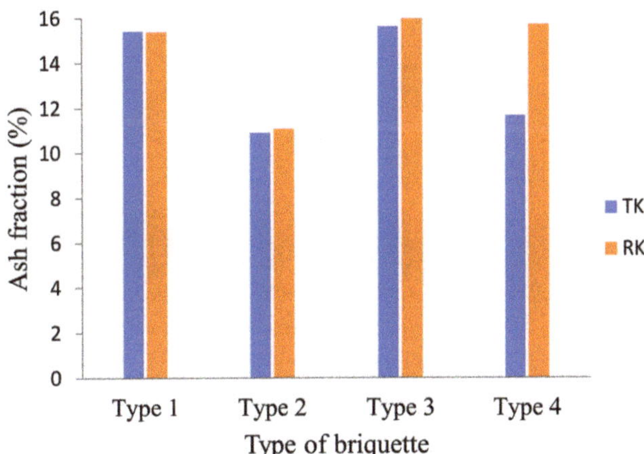

Figure 4. Ash fraction in the rotary kiln (RK) and the tubular kiln (TK).

This fact can be explained by paper manufacturing process which reduces considerably the lignin and hemicellulose content of the biomass and added the mineral matters to enhance the paper properties. However, the ashes content are lower than those obtained by Sorum et al. [5] since a part of the ashes may be evaporated or entrained by the flowing flue gas during the combustion as mentioned by Jenkins et al. [8]. The fly ash was not accounted in the present study contrary to the protocol followed for the ash content determination. In addition, the differences in the ash content are also due to the heterogeneities of the waste paper/cardboard.

3.2. Heavy Metals in the Flue Gas and Ash

The heavy metals content (reported to a kg of fuel) in the flue gas emitted in the tubular furnace and in the collected ash in the rotary kiln were presented in the Table 3. The heavy metals content in the ash (reported to a kg of ash) were presented in Table 4. The tests in the tubular furnace were carried out at the average temperatures obtained in the rotary kiln, thus the obtained results were compared. Indeed, the combustion temperature and the fuel moister content are the most influencing factors of the heavy metal emission during the biomass combustion [21,22]. According to Wei et al. [22] the importance of the moisture content is due to its impact on combustion temperature.

Due to the high temperature of the combustion (750 °C to 950 °C), the heavy metal contained in the raw paper waste were predominantly transferred into the flue gas than into the ash as shown in the Table 3. According to Kovacs et al. [15], high percent of the heavy metals may be content in the flue gas even at relatively low temperature of 250 °C. Indeed, the temperature increases the heavy metal content in flue gas at the expense of ash [21]. The low volatilization temperature of heavy metal, particularly Hg, Cd and Pb, explained their large content in the flue gas comparatively to the bottom ash as shown in the literature [21,23]. However, the Mn was more present in the ash than in the flue gas. The low volatility of Mn explained this fact as it was confirmed by Yanjun et al. [21].

Table 3. Heavy metals contained in the flue gas and in the bottom ashes in mg_{metal}/kg_{fuel}.

Briquettes	As	Cd	Hg	Mn	Pb
In the flue gas					
Type 1	2.58	0.82	7.07	68.32	42.78
Type 2	7.83	1.32	6.01	205.01	53.58
Type 3	2.09	0.97	6.16	74.96	38.93
Type 4	2.92	0.84	5.31	87.62	37.97
In the bottom ashes					
Type 1	1.78	0.22	0.20	91.32	15.24
Type 2	1.73	0.14	0.36	77.85	5.25
Type 3	1.06	0.14	0.26	76.44	12.60
Type 4	0.92	0.25	0.75	90.96	10.73

The briquette of type 2, made of office paper, had the higher emission of heavy metal comparatively to the others type of briquettes since its combustion produced a highest mass of heavy metal which were concentrated in the flue gas. This fact may be explained by the presence of considerable heavy metals in the office paper due to the use of this type of paper for printing and writing. However, the mercury and cadmium emission were similar for all type of briquettes. Otherwise, the differences between the emissions of heavy metal in the flue gas were low for the other types of briquettes. For the ash, the heavy metal content was higher for the briquette of type 1, made of corrugated cardboard, comparatively to the other type of briquettes (Tables 3 and 4). Even though the highest temperature was achieved during the combustion of briquette of type 1, their high moisture content may favor the concentration of heavy metal in the collected ash. According to Yanjun et al. [21], the moisture content reduces the transfer of heavy metal in the flue gas. In addition, the low ash content of the briquette of Type 1 (Figure 3) increased the concentration of heavy metal in the ash (Table 4).

Note that the Pb, Hg and Mn contents in the flue gas and collected ash were higher than that reported in the raw biomass by the literature [15,24]. In contrast, the Cd and As content were similar to those reported in the raw biomass by Kovacs et al. [15]. Thus, the excess of Pb, Hg and Mn may be resulted to the impurity of waste paper mainly the ink and the additives of the paper manufacturing process. The heavy metal content in the ash is also important for analyzing their impact on soil quality since ash could be used to amend the soil. In this purpose, the toxicity of heavy metals in the ashes should be evaluated before considering any use as fertilizer. These heavy metals are indeed taken into account in the definition of standards for the spreading of mineral fertilizer on agricultural land. Note that, the heavy metals content obtained in the present study are largely lower than those found in the municipal waste by Xiao et al. [25]. In addition, the contents of heavy metals were compared to the limit values reported in the law of Burkina Faso [26] and in the Australian standard [27] as shown by Table 4.

Table 4. The heavy metals content in the ash and the limit value in heavy metals for application of ash in soil (mg_{metal}/kg_{ash}).

Heavy Metals	As	Cd	Hg	Mn	Pb
Heavy metals content in ash					
Type 1 (corrugated cardboard)	16.08	2.02	1.80	825.28	137.69
Type 2 (office paper)	11.21	0.93	2.35	505.80	34.13
Type 3 (mixture of paper waste)	6.65	0.89	1.61	478.80	78.96
Type 4 (mixture of paper and green waste)	5.86	1.56	4.80	579.45	68.37
Limit value of heavy metals					
Law of Burkina Faso [26]	-	40	25	-	1200
Narodoslawsky and Obernberger, [27]	20	10	-	-	500

For all the heavy metals analyzed, the ash obtained for all briquettes types can be used for soil amendment since the heavy metal content were under the limits values given by Table 4. However, the agronomic value must be determined by considering the fertilizing substances (NPK) content in the ash. This issue was addressed in the next section.

3.3. Ash Proportion and NPK Content

The ash could be valorized in the fields to amend soils, as said previously. The value of ash as fertilizer is mainly evaluated through it content of NPK. Thus, the contents of NPK have been determined and given in Table 5. The obtained fractions of NPK are comparable to those obtained for waste paper by other authors [8]. However, the obtained fractions of NPK are considerably lower than those obtained for the ash coming from the most biomass species as shown in the literature [8]. The manufacturing process of the paper may reduce the NPK fractions since these fractions are lower than those obtained for the parent wood used for paper production.

Table 5. NPK contents in the ash and minimum content of NPK for the ash qualification to fertilizers.

NPK	Type 1	Type 2	Type 3	Type 4	Minimum Content [28]
N (%)	0.003	0.010	0.003	0.011	1
P_2O_5 (%)	0.136	0.179	0.090	0.224	2
K_2O (%)	0.494	1.198	0.353	1.408	5
$N+P_2O_5 + K_2O$	0.633	1.387	0.446	1.644	7

The standard NF U 42 001 [28] gave the minimum levels of nitrogen, phosphorus and potassium (NPK) to qualify ash as fertilizer. The NPK contents obtained for the four types of briquettes were below the minimum value required by the standard NF U 42001, [28]. Thus, ash collected cannot be standardized as fertilizer according to the standard. However, in the context of a use in co-composting, the mixture of ash with other elements, may allow to obtain a product whose fertilizer standardization is possible.

4. Conclusions

Briquettes made of waste paper produced in Ouagadougou were combusted in the present work. The temperature variation, the heavy metal in the flue gas and the produced ash and NPK content in the produced ash were studied in the present work. The combustion of the briquette made of corrugated cardboard achieved the highest mean temperature of 950 °C and lowest duration of 10.5 min during the established phase of the combustion. The lowest mean temperature of the established phase was achieved for the combustion of briquettes of type 2 due to their low HHV. The highest combustion time was obtained for the combustion of briquettes made of the mixture of all paper waste due to flame retardant which may be present in the mixture of waste paper and cardboard. The combustion produced considerable residual ash which proportion varied from 11% to 16% of the raw briquettes mass. The analysis of the ash and the flue gas shown that large proportions of the heavy metal were transferred predominantly in the flue gas than in the ash. In addition, the manufacturing process and the usage of paper had increased some heavy metal content (Pb, Hg and Mn) comparatively to the parent wood. The heavy metal contents in the ash were under the limit content fixed by the standard in the context of soils amendment. Thus, the ash obtained after the briquettes combustion can be used to amend soils. For this purpose, the ash may be mixed with other element in order to improve fertilizer content since the ash fraction of NPK are lower than the minimum required by the standard.

Author Contributions: H.G.I. realized the experiments and wrote the paper. S.K.O. and A.Y. conceived and supervised the study. O.S. contributed in the data analysis. The paper presentation and formulation was improved by T.D., J.K. had made a critical review which improved the scientific quality of the paper.

Funding: This research received no external funding.

Acknowledgments: The authors want to acknowledge Dissa Alfa Omar and Sanon Pierre for their assistance with the present work.

Conflicts of Interest: The authors declare no conflict of interest.

References

1. Tezanou, J. Evaluation Environnementale et Technique de la Gestion des Déchets Ménagers de Ouagadougou: Schémas de Gestion et Expérimentation de Traitement Thermique. Ph.D. Thesis, Université de Poitiers, Poitiers, France, 2003; p. 27.
2. Karna, A.; Engstrom, J.; Kutnlahtim, T. Life cycle analysis of newsprint. In Proceedings of the 2nd Research Forum on Recycling, Canadian Pulp and Paper Association, Montreal, QC, Canada, 5–7 October 1993; pp. 171–178.
3. Merrild, H.; Anders, D.; Christensen, T.H. Life cycle assessment of waste paper management: The importance of technology data and system boundaries in assessing recycling and incineration. *Resour. Conserv. Recycl.* **2008**, *52*, 1391–1398. [CrossRef]
4. Harouna, I.G.; Ouiminga, S.K.; Daho, T.; Arsène, H.Y.; Sougoti, M.; Koulidiati, J. Characterization of Briquettes Coming From Compaction of Paper and Cardboard Waste at Low and Medium Pressures. *Waste Biomass Valoriz.* **2014**, *5*, 725–731.
5. Sorum, L.; Gronli, M.G.; Hustad, J.E. Pyrolysis characteristics and kinetics of municipal solid waste. *Fuel* **2001**, *80*, 1217–1227. [CrossRef]
6. Von Wilpert, K.; Bosch, B.; Puhlmann, H.; Zirlewagen, D. Wood ash recycling—An appropriate measure to close nutrient cycles in forests. In Proceedings of the 4th Central European Biomass Conference, Graz, Austria, 15–18 January 2014.
7. Hallenbarter, D.; Landolt, W.; Bucher, J.B.; Schutz, J.P. Effects of wood ash and liquid fertilization on the nutritional status and growth of Norway spruce (*Picea abies* (L.) Karst). *Forstwiss. Centralbl.* **2002**, *121*, 240–249. [CrossRef]
8. Jenkins, B.M.; Baxter, L.L.; Miles, T.R., Jr.; Miles, T.R. Combustion properties of biomass. *Fuel Process. Technol.* **1998**, *54*, 17–46. [CrossRef]
9. Jokinen, K.; Siren, K.; Osmonen, R.M. Quality aspects of recycled fibre. *World Pap.* **1995**, *220*, 31–33.
10. Vassilev, S.V.; Baxter, D.; Andersen, L.K.; Vassileva, C.G.; Morgan, T.J. An overview of the organic and inorganic phase composition of biomass. *Fuel* **2012**, *94*, 1–33. [CrossRef]
11. Ginebreda, A.; Guillen, D.; Barcelo, D.; Darbra, R.M. Additives in the paper industry. In *Global risk-based management of chemical additives I: Production, Usage and Environmental Occurrence*, 1st ed.; Bilitewski, B., Darbra, R.M., Barcelo, Eds.; Springer-Verlag: Heildenberg, Germany, 2012; Volume 18, pp. 11–34. ISBN 978-3-642-24876-4.
12. Pekarovicova, A.; Wu, Y.J.; Fleming, P. Quality analysis of gravure spot color reproduction with an ink jet printer. *J. Imaging Sci. Technol.* **2008**, *52*, 60501–60509.

13. Conti, M.E. Heavy metals in food packagings. In *Mineral Components in Food*, 1st ed.; Nriagu, J., Szefer, P., Eds.; CRC Press: Boca Raton, FL, USA, 2006; pp. 339–362. ISBN 978-1-420-00398-7.

14. Zalewski, S. Design, graphics arts, and environment. Thesis, RIT Roschester Institute of Technology, Rochester, NY, USA, 1994.

15. Kovacs, H.; Katalin, S.; Tamás, K. Theoretical and experimental metals flow calculations during biomass combustion. *Fuel* **2016**, *185*, 524–531. [CrossRef]

16. Lanzerstorfer, C. Chemical composition and physical properties of filter fly ashes from eight grate-fired biomass combustion plants. *J. Environ. Sci.* **2015**, *30*, 191–197. [CrossRef] [PubMed]

17. Nurmesniemi, H.; Makela, M.; Poykio, R.; Manskinen, K.; Dahl, O. Comparison of the forest fertilizer properties of ash fractions from two power plants of pulp and paper mills incinerating biomass-based fuels. *Fuel Process. Technol.* **2012**, *104*, 1–6. [CrossRef]

18. ISO 16968. *Solid Biofuels—Determination of Minor Elements*; ISO: Geneva, Switzerland, 2015.

19. Olsen, S.R.; Cole, C.V.; Watanabe, F.S.; Dean, L.A. *Estimation of Available Phosphorus in Soils by Extraction with Sodium Bicarbonate*; Circular 939; United States Department Of Agriculture: Washington, DC, USA, 1954; Volume 939, pp. 1–19.

20. ISO 16967. *Solid Biofuels—Determination of Major Elements—Al, Ca, Fe, Mg, P, K, Si, Na and Ti*; ISO: Geneva, Switzerland, 2015.

21. Yanjun, H.; Jiubing, W.; Kai, D.; Jianli, R. Characterization on Heavy Metals Transferring into flue gas during Sewage Sludge Combustion. *Energy Procedia* **2014**, *61*, 2867–2870.

22. Wei, Z.; Yindong, T.; Huanhuan, W.; Long, C.; Langbo, O.; Xuejun, W.; Guohua, L.; Yan, Z. Emission of Metals from Pelletized and Uncompressed Biomass Fuels Combustion in Rural Household Stoves in China. *Sci. Rep.* **2014**, *4*, 5611.

23. Van de Velden, M.; Dewil, R.; Baeyens, J.; Josson, L.; Lanssens, P. The distribution of heavy metals during fluidized bed combustion of sludge (FBSC). *J. Hazard. Mater.* **2008**, *151*, 96–102. [CrossRef] [PubMed]

24. Demirbas, A. Influence of Gas and Detrimental Metal Emissions from Biomass Firing and Co-Firing on Environmental Impact. *Energy Sources* **2005**, *27*, 1419–1428. [CrossRef]

25. Xiao, Z.; Xingzhong, Y.; Hui, L.; Longbo, J.; Lijian, L.; Xiaohong, C.; Guangming, Z.; Fei, L.; Liang, C. Chemical speciation, mobility and phyto-accessibility of heavy metals in fly ash and slag from combustion of pelletized municipal sewage sludge. *Sci. Total Environ.* **2015**, *536*, 774–783. [CrossRef] [PubMed]

26. Décret. *Le Décret N 2001-185/PRES/PM/MEE du 07 mai 2001*; Portant Fixation des Normes de Rejets de Polluants Dans L'air, L'eau et le sol; Journal officiel: Ouagadougou, Burkina Faso, 2001.

27. Narodoslawsky, M.; Obernberger, I. From waste to raw material–the route from biomass to wood ash for cadmium and other heavy metals. *J. Hazard. Mater.* **1996**, *50*, 157–168. [CrossRef]

28. NF U42-001. Engrais–Dénominations et Spécifications; 1981. Available online: https://www.boutique.afnor.org/norme/nf-u42-001/fertilizers-types-and-specifications/article/743864/fa032284 (accessed on 8 July 2018).

Review

Microwave Technologies: An Emerging Tool for Inactivation of Biohazardous Material in Developing Countries

Klaus Zimmermann

Danube BioSolutions, Hauptstrasse 119/2/2, 3021 Pressbaum, Austria; klaus.zimmermann@live.at

Received: 6 July 2018; Accepted: 1 August 2018; Published: 2 August 2018

Abstract: Inappropriate treatment and disposal of waste containing biohazardous materials occurs especially in developing countries and can lead to adverse effects on public and occupational health and safety, as well as on the environment. For the treatment of biohazardous waste, microwave irradiation is an emerging tool. It is a misbelief that microwave devices cannot be used for inactivation of solid biohazardous waste; however, the inactivation process, and especially the moisture content, has to be strictly controlled, particularly if water is required to be added to the process. Appropriate control allows also inactivation of waste containing inhomogeneous compositions of material with low fluid/moisture content. Where appropriate, especially where control of transport of waste cannot be guaranteed, the waste should be inactivated directly at the place of generation, preferably with a closed waste collection system. In waste containing sufficient moisture, there are direct useful applications, for example the treatment of sewage sludge or human feces. A number of examples of microwave applications with impacts for developing countries are presented in this review. In respect to energy costs and environmental aspects, microwave devices have clear advantages in comparison to autoclaves.

Keywords: microwave; inactivation; disinfection; biohazardous waste; healthcare waste; sludge; carbon footprint; developing countries

---

1. Introduction

In developing countries, a relatively low number of research facilities or commercial companies are working with microorganisms in higher biosafety level (BSL) containments, and thus a large part of biohazardous waste is generated in hospitals. Sewage sludge is also one potential source of biohazardous waste and its treatment has considerable interest for these countries [1].

Usually 5–90% of hospital waste is general waste (similar to household waste) and about 10% is classified as biohazardous [2,3]. Other types of dangerous wastes are toxic and radioactive. Segregation of waste prior to decontamination and/or disposal could be an important factor to save costs, which is especially relevant for low-income countries. The transport from waste generating facilities to the disposal site bears considerable risks [4], but in reality, also uncontrolled transport ways within facilities or hospitals from the site of generation to the treatment site should not be disregarded. The final disposal of untreated contaminated material on the ground should be avoided and may lead especially in developing countries to adverse impacts [5].

A variety of waste treatment technologies are available, and there is not one which is optimal for every need. Relevant factors for using a specific technology including environmental impacts are compiled in the "United Nations Environment Programme (UNEP) compendium of technologies for treatment/destruction of healthcare waste" [2], which gives an excellent introduction into this topic.

From a variety of different technologies to treat biohazardous waste, microwave irradiation is emerging and may especially be helpful to solve specific issues in developing countries [6]. However, in peer-reviewed literature, only limited information about these topics is available and sometimes not all necessary aspects for appropriate use of microwave technology are taken into account. In this review, the aforementioned aspects are highlighted in the context of whether microwave technologies could be an alternative tool for improving the management and treatment of biohazardous waste with a focus on developing countries.

2. Treatment Technologies and Challenges

There are four basic processes for the treatment of biohazardous components in waste (i.e., thermal, chemical, irradiative, and biological). From the thermal processes, incineration and autoclave treatment are most widely used; however, microwave irradiation is an emerging tool.

Inappropriate treatment and final disposal of wastes containing biohazardous materials buries especially in developing countries a variety of issues potentially leading to adverse impacts to public and occupational health and safety, as well as to the environment [5]. Accepted treatment options and processes are listed for example by the German Robert Koch Institute [7].

Not unexpectedly, there are differences in the management of healthcare waste especially between low, middle, and high-income countries [8]. When the content of waste containers was observed in a hospital in El Salvador, 61% of biohazardous waste was actually common waste, suggesting that the staff was possibly unaware of the requirements or just neglected them [9].

Waste management practices in three government hospitals of Agra, India indicated also a lack of knowledge and awareness regarding legislations on bio-medical waste management [10]. Other examples of suboptimal waste management practices are presented by Nandwani [11] and Zhang et al. [12].

Inappropriate transport is also a major challenge. Biosafety guidelines demand that transport of inactivated biohazardous material outside the facility is forbidden if not using specific precautions. Though healthcare facilities are usually exempted from these guidelines, the hospital management should nevertheless consider the risk [3]. In order to avoid infections of humans and environmental contamination, a disinfection system minimizing the risk should be used. A good example (for a microwave device) is a closed waste collection system with different container volumes in which the material is subsequently inactivated (Figure 1, with kind permission of Meteka, Judenburg, Austria). Such technology is especially useful for developing countries where appropriate control of transport of waste cannot be guaranteed.

In respect to microwave technologies, Tonuci et al. [13] showed that if the operational conditions of the equipment are not adequately controlled, the inactivation treatment is probably ineffective. Clearly, operational conditions are the most important factor and without tight control of parameters, especially moisture, a complete inactivation can never be guaranteed with microwave technologies.

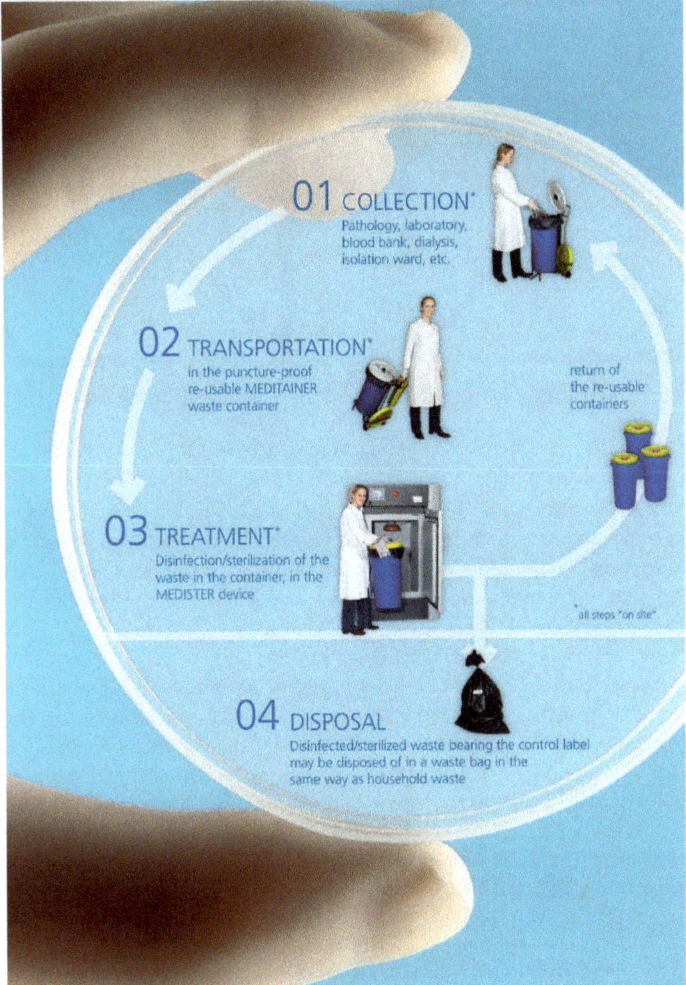

Figure 1. Example for closed waste disinfection system.

3. Microwave: A Variety of Applications

The main benefit of microwave energy is the direct delivery of energy to microwave-absorbing materials. As long as a material contains dipolar molecules (i.e., water), a complete heating of samples from outside and inside is guaranteed. Issues such as long heating periods, thermal gradients, and energy loss to the environment can thus be minimized [14]. Disinfection with microwave irradiation occurs mainly through the combination of moisture and (low) heat. In contrast to microwaves, the heat for autoclaves is generated outside of the waste material, and thus the material to be treated is heated starting from the surface [2]. Every type of material is inactivated; however, to be on the safe side, extended inactivation times should be used for waste with a large volume. A schematic drawing of the two principles is shown in Figure 2.

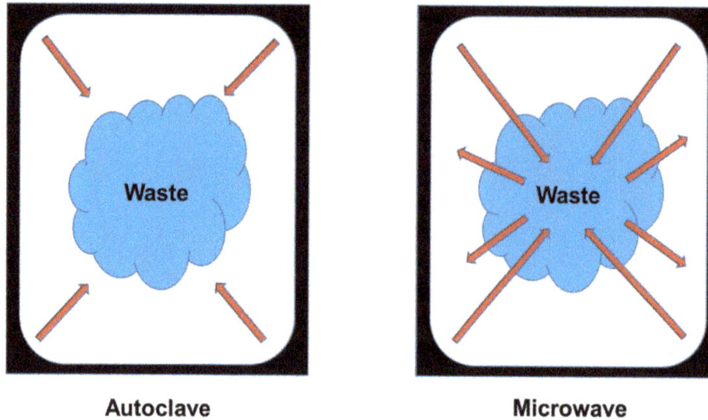

Figure 2. Heat generation in microwaves.

The majority knows microwave ovens just as a tool for cooking at home. It is not widely known, but even at home, microwave radiation can be used for microbial inactivation. When treating kitchen sponges, scrubbing pads, and syringes with microwaves at 100 percent power level, the total bacterial count can be reduced by more that 99 percent within 1 to 2 min, and complete inactivation can be achieved over longer exposures [15].

In recent years, microwave technologies proved to be a very attractive alternative for industrial conventional processing methods, and found an astonishing number of applications in different areas. In respect of processing food or beverages, one of its applications is drying of green tea [16]. However, main applications for food or beverages are inactivation of microorganisms, for example in or on peanut butter [17], beef [18], or catfish filet [19].

Regarding therapeutic products, microwaves are useful for freeze-thaw treatment of injectable drugs [20] or hold potential for vaccine production [21]. Microwave assisted extraction is a possibility for fragrance production [22], and became a popular method for extracting natural products and active ingredients from plants [23,24]. Another potential application of microwaves could be the preparation of safe drinking water [25].

4. Efficiency of Microwaves for Treating Biohazardous Waste

The destruction of all microbial life—sterilization—is usually not required for inactivation of healthcare waste [26], whereas disinfection—the prevention of any potential for transmission—is regarded as sufficient.

For many years, there seem to be no doubt that microwave systems are able to destroy pathogens. Already in the 1960s of the last century, suspensions of *E. coli* and *Bacillus subtilis* spores were exposed to a conventional microwave at 2450 MHz and approximately a 6-log cycle reduction in viability was encountered for both microorganisms [27]. Another early study supporting the efficient inactivation of microorganisms with "ordinary" microwaves was conducted by Souhrada [28]. Hoffman et al. [29] already used a more sophisticated microwave system with a self-generated steam decontamination cycle for inactivation of bacteria.

It became clear that the operational conditions have to be strictly controlled for microwave inactivation of microorganisms. Pre-sterilized public healthcare wastes, which were inoculated with 5×10^5 vegetative *E. coli* bacteria and then treated with a microwave device [13] showed that not only radiation exposure time and power per waste mass unit were important for the percentage of inactivation of the microorganisms, but also the incoming waste moisture had an significant influence.

If properly controlled, the misbelieving that microwave systems cannot be used for inactivation of solid biohazardous waste is thus contradicted. However, the inactivation process, and especially the moisture content has to be strictly controlled.

5. Advanced Microwave Technologies

In principal there are two system designs for advanced microwave technologies: batch processes and semi-continuous microwave systems [3]. The Sanitec waste disposal system is an example for continuous microwave technology including a shredding system [30]. This system is intended rather for the treatment of large amounts of biomedical waste, which inherently leads to collection of the waste at sites where it is generated, transport, and inactivation at one single site. For smaller amounts of waste, other microwave technologies combined with shredding are available as well [31]. Among the commercial suppliers is the company Bertin.

The Meteka batch microwave technology guarantees a controlled even heating of waste including also inhomogeneous compositions of material [32,33]. The environmental performance of this system is proven through environmental product declarations (EPDs). The microwave device automatically adds water and controls moisture air, and heats up and inactivates the waste. Table 1 gives a short summary of major advantages and disadvantages of conventional and sophisticated microwave technologies [34].

Table 1. Advantages/disadvantages of conventional and sophisticated microwave technologies.

	Conventional ("Household") Microwave	Sophisticated Microwave (Controlled Heat and Moisture)
Cost for device	Low	High
Energy consumption	Low	Low
Water consumption	None	Low
Control of inactivation process	Difficult	Very good

6. Comparison of Technologies for Treatment of Biohazardous Waste

For decontamination of infectious waste autoclaves are widely used. However, these devices have the disadvantage that they have a high energy consumption and a long runtime per disinfection cycle. Due to the operating conditions—overpressure and high temperatures—autoclaves demand more technical handling, have a high service requirement, and a comparatively limited durability.

Another widely used option is disposal and/or incineration of special waste. However, this option requires transport, mainly on the road. The transport risks are directly related to the operation safety of the dangerous goods transport enterprises and can be mathematically calculated [4]. The validity of the calculation was proven in a case analysis of five dangerous goods transport enterprises in the Inner Mongolia Autonomous Region.

The pros and cons of different waste treatment technologies must thus be objectively analyzed. An analysis on technologies such as autoclave, microwave, chemical disinfection, combustion, and disposal on the ground was carried out by Diaz et al. [5]. One large study comparing technologies was also conducted in China. From 272 modern, high-standard, centralized medical waste disposal facilities there were about 50% non-incineration treatment facilities, including the technologies of high temperature steam, chemical disinfection, and microwave [35], and every technology was found to have its advantages and disadvantages.

In a study in Istanbul, Turkey, it was concluded that the method of choice for the healthcare waste for this city would be incineration [36]. However, another study in the same country led to different results. When five different healthcare waste treatment/disposal alternatives including incineration, microwaving, autoclaving on-site, autoclaving off-site, and landfill were evaluated, the off-site autoclaving was found to be the most appropriate solution for the specific requirements [37].

A systematic approach for analyzing all factors including costs can be found in the UNEP compendium of technologies for treatment/destruction of healthcare waste [2]. The UNEP compendium advises to calculate scores from all aspects of environmental and occupational safety, operation costs, capacities, volume reduction, efficacy of inactivation, and installation requirements. In this study, costs stated for autoclaves were calculated to be between 0.14 and 0.33 USD/kg and for batch microwaves about 0.13 USD/kg.

Depending on the type of analysis, cost calculations do not always deliver the same result. Soares et al. [38] conducted a systematic approach for analyzing costs of small generators of healthcare waste for three disinfection techniques (microwave, autoclave, and lime) followed by transportation and landfilling using a life-cycle assessment. Microwaving had the lowest environmental impact (12.64 Pt) followed by autoclaving (48.46 Pt). Cost analyses revealed values of USD 0.12/kg for microwaves and USD 1.10/kg autoclaves. The conclusion was that microwave disinfection had the best eco-efficiency performance. It has to be noted that an ordinary household microwave was used for the inactivation experiments. As these microwave instruments are relatively cheap, but lack options for controlling the efficacy (heat and moisture), the costs need a recalculation in case advanced microwave technologies would be applied (prices starting at about USD 20,000 for smaller units).

With regard to the many differences in technologies and features of commercially available devices, a fair price comparison is hardly possible. Clearly, a comparison of costs of microwaves and autoclaves cannot simply be reduced to the price of the device. However, in respect of energy costs and additional costs such as maintenance there is a clear difference. An example was calculated for 150 kg of a typical mixture of solid biohazardous waste per day using a Medister 160 microwave device with 6.5 kW power input [34]. Inactivation of 150 kg waste needs approximately 12 runs with 60 L containers. The overall energy consumption is 40.9 kWh/day. In comparison, a typical autoclave with 110 L chamber volume and 17 kW power input has an overall energy consumption of 142.4 kWh/day. With a price example of 0.2 €/kWh, the cost save accumulates to an astonishing 20 €/day. In addition, the environmental factor should not be neglected. Based on a daily operation, the difference in energy consumption would be 37,000 kWh yearly. Assuming 0.583 kg CO_2 (International Energy Agency 2014) for one kWh the reduction of carbon foot print is 21.6 tons CO_2/year.

7. Examples for Use of Microwaves in Developing Countries

Microwaves hold great potential to treat biohazardous waste and may especially be helpful to solve specific issues with waste in developing countries [6]. Challenges and consequences of poor sanitation, especially in developing economies, demand the exploration of new sustainable sanitation technologies [1]. The treatment of sewage sludge is one of the widely discussed applications.

7.1. Sewage Sludge

The treatment of sewage sludge is a widely discussed application of microwave technologies and has generated considerable interest in developing countries. Sewage is an organic-rich resource that is typically high in moisture (up to 97%), making it a suitable candidate for microwave irradiation [1], without needing advanced technologies.

Not all studies opt for microwave technologies. When solid waste landfill leachate and sewage sludge samples were inactivated with different technologies and tested for several spores, microwaving was ineffective against the spores of *E. bieneusi* and *E. intestinalis* [39]. In addition, when a range of ultrasonication and microwave pretreatments in thickened sewage sludges were examined, the improvements for microwave pretreated sludge were relatively small [40].

In contrast to these studies, effective inactivation of Gram-negative microorganisms was achieved by microwaves in municipal secondary sludge [41]. Other positive effects for the treatment of sludge were observed as well [42,43]. When microwave treatment of fecal sludge from toilets in the slums of Nairobi, Kenya was studied [44], it was demonstrated that the microwave technology efficiently inactivated *E. coli* and *Ascaris lumbricoides* eggs, and it was concluded that the technology can be applied under real field conditions.

Similarly, treatment of human feces with microwave radiation showed efficient pathogen eradiation performances of six log units or more within a high range of microwave powers. In addition, enhanced moisture removal and volume reduction was achieved. The product was suggested to be used also as compost [45]. Microwaving human fecal sludge represents a thermally effective approach that not only destroys pathogens, but also eradicates the foul odor associated with human fecal sludge, improves de-waterability and heavy metals recovery, and reduces emissions [1].

Microwave pretreatment also significantly improved the dehydration and hydrogen production of sludge subjected to anaerobic digestion [46]. Furthermore, this study may provide theoretical and experimental basis for the development of a continuous microwave sludge-conditioning system. Microwave-H_2O_2 pretreatment on concentrated sludge anaerobic digestion showed in a study in China that a mixture of activated sludge and pretreated sludge at mass ratio of 1:1 was efficient for enhancing anaerobic digestion and methane production [47]. In a study in India, microwave irradiation has been used to disintegrate sludge biomass by de-agglomerating it with an ultra-sonicator and a net profit of 2.67 USD/t was calculated for this procedure [48].

Microwaves in combination with other methods hold also potential for the reduction of antibiotic resistant bacteria and antibiotic resistance genes during sludge treatment [49].

7.2. Healthcare Waste

The most widely used application of microwave irradiation is in the field of healthcare waste. Unfortunately, most economically developing countries suffer a variety of constraints to adequately manage healthcare wastes. Usually, only a few individuals in the staff of a healthcare facility are familiar with a proper waste management program [5]. How to perform an assessment of healthcare waste disposal alternatives including microwaves was illustrated for example with a case study in Shanghai [50]. It has to be decided case by case how to best meet the local biohazardous waste management requirements while minimizing the impact on the environment and public health.

Not unexpectedly, a major constraint is equipment itself. For example, when waste management practices in three government hospitals in Agra, India were studied, it was revealed that none of these hospitals were equipped with higher technological options such as incinerator, autoclave, and microwave. Furthermore, facilities to treat liquid waste generated inside the hospital [10] were not available.

There are number of successful practical uses of microwave technologies in developing countries. When studying 272 modern, high standard, centralized medical waste disposal facilities operating in various cities in China, the application of non-incineration technologies including microwaves was recommended [35]. Soares et al. [38] pointed at two different profiles of waste generators: (i) hospitals, which produce large quantities of healthcare waste; and (ii) small establishments, such as clinics, pharmacies, and other sources that generate dispersed quantities of healthcare waste and are scattered throughout the city. The microwave disinfection presented the best eco-efficiency performance of several studied technologies for small generators of healthcare waste. In Thailand, the influence of microwave irradiation in addition to conductive heating was studied for inactivation of 15 different *C. difficile* spores in aqueous suspension, and microwave proved as a simple and time-efficient tool to inactivate the spores [51].

Oliveira et al. [52] inoculated pre-sterilized public healthcare wastes from the region of Ribeirão Preto, Brazil with spores of *Bacillus atrophaeus* and inactivated the samples with microwave irradiation. The influence of waste moisture, presence of surfactant, power per unit mass of waste, and radiation exposure time was investigated. Microwave irradiation was successful and optimal conditions for inactivation of the *B. atrophaeus* spores in typical healthcare waste were demonstrated. Previously, in a similar study in Ribeirão Preto, *Escherichia coli* in vegetative form were processed in microwaves [13]. Under the operational conditions of the equipment employed in this study, the process of inactivation was ineffective, because the exposure time to radiation average power of approximately was probably inadequate—again pointing to control the inactivation conditions.

Further future potential applications of microwave technologies with impact for developing countries could be in the destruction of dioxins in the froth product after flotation of hospital solid waste incinerator fly ash [53].

8. Conclusions

Microwave irradiation is without doubt an effective tool for inactivation of solid biohazardous waste. However, the inactivation process, and especially the moisture content is not always strictly controlled. In the future only sophisticated microwave technologies allowing appropriate control of heat and fluid/moisture content should be utilized for treatment of in-homogenous solid waste. Otherwise, complete inactivation of biohazardous waste cannot be guaranteed. If following this recommendation, microwave technologies having benefits in contrast to autoclaves would be more widely accepted. Another major current challenge is that biohazardous waste is often generated at many different places within one facility and then transported to the place of inactivation. Because the transport bears risks, biohazardous waste should be preferably inactivated either directly at the place where it is generated or transported in appropriate closed systems. Decentralized inactivation by the relatively simple microwave technology would be especially useful for developing countries where appropriate control of transport of waste is sometimes a challenge. Finally, the aspect of costs should also not be neglected. In comparison to the more widely used autoclave technologies, microwave irradiation is a possibility to save energy costs and has the underlying effect of a reduced carbon footprint.

Conflicts of Interest: The author declares no conflict of interest.

References

1. Afolabi, O.O.D.; Sohail, M. Microwaving human faecal sludge as a viable sanitation technology option for treatment and value recovery—A critical review. *J. Environ. Manag.* **2017**, *1*, 401–415. [CrossRef] [PubMed]
2. UNEP. *Compendium of Technologies for Treatment/Destruction of Healthcare Waste*; United Nations Environment Programme (UNEP), UNEP DTIE International Environmental Technology Centre (IETC): Nairobi, Kenya, 2012.
3. World Health Organization (WHO). *Safe Management of Wastes from Health-Care Activities*, 2nd ed.; Chartier, Y., Emmanuel, J., Pieper, U., Prüss, A., Rushbrook, P., Stringer, R., Townend, W., Susan Wilburn, S., Zghondi, R., Eds.; World Health Organization: Geneva, Switzerland, 2014.
4. Wu, J.; Li, C.; Huo, Y. Safety assessment of dangerous goods transport enterprise based on the relative entropy aggregation in group decision making model. *Comput. Intell. Neurosci.* **2014**, *571058*. [CrossRef] [PubMed]
5. Diaz, L.F.; Savage, G.M.; Eggerth, L.L. Alternatives for the treatment and disposal of healthcare wastes in developing countries. *Waste Manag.* **2005**, *25*, 626–637. [CrossRef] [PubMed]
6. De Titto, E.; Savino, A.A.; Townend, W.K. Healthcare waste management: The current issues in developing countries. *Waste Manag. Res.* **2012**, *30*, 559–561. [CrossRef] [PubMed]
7. RKI (Robert-Koch-Institut). Liste der vom RKI geprüften und anerkannten Desinfektionsmittel und-verfahren. *Bundesgesundheitsblatt-Gesundheitsforschung-Gesundheitsschutz* **2013**, *56*, 1706–1728. [CrossRef]
8. Caniato, M.; Tudor, T.; Vaccari, M. International governance structures for health-care waste management: A systematic review of scientific literature. *J. Environ. Manag.* **2015**, *153*, 93–107. [CrossRef] [PubMed]
9. Johnson, K.M.; González, M.L.; Dueñas, L.; Gamero, M.; Relyea, G.; Luque, L.E.; Caniza, M.A. Improving waste segregation while reducing costs in a tertiary-care hospital in a lower-middle-income country in Central America. *Waste Manag. Res.* **2013**, *31*, 733–738. [CrossRef] [PubMed]
10. Sharma, S.; Chauhan, S.V. Assessment of bio-medical waste management in three apex government hospitals of Agra. *J. Environ. Biol.* **2008**, *29*, 159–162. [PubMed]

11. Nandwani, S. Study of biomedical waste management practices in a private hospital and evaluation of the benefits after implementing remedial measures for the same. *J. Commun. Dis.* **2010**, *42*, 39–44. [PubMed]

12. Zhang, H.J.; Zhang, Y.H.; Wang, Y.; Yang, Y.H.; Zhang, J.; Wang, Y.L.; Wang, J.L. Investigation of medical waste management in Gansu Province, China. *Waste Manag. Res.* **2013**, *31*, 655–659. [CrossRef] [PubMed]

13. Tonuci, L.R.; Paschoalatto, C.F.; Pisani, R., Jr. Microwave inactivation of *Escherichia coli* in healthcare waste. *Waste Manag.* **2008**, *28*, 840–848. [CrossRef] [PubMed]

14. Bélanger, J.M.; Paré, J.R.; Poon, O.; Fairbridge, C.; Ng, S.; Mutyala, S.; Hawkins, R. Remarks on various applications of microwave energy. *J. Microw. Power Electromagn. Energy* **2008**, *42*, 24–44. [CrossRef] [PubMed]

15. Park, D.K.; Bitton, G.; Melker, R. Microbial inactivation by microwave radiation in the home environment. *J. Environ. Health* **2006**, *69*, 17. [PubMed]

16. Gulati, A.; Rawat, R.; Singh, B.; Ravindranath, S.D. Application of microwave energy in the manufacture of enhanced-quality green tea. *J. Agric. Food Chem.* **2003**, *51*, 4764–4768. [CrossRef] [PubMed]

17. Song, W.J.; Kang, D.H. Inactivation of *Salmonella* Senftenberg, *Salmonella* Typhimurium and *Salmonella* Tennessee in peanut butter by 915 MHz microwave heating. *Food Microbiol.* **2016**, *53*, 48–52. [CrossRef] [PubMed]

18. Huang, L.; Sites, J. New automated microwave heating process for cooking and pasteurization of microwaveable foods containing raw meats. *J. Food Sci.* **2010**, *75*, E110–E115. [CrossRef] [PubMed]

19. Sheen, S.; Huang, L.; Sommers, C. Survival of *Listeria monocytogenes*, *Escherichia coli* O157:H7, and *Salmonella* spp. on catfish fillets exposed to microwave heating in a continuous mode. *J. Food Sci.* **2012**, *77*, E209–E214. [CrossRef] [PubMed]

20. Hecq, D.; Jamart, J.; Galanti, L. Microwave freeze-thaw treatment of dose-banded cytotoxics injectable drugs: A review of the literature from 1980 to 2011. *Ann. Pharm. Fr.* **2012**, *70*, 227–235. [CrossRef] [PubMed]

21. Craciun, G.; Martin, D.; Togoe, I.; Tudor, L.; Manaila, E.; Ighigeanu, D.; Matei, C. Vaccine preparation by radiation processing. *J. Microw. Power Electromagn. Energy* **2009**, *43*, 65–70. [CrossRef] [PubMed]

22. Kokolakis, A.K.; Golfinopoulos, S.K. Microwave-assisted techniques (MATs); a quick way to extract a fragrance: A review. *Natl. Prod. Commun.* **2013**, *8*, 1493–1504.

23. Delazar, A.; Nahar, L.; Hamedeyazdan, S.; Sarker, S.D. Microwave-assisted extraction in natural products isolation. *Methods Mol. Biol.* **2012**, *864*, 89–115. [CrossRef] [PubMed]

24. Chan, C.H.; Yusoff, R.; Ngoh, G.C.; Kung, F.W. Microwave-assisted extractions of active ingredients from plants. *J. Chromatogr. A* **2011**, *1218*, 6213–6225. [CrossRef] [PubMed]

25. Al-Hakami, S.M.; Khalil, A.B.; Laoui, T.; Atieh, M.A. Fast Disinfection of *Escherichia coli* Bacteria Using Carbon Nanotubes Interaction with Microwave Radiation. *Bioinorg. Chem. Appl.* **2013**, *458943*. [CrossRef]

26. *Guideline Hospital Waste Decontamination Equipment*; VROM, Ministry of Housing, Spatial Planning and the Environment: Hague, The Netherlands, 2006.

27. Goldblith, S.A.; Wang, D.I. Effect of Microwaves on *Escherichia coli* and *Bacillus subtilis*. *Appl. Microbiol.* **1967**, *15*, 1371–1375. [PubMed]

28. Souhrada, L. Sterilization wave of the future: Microwaves. *Hospitals* **1989**, *63*, 44. [PubMed]

29. Hoffman, P.N.; Hanley, M.J. Assessment of microwave-based clinical waste decontamination unit. *J. Appl. Bacteriol.* **1994**, *77*, 607–612. [CrossRef] [PubMed]

30. Edlich, R.F.; Borel, L.; Jensen, H.G.; Winters, K.L.; Long, W.B., III; Gubler, K.D.; Buschbacher, R.M.; Becker, D.G.; Chang, D.E.; Korngold, J.; et al. Revolutionary advances in medical waste management. The Sanitec system. *J. Long-Term Eff. Med. Implants* **2006**, *16*, 9–18. [CrossRef] [PubMed]

31. Veronesi, P.; Leonelli, C.; Moscato, U.; Cappi, A.; Figurelli, O. Non-incineration microwave assisted sterilization of medical waste. *J. Microw. Power Electromagn. Energy* **2007**, *40*, 211–218. [CrossRef] [PubMed]

32. Katschnig, H. Integriertes, kostengünstiges und umweltschonendes Sicherheitsabfallentsorgungskonzept für die Dialysestation. *Diatra J.* **1993**, *4*, 19–25.

33. Mucha, H. Desinfektion von infektiösen Gütern mittels Hochfrequenzverfahren. *Aseptica* **2001**, *1*, 18–20.

34. Zimmermann, K. Microwave as an emerging technology for the treatment of biohazardous waste: A mini-review. *Waste Manag. Res.* **2017**, *35*, 471–479. [CrossRef] [PubMed]

35. Chen, Y.; Ding, Q.; Yang, X.; Peng, Z.; Xu, D.; Feng, Q. Application countermeasures of non-incineration technologies for medical waste treatment in China. *Waste Manag. Res.* **2013**, *31*, 1237–1244. [CrossRef] [PubMed]

36. Alagöz, B.A.; Kocasoy, G. Treatment and disposal alternatives for health-care waste in developing countries—A case study in Istanbul, Turkey. *Waste Manag. Res.* **2007**, *25*, 83–89. [CrossRef] [PubMed]

37. Özkan, A. Evaluation of healthcare waste treatment/disposal alternatives by using multi-criteria decision-making techniques. *Waste Manag. Res.* **2013**, *31*, 141–149. [CrossRef] [PubMed]

38. Soares, S.R.; Finotti, A.R.; da Silva, V.P.; Alvarenga, R.A. Applications of life cycle assessment and cost analysis in health care waste management. *Waste Manag.* **2013**, *33*, 175–183. [CrossRef] [PubMed]

39. Graczyk, T.K.; Kacprzak, M.; Neczaj, E.; Tamang, L.; Graczyk, H.; Lucy, F.E.; Girouard, A.S. Human-virulent microsporidian spores in solid waste landfill leachate and sewage sludge, and effects of sanitization treatments on their inactivation. *Parasitol. Res.* **2007**, *101*, 569–575. [CrossRef] [PubMed]

40. Cella, M.A.; Akgul, D.; Eskicioglu, C. Assessment of microbial viability in municipal sludge following ultrasound and microwave pretreatments and resulting impacts on the efficiency of anaerobic sludge digestion. *Appl. Microbiol. Biotechnol.* **2016**, *100*, 2855–2868. [CrossRef] [PubMed]

41. Zhou, B.W.; Shin, S.G.; Hwang, K.; Ahn, J.H.; Hwang, S. Effect of microwave irradiation on cellular disintegration of Gram positive and negative cells. *Appl. Microbiol. Biotechnol.* **2010**, *87*, 765–770. [CrossRef] [PubMed]

42. Pino-Jelcic, S.A.; Hong, S.M.; Park, J.K. Enhanced anaerobic biodegradability and inactivation of fecal coliforms and *Salmonella* spp. in wastewater sludge by using microwaves. *Water Environ. Res.* **2006**, *78*, 209–216. [CrossRef] [PubMed]

43. Hong, S.M.; Park, J.K.; Teeradej, N.; Lee, Y.O.; Cho, Y.K.; Park, C.H. Pretreatment of sludge with microwaves for pathogen destruction and improved anaerobic digestion performance. *Water Environ. Res.* **2006**, *78*, 76–83. [CrossRef] [PubMed]

44. Mawioo, P.M.; Hooijmans, C.M.; Garcia, H.A.; Brdjanovic, D. Microwave treatment of faecal sludge from intensively used toilets in the slums of Nairobi, Kenya. *J. Environ. Manag.* **2016**, *184*, 575–584. [CrossRef] [PubMed]

45. Nguyen, T.A.; Babel, S.; Boonyarattanakalin, S.; Koottatep, T. Rapid and Decentralized Human Waste Treatment by Microwave Radiation. *Water Environ. Res.* **2017**, *89*, 652–662. [CrossRef] [PubMed]

46. Zhou, C.; Huang, X.; Zeng, M. Experimental continuous sludge microwave system to enhance dehydration ability and hydrogen production from anaerobic digestion of sludge. *J. Environ. Sci. (China)* **2018**, *67*, 145–153. [CrossRef] [PubMed]

47. Liu, J.; Yang, M.; Zhang, J.; Zheng, J.; Xu, H.; Wang, Y.; Wei, Y. A comprehensive insight into the effects of microwave-H_2O_2 pretreatment on concentrated sewage sludge anaerobic digestion based on semi-continuous operation. *Bioresour. Technol.* **2018**, *256*, 118–127. [CrossRef] [PubMed]

48. Kavitha, S.; Banu, J.R.; Kumar, G.; Kaliappan, S.; Yeom, I.T. Profitable ultrasonic assisted microwave disintegration of sludge biomass: Modelling of biomethanation and energy parameter analysis. *Bioresour. Technol.* **2018**, *254*, 203–213. [CrossRef] [PubMed]

49. Tong, J.; Liu, J.; Zheng, X.; Zhang, J.; Ni, X.; Chen, M.; Wei, Y. Fate of antibiotic resistance bacteria and genes during enhanced anaerobic digestion of sewage sludge by microwave pretreatment. *Bioresour. Technol.* **2016**, *217*, 37–43. [CrossRef] [PubMed]

50. Liu, H.C.; Wu, J.; Li, P. Assessment of health-care waste disposal methods using a VIKOR-based fuzzy multi-criteria decision making method. *Waste Manag.* **2013**, *33*, 2744–2751. [CrossRef] [PubMed]

51. Ojha, S.C.; Chankhamhaengdecha, S.; Singhakaew, S.; Ounjai, P.; Janvilisri, T. Inactivation of Clostridium difficile spores by microwave irradiation. *Anaerobe* **2015**, *38*, 14–20. [CrossRef] [PubMed]

52. Oliveira, E.A.; Nogueira, N.G.; Innocentini, M.D.; Pisani, R., Jr. Microwave inactivation of Bacillus atrophaeus spores in healthcare waste. *Waste Manag.* **2010**, *30*, 2327–2335. [CrossRef] [PubMed]

53. Wei, G.X.; Liu, H.Q.; Zhang, R.; Zhu, Y.W.; Xu, X.; Zang, D.D. Application of microwave energy in the destruction of dioxins in the froth product after flotation of hospital solid waste incinerator fly ash. *J. Hazard. Mater.* **2017**, *325*, 230–238. [CrossRef] [PubMed]

Review

Nomenclature for Healthcare Waste in the Healthcare Sector and Its Alignment with the Provisions Made by The World Health Organization's Manual for Healthcare Waste Management: A Scoping Review

Lydia Hangulu

Discipline of Psychology, Health Promotion Programme, University of KwaZulu-Natal, Durban 4041, South Africa; lydiahangulu@yahoo.com

Received: 10 July 2018; Accepted: 5 November 2018; Published: 12 November 2018

Abstract: There is lack of uniform nomenclature for healthcare waste (HCW) globally, which could undermine efforts to develop and implement appropriate policies relating to healthcare waste management (HCWM) in developing countries. This study sought to understand the terminologies used to describe HCW, including their definitions, categories, classification, and how they align with those that are provided by the World Health Organization (WHO)'s global manual for HCWM from healthcare facilities. The study first identified terms from the existing literature; then, it conceptually mapped the literature, and identified gaps and areas of further inquiry. Six electronic databases—EBSCOhost, Open Access, ProQuest, PubMed, Web of Science, and Google Scholar were used to search for literature. A total of 112 studies were included in the study. Despite having various nomenclature for HCW globally that align with those provided by the WHO manual, the use of varying nomenclature could create confusion among healthcare workers in the quest of managing HCW properly, especially in low and middle-income countries (LMICs). Further studies must be conducted to determine how these terminologies are interpreted and implemented in practice by healthcare workers. This will help to understand how their implementation aligns with the recommendations provided by the WHO manual.

Keywords: healthcare waste; healthcare workers; health sector; scoping review; WHO

---

1. Introduction

Healthcare activities such as operative and diagnostic procedures that involve administering injections, medications, drips, and surgery improve the health and quality of life of individuals [1]. However, the healthcare waste (HCW) that is generated during these healthcare activities could have adverse effects on both the people and the environment if it is poorly managed [2,3]. The literature regarding healthcare waste management (HCWM) shows that large volumes of HCW is generated from healthcare facilities globally [4], and it is poorly managed, especially in low and middle-income countries (LMICs) [5]. The main factors attributed to poor HCWM practices in LMICs include: lack of financial investment and clear policies to manage HCW by most governments [3–6], low levels of knowledge by healthcare staff on how to handle HCW properly [7,8], poor segregation practices among healthcare workers [9], and inappropriate transport for transporting HCW, which is driven by untrained drivers who are also not registered to handle HCW [10]. More so, there are inadequate technologies for treating HCW [11–14]; as a result, HCW is often illegally dumped, openly burnt, and buried in poorly managed dumpsites [10–17].

Poor management of HCW exposes healthcare workers, waste handlers, and the community members to toxins, injuries, and infections [2]. For example, in a study conducted in Tripoli,

Libya, it was found that exposure to HCW among waste handlers caused 5% of them to develop hepatitis B virus, and 0.3% had hepatitis C virus [18]. Exposure to HCW can cause tuberculosis (TB) infections [19], and damage the respiratory, nervous, and reproductive systems of the patients, family members, caregivers, and the public. HCW has mutagenic, teratogenic, and carcinogenic effects [20]. Exposure to HCW can also cause diseases such as diarrhea, leptospirosis, typhoid, cholera, and HIV [21]. The disposal of HCW into unprotected dumpsites promotes scavenging for reusable items for reselling. For example, one study in India revealed that, in almost 10% of the healthcare facilities in the country, more than 30% of the three to six billion injections that were administered every year were done with used equipment [22]. Similarly, in 2009, 240 people in the state of Gujarat in India contracted hepatitis B, because medical care was delivered with previously used syringes that were acquired through the black market [23].

While the environmental and health impacts of HCW have been well documented, having an operationalised nomenclature 'terminology' for HCW is important. The World Health Organization (WHO), in its global manual for healthcare waste management, provides guidelines for all of the issues relating to the proper management of HCW from healthcare settings. The manual uses the nomenclature 'healthcare waste' to mean all of the waste that is generated as a result of healthcare activities, and further classifies HCW into non-hazardous and hazardous waste [2]. Despite having a global manual, different nomenclatures are used to describe HCW by various authors from high-income countries (HICs) and LMICs. For example, some have used 'medical waste' [10,12,18,19,21,23,24], 'biomedical waste' [1,3], 'hazardous waste' [20,25], 'hospital waste' [6,11,13], or 'yellow bag waste' [26,27]. Consequently, a study on HCWM practices in healthcare facilities in Botswana [16] found that the use of the nomenclature 'clinical waste' to mean HCW confused healthcare workers and the public. Both the healthcare workers and ordinary people correctly defined clinical waste as any waste from healthcare facilities, but failed to consider that HCW is further categorized as non-hazardous and hazardous waste. Failure to classify HCW into these categories resulted in the improper segregation of HCW [16]. In other contexts such as clinical practice, the consistence use of specific nomenclature in describing suicidal behavior is critical for case development among healthcare providers [28]. Furthermore, it is argued that consistent and specific nomenclature allows for appropriate diagnosis, treatment, and the subsequent creation of public policy [29]. Similarly, considering that HCWM practices involve cognitive behavior with a triadic relationship which involves peoples' perceptions, feelings, and actions, one can argue that using various nomenclatures to describe HCW could be confusing to waste generators and handlers, thereby affecting the HCWM practices. Despite having various nomenclatures to describe HCW, is not clear how they align with the definitions that are provided by the WHO's manual on HCWM. Furthermore, there is no scoping review that describes, defines, and characterizes HCW in comparison with the provisions made by the WHO manual.

This scoping review endeavors to determine the common nomenclatures for HCW from HICs and LMICs, including their definition, classification, and categorization in comparison to those provided by the WHO global manual for HCWM. This scoping review will answer three specific research questions: (1) What are the various nomenclatures that are used to describe HCW in HICs and LMICs? (2) To what extent do such nomenclatures align with the provisions made by the WHO manual on HCWM? (3) What are the gaps that exists in the literature? Findings of the review will add scientific knowledge to the body of literature on HCWM, and will help determine any inconsistencies that exist that will help make recommendations aimed at improving HCWM by healthcare workers and other policy implementers, especially in LMICs.

2. Methods

Unlike systematic reviews that aim at combining, summarizing, and synthesizing the findings of particular research [30], scoping reviews are conducted for the purpose of mapping the key concepts underpinning a research area, and the main sources and types of evidence that are available [31–33]. Scoping reviews can be undertaken as standalone projects in their own right, especially where an area is complex or has previously not been reviewed comprehensively [30]. This scoping review was conducted because to our knowledge, no such study has been conducted to explore and map the nomenclature that is used to describe, define, and categorize HCW in HICs and LMICs, in comparison with those provided in the WHO manual for HCW. Both deductive and inductive approaches were applied when conducting this review. First, search terms from the literature were identified. Secondly, the search terms that were identified in the literature were used to search for literature from various databases. Thirdly, the relevant literature was reviewed and selected, and finally, all of the selected literature was mapped.

2.1. Identifying Search Terms

We used an iterative process to conduct the searches. First, the WHO manual was read in order to derive the first term 'healthcare waste' for the search. The term healthcare waste was used to search and conduct a broad but rapid review of the literature. In order to identify the terms used by ordinary people, policy-makers, and stakeholders, news stories from the 20 newspapers retrieved from the South African print media database were searched. The search was limited to the South African print media, because it was the only available database for news stories within the university at the time of research. The characteristics of the newspapers are summarized in Supplementary Table S1. The various searches yielded the following terms: "healthcare waste", "medical waste", "clinical waste", "biomedical waste", and "hospital waste".

2.2. Literature Search

The initial search was conducted in September 2015 by two student assessors (LH and SM) under the supervision of (OA). Six electronic databases: EBSCOhost, Open Access, ProQuest, PubMed, Web of Science, and Google Scholar were used (see Supplementary Table S2 for the table of all of the databases that were used in the study). These databases were those available at the University of KwaZulu-Natal, Durban, South Africa. We chose both gray and peer-reviewed literature in order to have a broader coverage of the literature. From the initial search, the results were 9735, and too broad. To limit the search, we developed an inclusion criteria to include: (1) only full texts of both grey and peer-reviewed literature that were available through the library at the University of KwaZulu-Natal, (2) literature published from 1990 to 2015, because this period had the highest hits, (3) only English literature, and (4) literature with key search terms in their title and/or their abstracts.

2.3. Review and Selection of Literature

After applying the inclusion criteria, 8468 studies were excluded, and 1267 remained. Thereafter, the two assessors (LH and SM) worked together to develop a set of explicit exclusion criteria. The exclusion criteria were applied independently by the two assessors, who met regularly to compare the assessments and resolve discrepancies. The criteria were applied as follows. Studies were removed if they: (1) were duplicates and (2) did not define, categorize, and classify HCW in their full texts. After applying the exclusion criteria, 107 duplicates were removed, and 1157 studies remained. A total of 1045 studies that did not define, categorize, or classify HCW in their full text were excluded. The remaining 112 studies (see supplementary table of all included studies Table S3) that met the criteria were mapped as summarized in the PRISMA flow chart in Figure 1.

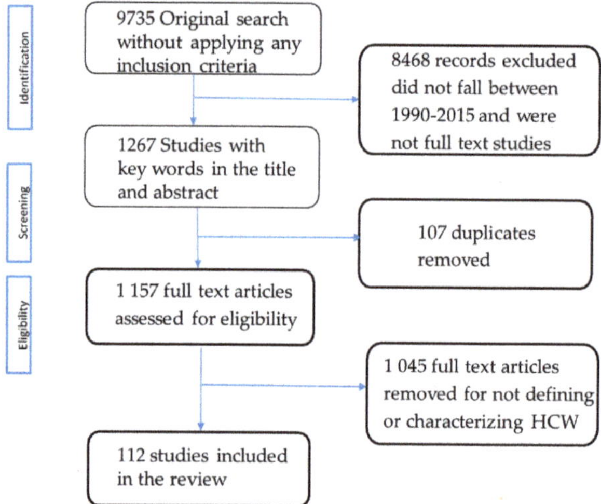

Figure 1. PRISMA flowchart of study selection process.

3. Results

The findings were coded using a coding framework which include the source of literature, country source, disciplines, topics covered, methods used and the common terminologies used all the details are discussed below (see the coding framework Supplementary Table S4).

3.1. Characteristics of the Mapped Literature

Source of literature: The majority of the studies N = 100 (89.3%), in this scoping review were published in journals, and 12 (10.7%) were student dissertations. There were more studies N = 100 (89.28%) from LMICs than from HICs N = 12 (10.71%). The studies from LMICs are divided into regions. For example, the Asian countries included: Bangladesh, China, India, Lao Republic, Malaysia, Pakistan, Philippines, Romania, Thailand, Kingdom of Bahrain, Nepal, Vietnam, Taiwan. African countries include, Algeria, Botswana, Egypt, Ethiopia, Ghana, Libya, Nigeria, Tanzania, and Zimbabwe. The Middle East countries included: the kingdom of Buhrain, Iran, Jordan, and Palestine; further, there was Brazil from South America and Turkey from Europe (Turkey). The minority of the studies from HICs were from Croatia, Japan, the United Kingdom, Greece, Portugal, the United States of America, and Italy. All of these are summarized in Table 1.

Table 1. A summary table for all low and middle-income countries (LMICs) and high-income countries (HICs) with various terminologies used.

Region	Country	Terminology	Reference	Number of Studies (N = 112)	Percentage
Asia	Bangladesh	Hospital waste Healthcare waste Medical waste	Akter et al., 1999, Alam and Hossain 2013, Rahman and Ali, 2000, Patwary et al., 2009	4	3.57
	China	Biomedical waste Clinical waste	Chen et al., 2013, Chung and Lo 2003, Tam, 1996	3	2.67
	India	Biomedical waste Hazardous waste	Chitnis et al., 2005, Nema et al., 2011, Ujwala, et al., 2012, Naik et al., 2012, Babu et al., 2009, Aravindan and Vasumathi, 2015, Bansal et al., 2013, Bhatt et al., 2013, Imtiaz, et al., 2014, Chaithra and Sadashivamurthy, 2014, Chaurasia et al., 2014, Chethana et al., 2014, Chowdhary and Slathia, 2014, Gupta et al., 2009, Sumi, 2010, Singh et al., 2014, Sehgal et al., 2015, Sanjeev et al., 2014, Rathod et al., 2012, Rajor et al., 2012, Prakash et al., 2015, Patil, 2015, Nayak, and Nayak, 2014, Kapoor et al., 2014, Komilis et al., 2012	25	22.32
	Lao republic	Hospital waste	Saad, 2013	1	0.89
	Malaysia	Healthcare waste	Shanmugasundaram, Soulalay and Chettiyappan, 2012	1	0.89
	Pakistan	Clinical waste	Ibrahim, 2005, Hossain et al., 2012	2	1.78
	Philippines	Biomedical waste	Ali, H.A. 2000, Ibrahim, Z.B. 2005, Hossain, et al., 2012, Ambali et al., 2013, Abdullah, et al., 2013, Chitralekha and Agrawal, 2010, Cruz et al., 2014.	6	5.35
	Japan	Biomedical waste	Miyazaki and Une 2005	1	0.89
	Thailand	Healthcare waste	Ananth et al., 2010, Manowan, 2009	2	1.78
	Nepal	Hazardous Healthcare waste	Sapkota et al., 2014	1	0.89
	Vietnam	Healthcare waste	Phengxay et al., 2005	1	0.89
	Taiwan	Biomedical, Medical waste	Liao and Ho, 2014, Cheng et al., 2009	2	1.78
Africa	Algeria	Healthcare waste	Bendjoudi et al., 2009	1	0.89
	Botswana	Clinical waste Biomedical waste Healthcare waste	Kang'ethe, 2008, Kudoma, 2013, Mbongwe et al., 2008, Mmereki et al., 2008, Mmereki et al., 2017	5	4.46
	Cameroon	Clinical waste	Mochungong, 2011, Manga et al., 2011	2	1.78
	Egypt	Hospital waste Biomedical waste	El-Salam, 2010, Soliman and Ahmed, 2007	2	1.78

Table 1. *Cont.*

	Ethiopia	Healthcare waste	Haylamicheal, et al., 2011, Tesfahun et al. 2014, Tadesse and Kumie, 2014	3	2.67
	Ghana	Healthcare waste	Abor, 2013, Asante et al., 2014	2	1.78
	Libya	Healthcare waste	Sawalem et al., 2009	1	0.89
	Nigeria	Biomedical waste Hospital waste Healthcare waste	Turaki, 2015, Chima et al., 2014, Joshua et al., 2014, Longe, 2012, Ngwuluka et al., 2009, Abor, 2007, Maseko, 2014	7	6.25
	South Africa	Medical waste Healthcare waste	Gabela and Knight, 2010, Maseko, 2014, Nemathaga et al., 2008	3	2.67
	Tanzania	Hospital waste	Nemathaga et al., 2008	1	0.89
	Zimbabwe	Medical waste	Mziray, 2009, Taru and Kuvarega, 2005	2	1.89
Middle East	Kingdom of Bahrain	Healthcare waste Hospital waste	Askarian et al., 2010, Mesdaghinia, 2009, Mohamed et al., 2009	3	2.67
	Iran	Healthcare waste Medical waste Hospital waste Infectious waste	Askarian et al., 2010, Bazrafshan et al., 2014, Taghipour and Mosaferi, 2009, Sabour et al., 2007, Oroei et al., 2014	5	4.46
	Jordan	Medical waste	Abu-Awwad 2008, Al-Khatib et al., 2011, Qdais et al., 2007	3	2.67
	Palestine	Healthcare waste Hospital waste	Al-khatib et al., 2009, Eleyan et al., 2013, Maroufi, and Javadi, 2012	3	2.67
American	Brazil	Hospital Biomedical waste Medical waste	Ferreira et al., 2003, Paiz et al., 2014, Silva et al., 2005	3	2.67
	San Salvador	Healthcare waste	Johnson et al., 2013	1	0.89
	Mexico	Medical waste	Spence, 2000	1	0.89
	United States of America	Healthcare waste	Klangsin, 1994	1	0.89
European	Turkey	Medical waste Hospital waste	Akbolat and Saglam, 2011, Eker et al., 2010, Uysal and Tinmaz, 2004, Altin et al., 2003	4	3.57
	Croatia	Hazardous medical waste	Marinković et al., 2008	1	0.89
	United Kingdom	Hazardous waste Healthcare waste Clinical waste Medical waste	Blenkharn, 2006, Akpieyi et al., 2015, Moritz, 1995, Pudussery, 2010	4	3.57
	Greece	Hospital waste Medical waste	Tsakona et al., 2007, Graikos et al., 2010	2	1.89
	Portugal	Medical waste	Botelho, 2012	1	0.89
	Italy	Medical waste	Giacchetta and Marchetti, 2013	1	0.89
	Romania	Medical waste	Bulucea et al., 2008	1	0.89

3.2. Disciplines Represented by the Various Studies

The 112 studies were from science-related disciplines ranging from administrative science and policy to process engineering. The studies covered various topics for example, 44 (39.3%) covered HCWM practices, 14 (12.5%) discussed the knowledge and attitudes of healthcare staff about HCWM, 13 (11.6%) addressed the segregation and quantification of HCW, 12 (10.7%) presented the risks associated with HCWM, 11 (9.8%) focused on HCW treatment and disposal options, nine (8.0%) reviewed existing policies on HCWM, and nine (8.0%) addressed models for HCWM. The majority 22 (19.6%) of the studies were cross-sectional studies followed by 16 mixed methods (14.3%), 13 literature reviews (11.6%), 13 quantitative surveys (11.6%), 11 case studies (9.8%), eight qualitative studies (7.1%), six experiments (5.4%), five document analyses (4.5%), two commentaries (1.8%) and three systematic reviews (2.8%).

3.3. The Common Terminologies Used to Describe Healthcare Waste by Various Countries

The WHO manual uses the term 'healthcare waste' to describe all of the waste resulting from healthcare activities. The common terminologies used by HICs are medical waste (4.46%), healthcare waste, and hospital waste (1.78% each), followed by clinical, biomedical, and hazardous waste, at 0.89% each. In the LMICs, biomedical waste was found to be the common terminology used with 32.14%, followed by healthcare waste (24.10%), medical waste (13.39%), hospital waste (10.71%), clinical waste (6.25%), and hazardous healthcare waste (2.67%), as summarized in Table 2.

Table 2. The common terminologies used by LMICs and HICs to describe healthcare waste.

Country Focus	Common Terminology	Number of Studies	Total Percentage
	Medical Waste	5	4.46
	Healthcare Waste	2	1.78
	Hospital Waste	2	1.78
High-Income Countries	Clinical Waste	1	0.89
	Biomedical Waste	1	0.89
	Hazardous Waste	1	0.89
			(89.28)
	Biomedical Waste	36	32.14
	Healthcare Waste	27	24.10
Low and Middle-Income	Medical Waste	15	13.39
Countries	Hospital Waste	12	10.71
	Clinical Waste	7	6.25
	Hazardous Healthcare Waste	3	2.67
Total		112	100

3.4. Categorization of Healthcare Waste

The WHO manual categorizes HCW into non-hazardous and hazardous waste. All of the categories presented by the eligible studies from LMICs conformed to the WHO's categories despite using terminologies such as biohazardous or pathological waste to mean all of the waste that is capable of transmitting microbes, and non-biohazardous/non-pathological waste as waste that does not pose harm upon contact, as presented in Table 3.

On the other hand, the HICs categorizes all waste into five groups/classes using letters of the alphabet. These groups are Group A waste, which is all waste with human tissue including blood, animal carcasses, tissue from veterinary centers, hospitals, and laboratories. Group B waste consists of all discarded syringes needles, cartridges, broken glass, and any other contaminated disposable sharp instruments. Group C waste includes all microbiological cultures and all potentially infected waste from pathology departments such as clinical or research laboratories and post-mortem rooms. Group D waste are all of the pharmaceutical products and chemical wastes. Group E waste includes all of the items that are used to dispose of urine, feces, body secretions, and excretions, as summarized in Table 4. Despite using letters of the alphabet to categorize HCW, the description of the HCW as well as the examples that are provided under each category align with those provided by the WHO manual.

Table 3. Classification of healthcare waste by low and middle-income countries.

Classification of Healthcare Waste	Examples of Waste
1. Biohazardous/Pathological Waste	
Infectious waste	Waste suspected to contain pathogens e.g., laboratory cultures; waste from isolation wards; tissues (swabs), materials, or equipment that have been in contact with infected patients; excreta
Pathological waste	Human tissues or fluids e.g., body parts; blood and other body fluids; fetuses
Sharp waste	This is sharp waste e.g., needles; infusion sets; scalpels; knives; blades; broken glass
Pharmaceutical waste	Waste containing pharmaceuticals e.g., pharmaceuticals that are expired or no longer needed; items contaminated by or containing pharmaceuticals (bottles, boxes)
Genotoxic waste	Waste containing substances with genotoxic properties e.g., waste containing cytostatic drugs (often used in cancer therapy); genotoxic chemicals
Chemical waste	Waste containing chemical substances e.g., laboratory reagents; film developer; disinfectants that are expired or no longer needed; solvents
Radioactive waste	Waste containing radioactive substances e.g., unused liquids from radiotherapy or laboratory research; contaminated glassware, packages, or absorbent paper; urine and excreta from patients treated or tested with unsealed radionuclides; sealed sources
2. Non-biohazardous/non pathological waste	Paper, plastic, cans, leftover food

Table 4. Classification of healthcare waste by high-income countries.

Classification of Healthcare Waste	Examples
Group A	All waste contaminated with human tissue, including blood, animal carcasses, tissue from veterinary centers, hospitals, and laboratories, including soiled surgical dressing, swabs, and other soiled waste from treatment areas
Group B	All discarded syringe needles, cartridges, broken glass, and any disposable sharp instruments
Group C	All microbiological cultures and all waste from pathology departments such as clinical or research laboratories and post-mortem rooms
Group D	All pharmaceutical products and chemical wastes e.g., all discarded medicines, cytotoxic drugs
Group E	All items that are used to dispose of urine, feces, body secretions, and excretions e.g., incontinence pads, disposable bedpans, urine containers

4. Discussion

The majority of the studies were from LMICs, which shows that HCWM is more of a concern in LMICs [2–26,30]. In order to address the issue of HCWM, the 112 studies in this review used various methods ranging from case studies to systematic reviews. The studies are spread across different disciplines, with the largest number coming from the public health discipline, followed by environmental engineering, environmental health, environmental management, waste management engineering, and lastly, community medicine. These findings show that HCW is a multidisciplinary issue [22,31] that has implications for the well-being of the people and the environment.

This review has also found that both HICs and LMICs use different terminologies to describe HCW, although the definition, categorization, and classification align with those provided by the WHO manual [2]. It is not clear why the use of nomenclature vary in countries, and yet, there is only one global manual. More so, the adverse consequences of the improper management of HCW have been documented extensively [2,18–23], and using different nomenclatures to describe HCW is one of the factors contributing to the improper segregation practices of HCW among healthcare workers [4]. For example, a study in Botswana on HCWM current practices in healthcare facilities [16] found that the term 'clinical waste' is known to mean all of the waste that is generated from healthcare facilities. Due to this definition, most healthcare workers and the general public ended up not segregating HCW into non-hazardous and hazardous waste. The health workers disposed of all of the unsegregated HCW into red bags. The consequence of this practice was the unnecessary use and wasting of red bags, and the overloading of resources needed for the transportation and storage of HCW. Although this particular study cannot be generalized, more of such studies are yet to be found. Varied nomenclatures could also cause confusion when it comes to developing HCWM policies by policy-makers, and can affect the practices of HCWM by healthcare workers who are generators and handlers of HCW. More importantly, a lesson learnt from Mbongwe's study is the need to have a standard nomenclature to describe, define, categorize, and classify HCW.

Conclusion and Recommendations

Considering that all of the nomenclatures that are used by various authors from HICs and LMICs are different, but align with those provided by the WHO manual, there is a need to adopt the terminology used by the manual. A uniform nomenclature could be beneficial for the healthcare workers who are HCW generators and handlers. A standard nomenclature could also be beneficial to policy-makers for designing HCWM policies that monitor appropriate HCWM practices.

5. Strengths and Limitations

The primary strength of this scoping review is its ability to answer all of the research questions with the use of transparent methods to conduct the review. Limitations to this scoping review are that only six databases were used to search for literature given the limited resources (at the University of KwaZulu-Natal). Furthermore, only 20 South African print media were used to identify search terms, because these were the only ones available at the time of the research. The search only used five key search terms (HCW, medical waste, clinical waste, biomedical waste, and hospital waste), because these are the most dominant terminologies found in the literature. Lastly, the review did not cover literature from 2016 onwards due to limited resources at the time of research.

6. Areas for Future Research

This scoping review did not explore the extent to which the WHO's guidelines have been adopted and implemented in practice by various countries. Such studies should be conducted to provide more insights into HCW and its management. More importantly, this review did not include literature from 2016 onwards. More studies should be conducted and include literature from 2016 onwards. This review did not explore search terms that are used in other media venues besides the ones from the South African print media. Other studies should be conducted to include search terms from other media venues to help yield more insights into the study on HCWM.

Supplementary Materials: The following are available online at http://www.mdpi.com/2313-4321/3/4/51/s1, Table S1: Characteristics of newspaper covered in the analysis; Table S2: Table showing databases used in the study; Table S3: A list of all references analysed for the scoping review. Table S4: Coding framework used for the studies included in the scoping review.

Funding: This research received no external funding.

Acknowledgments: Lydia Hangulu (L.H.) would like to acknowledge the assistance provided by the Samantha Moodley (S.M.) for working as an assessor in the data collection phase. More importantly, Olagoke Akintola (O.A.) who supervised the research. The writing of the manuscript was supported by the postdoctoral fellowship grant offered to Lydia Hangulu by the National Research Foundation (NRF) of South Africa. However, all views expressed in this manuscript are of the author and should and not necessarily to be attributed to NRF.

Conflicts of Interest: The author declares no conflict of interest.

References

1. Verma, L.K.; Mani, S.; Sinha, N.; Rana, S. Biomedical waste management in nursing homes and smaller hospitals in Delhi. *Waste Manag.* **2008**, *28*, 2723–2734. [CrossRef] [PubMed]

2. Chartier, Y.; Emmanuel, J.; Pieper, U.; Prüss, A.; Rushbrook, P.; Stringer, R.; Townend, W.; Wilburn, S.; Zghondi, R. *Safe Management of Wastes from Health-Care Activities*, 2nd ed.; World Health Organization: Geneva, Switzerland, 2014.

3. Mathur, P.; Patan, S.; Shobhawat, A.S. Need of biomedical waste management system in hospitals-An emerging issue-a review. *Curr. World Environ.* **2017**, *7*, 117–124. [CrossRef]

4. Ali, M.; Wang, W.; Chaudhry, N.; Geng, Y. Hospital waste management in developing countries: A mini review. *Waste Manag. Res.* **2017**, *35*, 581–592. [CrossRef] [PubMed]

5. Delmonico, D.V.; Santos, H.H.; Pinheiro, M.A.; de Castro, R.; de Souza, R.M. Waste management barriers in developing country hospitals: Case study and AHP analysis. *Waste Manag. Res.* **2018**, *36*, 48–58. [CrossRef] [PubMed]

6. Sawalem, M.; Selic, E.; Herbell, J.D. Hospital waste management in Libya: A. case study. *Waste Manag.* **2009**, *29*, 1370–1375. [CrossRef] [PubMed]

7. Kumar, R.; Khan, E.A.; Ahmed, J.; Khan, Z.; Magan, M.; Mughal, M.I. Healthcare waste management in Pakistan: Current situation and training options. *J. Ayub Med. Coll. Abbottabad* **2010**, *22*, 101–105. [PubMed]

8. Mmereki, D.; Baldwin, A.; Li, B.; Liu, M. Healthcare waste management in Botswana: Storage, collection, treatment and disposal system. *J. Mater. Cycles Waste Manag.* **2017**, *19*, 351–365. [CrossRef]

9. Ferreira, V.; Teixeira, M.R. Healthcare waste management practices and risk perceptions: Findings from hospitals in the Algarve region, Portugal. *Waste Manag.* **2010**, *30*, 2657–2663. [CrossRef] [PubMed]

10. Komilis, D.P. Issues on medical waste management research. *Waste Manag.* **2016**, *48*, 1–2. [CrossRef] [PubMed]

11. Nemathaga, F.; Maringa, S.; Chimuka, L. Hospital solid waste management practices in Limpopo Province, South Africa: A case study of two hospitals. *Waste Manag.* **2008**, *28*, 1236–1245. [CrossRef] [PubMed]

12. Hassan, M.M.; Ahmed, S.A.; Rahman, K.A.; Biswas, T.K. Pattern of medical waste management: Existing scenario in Dhaka City, Bangladesh. *BMC Public Health* **2008**, *8*, 36. [CrossRef] [PubMed]

13. El-Salam, M.M.A. Hospital waste management in El-Beheira Governorate, Egypt. *J. Environ. Manag.* **2010**, *91*, 618–629. [CrossRef] [PubMed]

14. Gabela, S.D.; Knight, S.E. Healthcare waste management in clinics in a rural health district in KwaZulu-Natal: Brief report. *S. Afr. J. Epidemiol. Infect.* **2010**, *25*, 19–21. [CrossRef]

15. Hangulu, L.; Akintola, O. Health care waste management in community-based care: Experiences of community health workers in low resource communities in South Africa. *BMC Public Health* **2017**, *17*, 448. [CrossRef] [PubMed]

16. Mbongwe, B.; Mmereki, B.T.; Magashula, A. Healthcare waste management: Current practices in selected healthcare facilities, Botswana. *Waste Manag.* **2008**, *28*, 226–233. [CrossRef] [PubMed]

17. Manga, V.E.; Forton, O.T.; Mofor, L.A.; Woodard, R. Health care waste management in Cameroon: A case study from the Southwestern Region. *Resour. Conserv. Recycl.* **2011**, *57*, 108–116. [CrossRef]

18. Franka, E.; El-Zoka, A.H.; Hussein, A.H.; Elbakosh, M.M.; Arafa, A.K.; Ghenghesh, K.S. Hepatitis B virus and hepatitis C virus in medical waste handlers in Tripoli, Libya. *J. Hosp. Infect.* **2009**, *72*, 258–261. [CrossRef] [PubMed]

19. Bdour, A.; Altrabsheh, B.; Hadadin, N.; Al-Shareif, M. Assessment of medical wastes management practice: A case study of the Northern part of Jordan. *Waste Manag.* **2007**, *27*, 746–759. [CrossRef] [PubMed]

20. Blackman, W.L., Jr. *Basic Hazardous Management*; Lewis: Boca Raton, FL, USA, 1993.

21. Mato, R.R.A.; Kassenga, G.R.; Mbuligwe, S.E. *Tanzania Environmental Profile*; Report prepared for Japan International Co-operation (JICA): Dar es Salaam, Tanzania, 1997.

22. Harhay, M.O.; Halpern, S.D.; Harhay, J.S.; Olliaro, P.L. Health care waste management: A neglected and growing public health problem worldwide. *Trop. Med. Int. Heal.* **2009**, *14*, 1414–1417. [CrossRef] [PubMed]

23. Solberg, K.E. Trade in medical waste causes deaths in India. *Lancet* **2009**, *373*, 1067. [CrossRef]

24. Abdulla, F.; Qdais, H.A.; Rabi, A. Site investigation on medical waste management practices in northern Jordan. *Waste Manag.* **2008**, *28*, 450–458. [CrossRef] [PubMed]

25. LaGrega, M.D.; Buckingham, P.L.; Evans, J.C. *Hazardous Waste Management*; Waveland Press: Long Grove, IL, USA, 2010.

26. Diaz, L.F.; Eggerth, L.L.; Enkhtsetseg, S.; Savage, G.M. Characteristics of healthcare wastes. *Waste Manag.* **2008**, *28*, 1219–1226. [CrossRef] [PubMed]

27. Ramokate, T. Knowledge and Practices of Doctors and Nurses about Management of Health Care Waste at Johannesburg Hospital in the Gauteng Province, South Africa. Doctoral Dissertation, University of Witwatersrand, Johannesburg, South Africa, 2008.

28. Bryan, C.J. (Ed.) *Cognitive Behavioral Therapy for Preventing Suicide Attempts: A Guide to Brief Treatments Across Clinical Settings*; Routledge: London, UK, 2015.

29. De Leo, D.; Burgis, S.; Bertolote, J.M.; Kerkhof, A.J.; Bille-Brahe, U. Definitions of suicidal behavior: Lessons learned from the WHO/EURO Multicentre Study. *Crisis* **2006**, *27*, 4–15. [CrossRef] [PubMed]

30. Chircop, A.; Bassett, R.; Taylor, E. Evidence on how to practice intersectoral collaboration for health equity: A scoping review. *Crit. Public Heal.* **2015**, *25*, 178–191. [CrossRef]

31. Mays, N.; Roberts, E.; Popay, J. *Synthesising Research Evidence: Studying the Organisation and Delivery of Health Services: Research Methods*; Routledge: London, UK, 2001; pp. 188–220.

32. Arksey, H.; O'Malley, L. Scoping studies: Towards a methodological framework. *Int. J. Soc. Res. Methodol.* **2005**, *8*, 19–32. [CrossRef]

33. Daudt, H.M.; Van Mossel, C.; Scott, S.J. Enhancing the scoping study methodology: A large, inter-professional team's experience with Arksey and O'Malley's framework. *Med. Res. Methodol.* **2013**, *13*, 48. [CrossRef] [PubMed]

MDPI
St. Alban-Anlage 66
4052 Basel
Switzerland
Tel. +41 61 683 77 34
Fax +41 61 302 89 18
www.mdpi.com

Recycling Editorial Office
E-mail: recycling@mdpi.com
www.mdpi.com/journal/recycling